A TEXTBOOK OF
DISCRETE MATHEMATICS

A TEXTBOOK OF
DISCRETE MATHEMATICS

[For B Tech. Third Year (CSE & IT) students of West Bengal University of Technology]

Guruprasad Samanta

PhD, PRS, Mouat Medallist

Professor

Department of Mathematics

Bengal Engineering and Science University, Shibpur

Howrah, West Bengal

NEW AGE
WBUT

An Imprint of

NEW AGE INTERNATIONAL (P) LIMITED, PUBLISHERS

New Delhi • Bangalore • Chennai • Cochin • Guwahati
Hyderabad • Kolkata • Lucknow • Mumbai
Visit us at www.newagepublishers.com

Published by New Age International (P) Ltd., Publishers
First Edition: 2014

BRANCHES

- **Bangalore** 37/10, 8th Cross (Near Hanuman Temple), Azad Nagar, Chamarajpet, Bangalore-560 018
Tel.: (080) 26756823, Telefax: 26756820, **E-mail: bangalore@newagepublishers.com**
- **Chennai** 26, Damodaran Street, T. Nagar, Chennai-600 017
Tel.: (044) 24353401, Telefax: 24351463, **E-mail: chennai@newagepublishers.com**
- **Cochin** CC-39/1016, Carrier Station Road, Ernakulam South, Cochin-682 016
Tel.: (0484) 2377004, Telefax: 4051303, **E-mail: cochin@newagepublishers.com**
- **Guwahati** Hemsen Complex, Mohd. Shah Road, Paltan Bazar, Near Starline Hotel, Guwahati-781 008
Tel.: (0361) 2513881, Telefax: 2543669, **E-mail: guwahati@newagepublishers.com**
- **Hyderabad** 105, 1st Floor, Madhiray Kaveri Tower, 3-2-19, Azam Jahi Road, Near Kumar Theater Nimboliadda, Kachiguda, Hyderabad-500 027, Tel.: (040) 24652456, Telefax: 24652457
E-mail: hyderabad@newagepublishers.com
- **Kolkata** RDB Chambers (Formerly Lotus Cinema) 106A, 1st Floor, S.N. Banerjee Road, Kolkata-700 014
Tel.: (033) 22273773, Telefax: 22275247, **E-mail: kolkata@newagepublishers.com**
- **Lucknow** 16-A, Jopling Road, Lucknow-226 001
Tel.: (0522) 2209578, 4045297, Telefax: 2204098, **E-mail:lucknow@newagepublishers.com**
- **Mumbai** 142C, Victor House, Ground Floor, N.M. Joshi Marg, Lower Parel, Mumbai-400 013
Tel.: (022) 24927869, Telefax: 24915415, **E-mail: mumbai@newagepublishers.com**
- **New Delhi** 22, Golden House, Daryaganj, New Delhi-110 002
Tel.: (011) 23262368, 23262370, Telefax: 43551305, **E-mail: sales@newagepublishers.com**

ISBN: 978-81-224-3680-8

₹ 199.00

C-13-11-7216

Printed in India at Mohanlal Printers, Delhi.
Typeset at Excellent Graphics, Delhi.

PUBLISHING GLOBALLY
NEW AGE INTERNATIONAL (P) LIMITED, PUBLISHERS
7/30 A, Daryaganj, New Delhi-110002
Visit us at **www.newagepublishers.com**

Dedicated to

My daughter, Debaleena Samanta

&

My son, Soham Samanta

Preface

It gives me great pleasure to introduce this textbook of Discrete Mathematics to the students pursuing the third year B.Tech. (CSE, fifth semester & IT, sixth semester) course offered by the Engineering or Technological Institutes affiliated to the West Bengal University of Technology (W.B.U.T.). Discrete Mathematics is one of the basic subjects of Computer Science and Engineering, Information Technology and Science. Discrete Mathematics is a systematic science of practical importance resting on relatively small number of basic concepts and involving powerful unifying methods. Students of engineering science and technology must have clear concepts of the mathematical logic and procedures to be applied to various engineering problems and should have more inclination towards the procedures rather than the underlying theories.

This textbook has been written according to the new revised syllabus of West Bengal University of Technology for the third year B.Tech. (CSE, fifth semester & IT, sixth semester) students. It consists of seven chapters. I think this book can also be treated as a supplementary book of Discrete Mathematics for B.E./B.Tech/B.Sc./BCA/M.Sc./ M.E./ M.Tech./MCA students of other Universities/Institutes. This book contains various types of problems along with worked out model problems which are essential to the Engineering/Technological/ Science students. It will help the students to get through their examination and at the same time will arouse greater intellectual curiosity in them. Multiple choice questions with answers have been provided in this textbook which will be helpful for GATE and other competitive examinations.

Any healthy suggestions and comments for the improvement of this book from students as well as from teachers will be heartily and gratefully acknowledged.

I am thankful to Sri Prosenjit Sen, Assistant Professor, Om Dayal Group of Institutions, Uluberia, Howrah for his help and suggestions during the preparation of the manuscript.

Finally, I express my heartiest gratitude to Mr. Saumya Gupta, Managing Director, Mr. T.K. Chatterjee, Regional Manager, Mr. Anup K. Kundu, Area Sales Manager, Mr. Mahesh Dhama and other members of New Age International (P) Limited, Publishers for their immense cooperation in bringing out this book in an excellent form within a very short span of time.

Guruprasad Samanta

Syllabus

WBUT Code CS-503 & IT-605A

Module I

Introduction to Propositional Calculus: Propositions, Logical Connectives, Conjunction, Disjunction, Negation and their truth table. Conditional Connectives, Implication, Converse, Contrapositive, Inverse, Biconditional statements with truth table, Logical Equivalence, Tautology, Normal forms – CNF, DNF; Predicates and Logical Quantifications of propositions and related examples. [10L]

Module II

Theory of Numbers: Well Ordering Principle, Divisibility theory and properties of divisibility; Fundamental theorem of Arithmetic; Euclidean Algorithm for finding G. C. D. and some basic properties of G. C. D. with simple examples; Congruences; Residue classes of integer modulo $n(Z_n)$ and its examples.

Order, Relation and Lattices: POSET, Hasse Diagram, Minimal, Maximal, Greatest and Least elements in a POSET, Lattices and its properties, Principle of Duality, Distributive and Complemented Lattices. [10L]

Module III

Counting Techniques: Permutations, Combinations, Binomial coefficients, Pigeon-hole Principle, Principles of inclusion and exclusions; Recurrence relations: Formulation/Modelling of different counting problems in terms of recurrence relations, Solution of linear recurrence relations with constant coefficients (upto second order) by (*i*) The iterative method (*ii*) Characteristic roots method (*iii*) Generating functions method. [10L]

Module IV

Graph Colouring: Chromatic Numbers and its bounds, Independence and Clique Numbers, Perfect Graphs—Definition and examples, Chromatic polynomial and its determination, Applications of Graph Colouring.

Matchings: Definitions and Examples of Perfect Matching, Maximal and Maximum Matching, Hall's Marriage Theorem (Statement only) and related problems. [6L]

Contents

Module II

Module III

Module III

Module IV

Module IV

Module I

CHAPTER 1

Introduction to Propositional Calculus

1.1 INTRODUCTION

A number of words making a complete grammatical structure having a sense and meaning is called a **sentence.** Sentences are usually classified as declarative, exclamatory, interrogative or imperative.

Proposition: A **proposition** or **statement** is a declarative sentence which is either true or false but not both. The truth or falsity of a proposition is called its **truth-value.**

These two values 'true' and 'false' are denoted by the symbols T and F respectively. Sometimes these are also denoted by the symbols 1 and 0 respectively.

Example 1: Consider the following sentences:

1. Delhi is the capital of India.
2. Kolkata is a country.
3. 5 is a prime number.
4. $2 + 3 = 4$.

These are propositions (or statements) because they are either true of false.

Next consider the following sentences:

5. How beautiful are you?
6. Wish you a happy new year
7. $x + y = z$
8. Take one book.

These are not propositions as they are not declarative in nature, that is, they do not declare a definite truth value T or F.

Propositional Calculus is also known as **statement calculus.** It is the branch of mathematics that is used to describe a logical system or structure. A logical system consists of (1) a universe of propositions, (2) truth tables (as axioms) for the logical operators and (3) definitions that explain equivalence and implication of propositions.

1.2 FIVE BASIC CONNECTIVES

The words or phrases or symbols which are used to make a proposition by two or more propositions are called **logical connectives** or **simply connectives.** There are five basic connectives called negation, conjunction, disjunction, conditional and biconditional.

Negation

The **negation** of a statement is generally formed by writing the word 'not' at a proper place in the statement (proposition) or by prefixing the statement with the phrase 'It is not the case that'. If p denotes a statement then the negation of p is written as $\urcorner p$ and read as 'not p'. If the truth value of p is T then the truth value of $\urcorner p$ is F. Also if the truth value of p is F then the truth value of $\urcorner p$ is T.

Table 1. Truth table for negation

p	$\urcorner p$
T	F
F	T

Example 2: Consider the statement p : Kolkata is a city. Then $\urcorner p$: Kolkata is not a city. or It is not the case that Kolkata is a city.

Although the two statements 'Kolkata is not a city' and 'It is not the case that Kolkata is a city' are not identical, we have translated both of them by $\urcorner p$. The reason is that both these statements have the same meaning.

Notes: (*i*) A given statement (proposition) in the object language is denoted by a symbol and it may correspond to several statements in English. This multiplicity happens since in a natural language one can express oneself in several manners.

(*ii*) $\urcorner p$ is also denoted as $\sim p$, $\bar{p}$ and p'.

Conjunction

The **conjunction** of two statements (or propositions) p and q is the statement $p \wedge q$ which is read as 'p and q'. The statement $p \wedge q$ has the truth value T whenever both p and q have the truth value T. Otherwise it has truth value F.

Table 2. Truth table for conjunction

p	q	$p \wedge q$
T	T	T
T	F	F
F	T	F
F	F	F

Example 3: Consider the following statements

p : It is raining today.

q : There are 10 chairs in the room.

Then $p \wedge q$: It is raining today and there are 10 chairs in the room.

Note: Usually, in our everyday language the conjunction 'and' is used between two statements which have some kind of relation. Thus a statement 'It is raining today and 1 + 1 = 2' sounds odd, but in logic it is a perfectly acceptable statement formed from the statements 'It is raining today' and '1 + 1 = 2'.

Example 4: *Translate the following statement:*

'Jack and Jill went up the hill' into symbolic form using conjunction.

Solution: Let p : Jack went up the hill,

q : Jill went up the hill.

Then the given statement can be written in symbolic form as $p \wedge q$.

Disjunction

The **disjunction** of two statements p and q is the statement $p \vee q$ which is read as 'p or q'. The statement $p \vee q$ has the truth value F only when both p and q have the truth value F. Otherwise it has truth value T.

Table 3. Truth table for disjunction

p	q	$p \vee q$
T	T	T
T	F	T
F	T	T
F	F	F

Example 5: Consider the following statements

p : I shall go to the game.

q : I shall watch the game on television.

Then $p \vee q$: I shall go to the game or watch the game on television.

Conditional proposition

If p and q are any two statements (or propositions) then the statement $p \rightarrow q$ which is read as, 'If p, then q' is called a **conditional statement** (or **proposition**) or **implication** and the connective is the **conditional connective.**

The conditional is defined by the following table:

Table 4. Truth table for conditional

p	q	$p \rightarrow q$
T	T	T
T	F	F
F	T	T
F	F	T

In this conditional statement, p is called the **hypothesis** or **premise** or **antecedent** and q is called the **consequence** or **conclusion.**

To understand better, this connective can be looked as a conditional promise. If the promise is violated (broken), the conditional (implication) is false. Otherwise it is true. For this reason, the only circumstances under which the conditional $p \rightarrow q$ is false is when p is true and q is false.

Example 6: *Translate the following statement:*

'The crop will be destroyed if there is a flood' into symbolic form using conditional connective.

Solution: Let c : the crop will be destroyed; f : there is a flood.

Let us rewrite the given statement as

'If there is a flood, then the crop will be destroyed'. So, the symbolic form of the given statement is $f \to c$.

Example 7: *Let p and q denote the statements:*

p : You drive over 70 kms per hour.

q : You get a speeding ticket.

Write the following statements into symbolic forms.

(i) You will get a speeding ticket if you drive over 70 kms per hour.

(ii) Driving over 70 kms per hour is sufficient for getting a speeding ticket.

(iii) If you do not drive over 70 kms per hour then you will not get a speeding ticket.

(iv) Whenever you get a speeding ticket, you drive over 70 kms per hour.

Solution: (i) $p \to q$ (ii) $p \to q$ (iii) $\neg p \to \neg q$ (iv) $q \to p$.

Notes: 1. In ordinary language, it is customary to assume some kind of relationship between the antecedent and the consequent in using the conditional. But in logic, the antecedent and the consequent in a conditional statement are not required to refer to the same subject matter. For example, the statement 'If I get sufficient money then I shall purchase a high-speed computer' sounds reasonable. On the other hand, a statement such as 'If I purchase a computer then this pen is red' does not make sense in our conventional language. But according to the definition of conditional, this proposition is perfectly acceptable and has a truth-value which depends on the truth-values of the component statements.

2. Some of the alternative terminologies used to express $p \to q$ (if p, then q) are the following:

(i) p implies q

(ii) p only if q ('If p, then q' formulation emphasizes the antecedent, whereas 'p only if q' formulation emphasizes the consequent. The difference is only stylistic.)

(iii) q if p, or q when p.

(iv) q follows from p, or q whenever p.

(v) p is sufficient for q, or a sufficient condition for q is p.

(vi) q is necessary for p, or a necessary condition for p is q.

(vii) q is consequence of p.

Converse proposition

The conditional proposition $q \to p$ is called the **converse proposition** of $p \to q$.

Contrapositive proposition

The conditional proposition $\neg q \to \neg p$ is called the **contrapositive proposition** of $p \to q$.

Inverse proposition

The conditional preposition $\neg p \to \neg q$ is called the **inverse proposition** of $p \to q$.

Table 5. Truth table for four propositions

				Conditional	*Converse*	*Contrapositive*	*Inverse*
p	q	$\neg p$	$\neg q$	$p \to q$	$q \to p$	$\neg q \to \neg p$	$\neg p \to \neg q$
T	*T*	*F*	*F*	*T*	*T*	*T*	*T*
T	*F*	*F*	*T*	*F*	*T*	*F*	*T*
F	*T*	*T*	*F*	*T*	*F*	*T*	*F*
F	*F*	*T*	*T*	*T*	*T*	*T*	*T*

Example 8: Consider the following statements.

p : It rains; q : The crops will be grown.

Conditional

$p \to q$: If it rains then the crops will be grown.

Converse

$q \to p$: If the crops grow then there has been rain.

Contrapositive

$\neg q \to \neg p$: If the crops do not grow then there has been no rain.

Inverse

$\neg p \to \neg q$: If there is no rain then the crops will not grow.

Example 9: *Construct the truth table for* $(p \to q) \wedge (q \to p)$.

Solution:

p	q	$p \to q$	$q \to p$	$(p \to q) \wedge (q \to p)$
T	T	T	T	T
T	F	F	T	F
F	T	T	F	F
F	F	T	T	T

Biconditional proposition

If p and q are any two statements (propositions), then the statement $p \rightleftarrows q$ which is read as 'p if and only if q' and abbreviated as 'p iff q' is called a **biconditional statement** and the connective is the **biconditional connective.**

The truth table of $p \rightleftarrows q$ is given by the following table:

Table 6. Truth table for biconditional

p	q	$p \rightleftarrows q$
T	T	T
T	F	F
F	T	F
F	F	T

It may be noted that $p \rightleftarrows q$ is true only when both p and q are true or when both p and q are false. Observe that $p \rightleftarrows q$ is true when both the conditionals $p \to q$ and $q \to p$ are true, *i.e.*, the truth-values of $(p \to q) \wedge (q \to p)$, given in Ex. 9, are identical to the truth-values of $p \rightleftarrows q$ defined here.

$p \rightleftarrows q$ can also be stated as p is a necessary and sufficient condition for q.

Note: The notation $p \leftrightarrow q$ is also used instead of $p \rightleftarrows q$.

Example 10: *Let p and q denote the following propositions:*

p : He is poor; q : He is happy.

Express each of the following propositions in symbolic form using p and q.

(*i*) *To be poor is to be unhappy.*

(*ii*) *He is poor if and only if he is happy.*

(*iii*) *Being rich is a necessary and sufficient condition to being happy.*

Solution:

(*i*) $p \rightleftarrows \urcorner q$ (*ii*) $p \not\rightleftarrows q$

(*iii*) $\urcorner p \rightleftarrows q$.

1.3 STATEMENT FORMULAS AND TRUTH TABLES

We have already defined the connectives $\urcorner, \wedge, \vee, \rightarrow$ and $\rightleftarrows$. We shall occasionally distinguish between two kinds of propositions (or statements) in our symbolic language. The statements which do not contain any connectives are called **atomic** or **primary** or **simple statements.** On the other hand, those statements which contain one or more primary statements and some connectives are known as **molecular** or **composite** or **compound statements.** Some of the compound statements formed by using any two statements p and q are

$$\urcorner p, p \vee q, (p \wedge q) \vee (\urcorner p), (p \vee q) \rightleftarrows p \wedge (\urcorner q).$$

These compound statements are statement formulas derived from the statement variables p and q. Therefore, p and q may be called the components of these statement formulas. It is to be noted that in addition to the connectives we have also used parentheses in some cases in order to make the formula unambiguous.

Our basic concern is to find the truth-value of a statement formula for each possible combination of the truth-values of the component statements. A table showing all such truth values is called the **truth table** of the given statement formula.

Example 11: *Construct the truth table for the statement formula* $p \vee \urcorner q$.

Solution:

p	q	$\urcorner q$	$p \vee \urcorner q$
T	T	F	T
T	F	T	T
F	T	F	F
F	F	T	T

Example 12: *Construct the truth table for* $(p \vee q) \vee \urcorner p$.

Solution:

p	q	$p \vee q$	$\urcorner p$	$(p \vee q) \vee \urcorner p$
T	T	T	F	T
T	F	T	F	T
F	T	T	T	T
F	F	F	T	T

1.4 TAUTOLOGY AND CONTRADICTION

Tautology: A statement formula which is true regardless of the truth values of the statements which replace the variables in it is called a **universally valid formula** or a **logical truth** or a **tautology.**

Contradiction: A statement formula which is false regardless of the truth values of the statements which replace the variables in it is said to be a **contradiction.**

Contingency: A statement formula which is neither a tautology nor a contradiction is known as a **contingency.**

Example 13: *Show that* $p \vee \neg p$ *is a tautology and* $p \wedge \neg p$ *is a contradiction.*

Solution:

p	$\neg p$	$p \vee \neg p$	$p \wedge \neg p$
T	F	T	F
F	T	T	F

From the truth table it is clear that $p \vee \neg p$ is a tautology and $p \wedge \neg p$ is a contradiction.

Example 14: *Using truth table, show that the statement formula* $\neg (p \wedge q) \rightleftarrows (\neg p \vee \neg q)$ *is a tautology.*

Solution:

p	q	$p \wedge q$	$\neg (p \wedge q)$	$\neg p$	$\neg q$	$\neg p \vee \neg q$	$\neg (p \wedge q) \rightleftarrows (\neg p \vee \neg q)$
T	T	T	F	F	F	F	T
T	F	F	T	F	T	T	T
F	T	F	T	T	F	T	T
F	F	F	T	T	T	T	T

Here the truth values of the given formula are T for all possible truth values of p and q. Hence $\neg (p \wedge q) \rightleftarrows (\neg p \vee \neg q)$ is a tautology.

Notes: (*i*) Obviously, the negation of a contradiction is a tautology and conversely.

(*ii*) A statement formula which is a tautology is **identically true** and a formula which is a contradiction is **identically false.**

(*iii*) Suppose $P(p, q, ...)$ is a tautology. Then it does not depend upon the particular truth values of its variables $p, q, ...$, so we can substitute a for p, b for q, ..., for any propositions $a, b, ...$ in the tautology $P(p, q, ...)$ and still have a tautology $P(a, b, ...)$. It is also true for contradiction.

1.5 EQUIVALENCE OF FORMULAS

Two statement formulas A and B are **equivalent** provided $A \rightleftarrows B$ is a tautology. Conversely, if $A \rightleftarrows B$ is a tautology then A and B are **equivalent.** We shall represent the equivalence of two statement formulas, say A and B, by writing '$A \Leftrightarrow B$' which is read as 'A is equivalent to B'.

Example 15: *Prove that* (*i*) $(p \vee \neg p) \Leftrightarrow (q \vee \neg q)$ (*ii*) $(p \to q) \Leftrightarrow (\neg p \vee q)$.

Solution: (*i*)

p	$\neg p$	$p \vee \neg p$	q	$\neg q$	$q \vee \neg q$	$(p \vee \neg p) \rightleftarrows (q \vee \neg q)$
T	F	T	T	F	T	T
T	F	T	T	F	T	T
F	T	T	F	T	T	T
F	T	T	F	T	T	T

$\therefore \quad (p \vee \neg p) \Leftrightarrow (q \vee \neg q)$ since $(p \vee \neg p) \rightleftarrows (q \vee \neg q)$ is a tautology.

(*ii*)

p	q	$p \to q$	$\urcorner p$	$\urcorner p \vee q$	$(p \to q) \rightleftarrows (\urcorner p \vee q)$
T	T	T	F	T	T
T	F	F	F	F	T
F	T	T	T	T	T
F	F	T	T	T	T

Observe that the truth values in the columns for $p \to q$ and $\urcorner p \vee q$ are identical and so the biconditional will have the truth values T.

$$\therefore \qquad (p \to q) \Leftrightarrow (\urcorner p \vee q).$$

1.6 EQUIVALENT FORMULAS OR LOGICAL EQUIVALENCES

1. (*i*) $p \vee p \Leftrightarrow p$ (*ii*) $p \wedge p \Leftrightarrow p$ (Idempotent laws)
2. (*i*) $p \vee q \Leftrightarrow q \vee p$ (*ii*) $p \wedge q \Leftrightarrow q \wedge p$ (Commutative laws)
3. (*i*) $p \vee (q \vee r) \Leftrightarrow (p \vee q) \vee r$ (*ii*) $p \wedge (q \wedge r) \Leftrightarrow (p \wedge q) \wedge r$ (Associative laws)
4. (*i*) $p \vee (q \wedge r) \Leftrightarrow (p \vee q) \wedge (p \vee r)$
   (*ii*) $p \wedge (q \vee r) \Leftrightarrow (p \wedge q) \vee (p \wedge r)$ (Distributive laws)
5. $p \to q \Leftrightarrow \urcorner p \vee q$ 6. $p \to q \Leftrightarrow \urcorner q \to \urcorner p$ (Law of contrapositive)
7. (*i*) $p \vee F \Leftrightarrow p$ (*ii*) $p \wedge T \Leftrightarrow p$
8. (*i*) $p \vee T \Leftrightarrow T$ (*ii*) $p \wedge F \Leftrightarrow F$
9. (*i*) $p \vee \urcorner p \Leftrightarrow T$ (*ii*) $p \wedge \urcorner p \Leftrightarrow F$
10. (*i*) $p \vee (p \wedge q) \Leftrightarrow p$ (*ii*) $p \wedge (p \vee q) \Leftrightarrow p$ (Absorption laws)
11. (*i*) $\urcorner (p \vee q) \Leftrightarrow \urcorner p \wedge \urcorner q$ (*ii*) $\urcorner (p \wedge q) \Leftrightarrow \urcorner p \vee \urcorner q$ (De Morgan's laws)
12. $p \rightleftarrows q \Leftrightarrow (p \to q) \wedge (q \to p)$

Note: Here most of the equivalences have straight forward intuitive interpretations and all of them can be verified by constructing truth tables.

Example 16: *Show that*

$$(\urcorner p \wedge (\urcorner q \wedge r)) \vee (q \wedge r) \vee (p \wedge r) \Leftrightarrow r.$$

Solution: $(\urcorner p \wedge (\urcorner q \wedge r)) \vee (q \wedge r) \vee (p \wedge r)$

$\Leftrightarrow$ $(\urcorner p \wedge (\urcorner q \wedge r)) \vee ((q \vee p) \wedge r)$ [By distributive law]

$\Leftrightarrow$ $((\urcorner p \wedge \urcorner q) \wedge r) \vee ((q \vee p) \wedge r)$ [By associative law]

$\Leftrightarrow$ $((\urcorner p \wedge \urcorner q) \vee (q \vee p)) \wedge r$ [By distributive law]

$\Leftrightarrow$ $(\urcorner (p \vee q) \vee (p \vee q)) \wedge r$ [By De Morgan's and commutative laws]

$\Leftrightarrow$ $T \wedge r \Leftrightarrow r.$ **(Proved)**

Example 17: *Prove that*

$$((p \vee q) \wedge \urcorner (\urcorner p \wedge (\urcorner q \vee \urcorner r))) \vee (\urcorner p \wedge \urcorner q) \vee (\urcorner p \wedge \urcorner r) \text{ is a tautology.}$$

Solution: Using De Morgan's laws:

$$\urcorner p \wedge \urcorner q \Leftrightarrow \urcorner (p \vee q), \urcorner p \wedge \urcorner r \Leftrightarrow \urcorner (p \vee r)$$

$$\therefore \qquad (\urcorner p \wedge \urcorner q) \vee (\urcorner p \wedge \urcorner r) \Leftrightarrow \urcorner (p \vee q) \vee \urcorner (p \vee r)$$

$\Leftrightarrow \quad \neg((p \vee q) \wedge (p \vee r))$

Also, $\neg(\neg p \wedge (\neg q \vee \neg r)) \Leftrightarrow \neg(\neg p \wedge \neg(q \wedge r))$

$\Leftrightarrow \quad \neg(\neg(p \vee (q \wedge r)))$

$\Leftrightarrow \quad p \vee (q \wedge r)$

$\Leftrightarrow \quad (p \vee q) \wedge (p \vee r)$ [By distributive law]

Therefore, the given expression becomes

$((p \vee q) \wedge (p \vee q) \wedge (p \vee r)) \vee \neg((p \vee q) \wedge (p \vee r))$

$\Leftrightarrow \quad ((p \vee q) \wedge (p \vee r)) \vee \neg((p \vee q) \wedge (p \vee r)) \Leftrightarrow T$

Hence the given statement formula is a tautology.

Example 18: *Show that* $r \rightleftarrows s \Leftrightarrow (r \wedge s) \vee (\neg r \wedge \neg s)$.

Solution: $r \rightleftarrows s \Leftrightarrow (r \to s) \wedge (s \to r)$.

$\Leftrightarrow \quad (\neg r \vee s) \wedge (\neg s \vee r)$ [$\because$ $(p \to q) \Leftrightarrow (\neg p \vee q)$, see Ex. 15(*ii*)]

$\Leftrightarrow \quad ((\neg r \vee s) \wedge \neg s) \vee ((\neg r \vee s) \wedge r)$ [By distributive law]

$\Leftrightarrow \quad ((\neg r \wedge \neg s) \vee (s \wedge \neg s)) \vee ((\neg r \wedge r) \vee (s \wedge r))$ [By distributive law]

$\Leftrightarrow \quad ((\neg r \wedge \neg s) \vee F) \vee (F \vee (s \wedge r))$ [$\because$ $p \wedge \neg p \Leftrightarrow F$ and By commutativity]

$\Leftrightarrow \quad (\neg r \wedge \neg s) \vee (s \wedge r)$ [$\because$ $p \vee F \Leftrightarrow p$ and By commutativity]

$\Leftrightarrow \quad (r \wedge s) \vee (\neg r \wedge \neg s)$ [By commutativity]

Example 19: *Prove that* $\neg(p \wedge q) \to (\neg p \vee (\neg p \vee q)) \Leftrightarrow \neg p \vee q$.

Solution: $\neg(p \wedge q) \to (\neg p \vee (\neg p \vee q)) \Leftrightarrow (\neg(\neg(p \wedge q))) \vee (\neg p \vee (\neg p \vee q))$

[$\because$ $(p \to q) \Leftrightarrow (\neg p \vee q)$, see Ex. 15(*ii*)]

$\Leftrightarrow \quad (p \wedge q) \vee ((\neg p \vee \neg p) \vee q)$

$\Leftrightarrow \quad (p \wedge q) \vee (\neg p \vee q)$ [$\because$ $p \vee p \Leftrightarrow p$]

$\Leftrightarrow \quad ((p \wedge q) \vee \neg p) \vee q$ [By associativity]

$\Leftrightarrow \quad ((p \vee \neg p) \wedge (q \vee \neg p)) \vee q$ (By distributive law)

$\Leftrightarrow \quad (T \wedge (q \vee \neg p)) \vee q$

$\Leftrightarrow \quad (q \vee \neg p) \vee q$

$\Leftrightarrow \quad \neg p \vee (q \vee q)$ (By commutative and associative laws)

$\Leftrightarrow \quad \neg p \vee q$. **(Proved)**

Example 20: *Prove that* $(p \wedge (p \rightleftarrows q)) \to q$ *is a tautology.*

Solution: $p \wedge (p \rightleftarrows q) \Leftrightarrow p \wedge (p \to q) \wedge (q \to p)$

$\Leftrightarrow \quad p \wedge (\neg p \vee q) \wedge (\neg q \vee p)$

$\Leftrightarrow \quad (p \wedge (p \vee \neg q)) \wedge (\neg p \vee q)$ [By commutativity and associativity]

$\Leftrightarrow \quad p \wedge (\neg p \vee q)$ [By absorption law]

$\Leftrightarrow \quad (p \wedge \neg p) \vee (p \wedge q)$ [By distributive law]

$\Leftrightarrow \quad F \vee (p \wedge q) \Leftrightarrow p \wedge q$.

$\therefore \quad (p \wedge (p \rightleftarrows q)) \to q \Leftrightarrow (p \wedge q) \to q$

$\Leftrightarrow \quad \neg(p \wedge q) \vee q$ [$\because$ $(a \to b) \Leftrightarrow (\neg a \vee b)$]

$\Leftrightarrow \quad (\neg p \vee \neg q) \vee q$ [By De Morgan's law]

$\Leftrightarrow \quad \neg p \vee (\neg q \vee q)$ [By associativity]

$\Leftrightarrow \quad \neg p \vee T \Leftrightarrow T$.

Hence $(p \wedge (p \rightleftarrows q)) \to q$ is a tautology. **(Proved)**

1.7 NORMAL FORMS : DNF AND CNF

It will be convenient to use the word **product** in place of **conjunction** ($\wedge$) and **sum** in place of **disjunction** ($\vee$) in the current discussion.

Elementary product: A product of the variables and their negations in a statement formula is called an **elementary product.**

Elementary sum: A sum of the variables and their negations in a statement formula is called an **elementary sum.**

Disjunctive normal form (DNF): A statement formula which is equivalent to a given formula and which consists of a sum of elementary products is called a **disjunctive normal form (DNF)** of the given formula.

Example 21: *Obtain disjunctive normal forms of*

(i) $p \wedge (p \rightarrow q)$ *(ii)* $\neg (p \vee q) \rightleftarrows (p \wedge q)$.

Solution:

(i) $p \wedge (p \rightarrow q) \Leftrightarrow p \wedge (\neg p \vee q)$ [See Ex. 15*(ii)*]

$\Leftrightarrow (p \wedge \neg p) \vee (p \wedge q)$ [By distributive law]

This is the required DNF.

(ii) $\neg (p \vee q) \rightleftarrows (p \wedge q) \Leftrightarrow (\neg (p \vee q) \wedge (p \wedge q)) \vee ((p \vee q) \wedge \neg (p \wedge q))$

[Using $r \rightleftarrows s \Leftrightarrow (r \wedge s) \vee (\neg r \wedge \neg s)$, see Ex.18]

$\Leftrightarrow (\neg p \wedge \neg q \wedge p \wedge q) \vee ((p \vee q) \wedge (\neg p \vee \neg q))$ (By De Morgan's laws)

$\Leftrightarrow (\neg p \wedge \neg q \wedge p \wedge q) \vee ((p \vee q) \wedge \neg p) \vee ((p \vee q) \wedge \neg q)$ (By distributive law)

$\therefore \neg (p \vee q) \rightleftarrows (p \wedge q)$

$\Leftrightarrow (\neg p \wedge \neg q \wedge p \wedge q) \vee (p \wedge \neg p) \vee (q \wedge \neg p) \vee (p \wedge \neg q) \vee (q \wedge \neg q)$

(By distributive law)

This is the required DNF.

Example 22: *Obtain a disjunctive normal form of*

$$p \rightarrow ((p \rightarrow q) \wedge \neg (\neg q \vee \neg p)).$$

Solution: $p \rightarrow ((p \rightarrow q) \wedge \neg (\neg q \vee \neg p))$

$\Leftrightarrow \neg p \vee ((p \rightarrow q) \wedge \neg (\neg q \vee \neg p))$ [$\because$ $r \rightarrow s \Leftrightarrow \neg r \vee s$]

$\Leftrightarrow \neg p \vee ((\neg p \vee q) \wedge \neg (\neg (q \wedge p)))$ [By De Morgan's law]

$\Leftrightarrow \neg p \vee ((\neg p \vee q) \wedge (q \wedge p))$

$\Leftrightarrow \neg p \vee ((\neg p \wedge (q \wedge p) \vee (q \wedge (q \wedge p))$ [By distributive law]

$\Leftrightarrow \neg p \vee (p \wedge \neg p \wedge q) \vee (q \wedge q \wedge p)$ [By commutative and associative laws]

$\Leftrightarrow \neg p \vee (F \wedge q) \vee (q \wedge p)$

$\Leftrightarrow \neg p \vee F \vee (q \wedge p) \Leftrightarrow \neg p \vee (q \wedge p)$

This is the required DNF.

Principal disjunctive normal form (PDNF):

Minterm: A product (or conjunction) in which each variable or its negation, but not both, occurs only once is called a **minterm.**

For two variables p and q, there are $2^2 = 4$ possible minterms : $p \wedge q$, $p \wedge \neg q$, $\neg p \wedge q$ and $\neg p \wedge \neg q$.

For three variables p, q and r, there are $2^3 = 8$ possible minterms : $p \wedge q \wedge r$, $\neg p \wedge q \wedge r$, $p \wedge \neg q \wedge r$, $p \wedge q \wedge \neg r$, $\neg p \wedge \neg q \wedge r$, $p \wedge \neg q \wedge \neg r$, $\neg p \wedge q \wedge \neg r$ and $\neg p \wedge \neg q \wedge \neg r$.

Note that there are 2^n minterms for n variables.

PDNF: A statement formula which is equivalent to a given formula and which consists of a sum (or disjunction) of minterms is called a **principal disjunctive normal form (PDNF)** of the given formula.

Note: If the PDNF of a formula A is known, the PDNF of $\neg A$ will consist of the sums (or disjunctions) of the remaining minterms which are not included in the PDNF of A.

Example 23: *Find PDNF of (i) $p \wedge (p \to q)$ (ii) $p \vee \neg q$.*

Solution: (*i*) $p \wedge (p \to q) \Leftrightarrow (p \wedge \neg p) \vee (p \wedge q)$ [See Example 21(*i*)]

$\Leftrightarrow$ $F \vee (p \wedge q) \Leftrightarrow p \wedge q$, this is the required *PDNF*.

(*ii*) $p \vee \neg q \Leftrightarrow p \wedge (q \vee \neg q) \vee \neg q \wedge (p \vee \neg p)$ [$\because$ $a \vee \neg a \Leftrightarrow T; a \wedge T \Leftrightarrow a$]

$\Leftrightarrow$ $(p \wedge q) \vee (p \wedge \neg q) \vee (\neg q \wedge p) \vee (\neg q \wedge \neg p)$ [By distributive law]

$\Leftrightarrow$ $(p \wedge q) \vee ((p \wedge \neg q) \vee (p \wedge \neg q)) \vee (\neg p \wedge \neg q)$ [By commutativity]

$\Leftrightarrow$ $(p \wedge q) \vee (p \wedge \neg q) \vee (\neg p \wedge \neg q)$ [By idempotent law]

This is the required PDNF.

Conjunctive normal form (CNF): A statement formula which is equivalent to a given formula and which consists of a product of elementary sums is called a **conjunctive normal form (CNF) of the given formula.**

Example 24: *Obtain conjunctive normal forms of (i) $p \wedge (p \to q)$ (ii) $q \vee (p \wedge \neg q) \vee (\neg p \wedge \neg q)$.*

Solution: (*i*) $p \wedge (p \to q) \Leftrightarrow p \wedge (\neg p \vee q)$ [see Ex. 15 (*ii*)]

Hence $p \wedge (\neg p \vee q)$ is the required CNF.

(*ii*) $q \vee (p \wedge \neg q) \vee (\neg p \wedge \neg q)$

$\Leftrightarrow$ $q \vee ((q \vee \neg p) \wedge \neg q)$ [By distributive law]

$\Leftrightarrow$ $(q \vee (p \vee \neg p)) \wedge (q \vee \neg q)$ [By distributive law]

$\Leftrightarrow$ $(q \vee p \vee \neg p) \wedge (q \vee \neg q)$ [By associativity] ...(1)

This is the required *CNF*.

Note: From (1): $q \vee (p \wedge \neg q) \vee (\neg p \wedge \neg q)$

$\Leftrightarrow$ $(q \vee T) \wedge T$ [$\because$ $r \vee \neg r \Leftrightarrow T$]

$\Leftrightarrow$ $T \wedge T \Leftrightarrow T$.

Therefore, $q \vee (p \wedge \neg q) \vee (\neg p \wedge \neg q)$ is a tautology.

Principal Conjunctive Normal Form (PCNF):

Maxterm: A sum (or disjunction) in which each variable or its negation, but not both, occurs only once is called a **maxterm.**

For two variables p and q, there are $2^2 = 4$ possible maxterms : $p \vee q$, $p \vee \neg q$, $\neg p \vee q$ and $\neg p \vee \neg q$.

For three variables p, q and r, there are $2^3 = 8$ possible maxterms : $p \vee q \vee r$, $\neg p \vee q \vee r$, $p \vee \neg q \vee r$, $p \vee q \vee \neg r$, $\neg p \vee \neg q \vee r$, $p \vee \neg q \vee \neg r$, $\neg p \vee q \vee \neg r$ and $\neg p \vee \neg q \vee \neg r$.

Note that there are 2^n maxterms for n variables.

PCNF: A statement formula which is equivalent to a given formula and which consists of a product (or conjunction) of maxterms is called a **principal conjunctive normal form (PCNF)** of the given formula.

Note: To obtain the PCNF of a statement formula A, we can use the fact that $A = \neg(\neg A)$ and apply De Morgan's laws to the PDNF of $\neg A$ repeatedly.

Example 25: *Obtain PCNF of (i) $p \rightleftarrows q$ (ii) $(p \wedge q) \vee (\neg p \wedge q) \vee (p \wedge \neg q)$.*

Solution: (i) $p \rightleftarrows q \Leftrightarrow (p \to q) \wedge (q \to p) \Leftrightarrow (\neg p \vee q) \wedge (\neg q \vee p)$

[$\because$ $(a \to b) \Leftrightarrow (\neg a \vee b)$]

This is the required *PCNF*.

(ii) Let $A \Leftrightarrow (p \wedge q) \vee (\neg p \wedge q) \vee (p \wedge \neg q)$

$\therefore$ $\neg A \Leftrightarrow (\neg p \wedge \neg q)$ [Remaining minterm]

So, $A \Leftrightarrow \neg(\neg A) \Leftrightarrow \neg(\neg p \wedge \neg q)$

$\Leftrightarrow$ $\neg(\neg p) \vee \neg(\neg q)$ [By De Morgan's law]

$\Leftrightarrow$ $p \vee q$

This is the required PCNF.

Determination of PDNF and PCNF of a statement formula using truth table

Working Rules:

1. If the given statement formula is not a contradiction then the disjunction (sum) of the minterms corresponding to the rows of the truth table having truth value T is the required PDNF, because it is equivalent to the given statement formula.

In the second row of the following example, the truth value T of the given statement corresponds to the truth values T, T, F for the variables p, q, r respectively. So, the corresponding minterm is $(p \wedge q \wedge \neg r)$.

2. If the given statement formula is not a tautology then the conjunction (product) of the maxterms corresponding to the rows of the truth table having truth value F is the required PCNF, because it is equivalent to the given statement formula.

In the third row of the following example, the truth value F of the given statement corresponds to the truth values T, F, T for the variables p, q, r respectively. So, the corresponding maxterm is $(\neg p \vee q \vee \neg r)$

Example 26: *Find the PDNF and PCNF of $(\neg p \to r) \wedge (p \rightleftarrows q)$ using truth table.*

Solution:

Row No.	p	q	r	$\neg p$	$\neg p \to r$	$p \rightleftarrows q$	$(\neg p \to r) \wedge (p \rightleftarrows q)$
1.	T	T	T	F	T	T	T
2.	T	T	F	F	T	T	T
3.	T	F	T	F	T	F	F
4.	T	F	F	F	T	F	F
5.	F	T	T	T	T	F	F
6.	F	T	F	T	F	F	F
7.	F	F	T	T	T	T	T
8.	F	F	F	T	F	T	F

The minterms corresponding to the three T values of the last column are given below:

Row 1 : $p \wedge q \wedge r$. Row 2: $p \wedge q \wedge \neg r$. Row 7: $\neg p \wedge \neg q \wedge r$

$\therefore$ $(\neg p \to r) \wedge (p \rightleftarrows q) \Leftrightarrow (p \wedge q \wedge r) \vee (p \wedge q \wedge \neg r) \vee (\neg p \wedge \neg q \wedge r)$

This is the required PDNF.

The maxterms corresponding to the five F values of the last column are given below:

Row 3: $\neg p \vee q \vee \neg r$. Row 4: $\neg p \vee q \vee r$. Row 5. $p \vee \neg q \vee \neg r$.

Row 6: $p \vee \neg q \vee r$. Row 8 : $p \vee q \vee r$.

$$\therefore \quad (\neg p \to r) \wedge (p \rightleftarrows q) \Leftrightarrow (\neg p \vee q \vee \neg r) \wedge (\neg p \vee q \vee r) \wedge (p \vee \neg q \vee \neg r) \wedge (p \vee \neg q \vee r) \wedge (p \vee q \vee r).$$

This is the required PCNF.

Notes: (*i*) PDNF (or PCNF) of a given statement formula is also a DNF (or CNF) of that statement formula. But the converse is not true in general.

(*ii*) In logic theory, it is very often required to find whether a given statement is a tautology or a contradiction. Let $P(p_1, p_2, ..., p_n)$ be a statement formula where $p_1, p_2, ..., p_n$ are the atomic variables. If we consider all possible assignments of truth values of $p_1, p_2, ..., p_n$ and obtain resulting truth values of the formula P, then we get truth table for P containing 2^n rows. If P has the truth value T for all possible assignments then P is said to be a tautology. If P has the truth value F for all possible assignments then P is said to be a contradiction. If P has the truth value T for at least one combination of truth values of $p_1, p_2, ..., p_n$ then P is said to be **satisfiable.**

The problem of finding whether a given statement formula is tautology or contradiction or satisfiable in a finite number of steps is known as decision problem. For decision problems, the construction of truth tables may not be practical always and therefore we consider alternate procedure known as reduction to normal forms.

1.8 TAUTOLOGICAL IMPLICATION

A compound statement (or proposition) $P(p_1, p_2, ..., p_n)$ is said to **tautologically imply** or **simply imply** the compound proposition $Q(p_1, p_2, ..., p_n)$ if Q is true whenever P is true or equivalently if and only if $P \to Q$ is a tautology. This is denoted by $P \Rightarrow Q$ and read as 'P implies Q'.

Note: $\Rightarrow$ is not a connective and $P \Rightarrow Q$ is not a statement (or proposition).

Example 27: *Prove that* $(p \to q) \Rightarrow (\neg q \to \neg p)$.

Solution:

p	q	$\neg p$	$\neg q$	$p \to q$	$\neg q \to \neg p$	$(p \to q) \to (\neg q \to \neg p)$
T	T	F	F	T	T	T
T	F	F	T	F	F	T
F	T	T	F	T	T	T
F	F	T	T	T	T	T

$\therefore \quad (p \to q) \Rightarrow (\neg q \to \neg p)$ since $(p \to q) \to (\neg q \to \neg p)$ is a tautology.

1.9 VALID ARGUMENTS

An **argument** is an assertion that a given set of propositions $p_1, p_2, ..., p_n$, called **premises** or **hypotheses,** yields (has a consequence) another proposition p, called the **conclusion.** Such an argument is denoted by $p_1, p_2, ..., p_n \vdash p$, or written as

$$\begin{array}{c} p_1 \\ p_2 \\ \vdots \\ p_n \\ \hline p \end{array}$$

Definition: An argument $p_1, p_2, ..., p_n \vdash p$ is said to be **valid** if p is true whenever all the premises $p_1, p_2, ... p_n$ are true. So, an argument $p_1, p_2, ..., p_n \vdash p$ is valid if and only if the proposition $(p_1 \wedge p_2 \wedge ... \wedge p_n) \rightarrow p$ is a tautology, *i.e.*, iff $p_1 \wedge p_2 \wedge ... \wedge p_n \Rightarrow p$.

Example 28: *Prove that*

$$\begin{array}{c} p \\ p \rightarrow q \\ \hline q \end{array}$$

is a valid argument. This rule of inference is known as ***modus ponens*** *or the* ***rule of detachment.***

Solution:

$$
\begin{array}{lll}
 & (p \wedge (p \rightarrow q)) \rightarrow q & \\
\Leftrightarrow & (p \wedge (\neg p \vee q)) \rightarrow q & \\
\Leftrightarrow & ((p \wedge \neg p) \vee (p \wedge q)) \rightarrow q & \text{(By distributive law)} \\
\Leftrightarrow & (F \vee (p \wedge q)) \rightarrow q & \\
\Leftrightarrow & (p \wedge q) \rightarrow q & \\
\Leftrightarrow & \neg (p \wedge q) \vee q & \\
\Leftrightarrow & (\neg p \vee \neg q) \vee q & \text{(By De Morgan's law)} \\
\Leftrightarrow & \neg p \vee (\neg q \vee q) & \text{(By associative law)} \\
\Leftrightarrow & \neg p \vee T \Leftrightarrow T. &
\end{array}
$$

Hence the given argument is valid. **(Proved)**

Alternative method (using truth table)

p	q	$p \rightarrow q$	$p \wedge (p \rightarrow q)$	$(p \wedge (p \rightarrow q)) \rightarrow q$
T	T	T	T	T
T	F	F	F	T
F	T	T	F	T
F	F	T	F	T

$\therefore \quad (p \wedge (p \rightarrow q)) \rightarrow q \Leftrightarrow T.$

Hence the given argument is valid.

Note: Modus ponens are Latin words meaning 'method of affirming'.

Example 29: *Represent the argument:*

You study hard

If you study hard, then you get A^+

You get A^+

symbolically and find whether the argument is valid.

Solution: Let p : You study hard; q : You get A^+. The given argument may be written symbolically as

$$\begin{array}{c} p \\ p \rightarrow q \\ \hline q \end{array}$$

Therefore, by modus ponens the given argument is valid.

Example 30: *Prove that*

$$\begin{array}{c} p \to q \\ q \to r \\ \hline p \to r \end{array}$$

is a valid argument. This rule of inference is known as the ***law of hypothetical syllogism*** *or the* ***transitive rule.***

Solution:

p	q	r	$p \to q$	$q \to r$	$p \to r$	$(p \to q) \wedge (q \to r) \Leftrightarrow a$	$a \to (p \to r)$
T	T	T	T	T	T	T	T
T	T	F	T	F	F	F	T
T	F	T	F	T	T	F	T
T	F	F	F	T	F	F	T
F	T	T	T	T	T	T	T
F	T	F	T	F	T	F	T
F	F	T	T	T	T	T	T
F	F	F	T	T	T	T	T

$\therefore \qquad ((p \to q) \wedge (q \to r)) \to (p \to r) \Leftrightarrow T.$

Hence the given argument is valid.

Example 31: *Represent the argument:*

If it rains today, then we will not meet today.
If we do not meet today, then we will meet tomorrow.
___
If it rains today, then we will meet tomorrow.

symbolically and find whether the argument is valid.

Solution: Let

p : It rains today.
q : We will not meet today.
r : We will meet tomorrow.

The given argument may be written symbolically as

$$\begin{array}{c} p \to q \\ q \to r \\ \hline p \to r \end{array}$$

Therefore, the given argument is valid by the rule of hypothetical syllogism.

Example 32: *Prove that*

$$\begin{array}{c} \neg p \\ p \vee q \\ \hline q \end{array}$$

is a valid argument. This rule of inference is known as the ***law of disjunctive syllogism.***

Solution:

$$(\neg p \wedge (p \vee q)) \rightarrow q$$

$\Leftrightarrow \quad \neg(\neg p \wedge (p \vee q)) \vee q$ [$\because$ $(a \rightarrow b) \Leftrightarrow (\neg a \vee b)$]

$\Leftrightarrow \quad p \vee \neg(p \vee q) \vee q$ [By De Morgan's law]

$\Leftrightarrow \quad p \vee (\neg p \wedge \neg q) \vee q$ [By De Morgan's law]

$\Leftrightarrow \quad ((p \vee \neg p) \wedge (p \vee \neg q)) \vee q$ [By distributive law]

$\Leftrightarrow \quad (T \wedge (p \vee \neg q)) \vee q$

$\Leftrightarrow \quad p \vee \neg q \vee q$

$\Leftrightarrow \quad p \vee (\neg q \vee q) \Leftrightarrow p \vee T \Leftrightarrow T.$

Hence the given argument is valid. **(Proved)**

Example 33: *Represent the argument:*

Mini is not an engineer.

Mini is an engineer or Soham is an engineer.

Soham is an engineer.

symbolically and find whether the argument is valid.

Solution: Let p : Mini is an engineer and

q : Soham is an engineer.

The given argument may be written symbolically as

$$\begin{array}{c} \neg p \\ \underline{p \vee q} \\ q \end{array}$$

Therefore, the given argument is valid by the rule of disjunctive syllogism.

Example 34: *Prove that*

$$\begin{array}{c} \neg q \\ \underline{p \rightarrow q} \\ \neg p \end{array}$$

is a valid argument. This rule of inference is called ***modus tollens.***

Solution:

$$(\neg q \wedge (p \rightarrow q)) \rightarrow \neg p$$

$\Leftrightarrow \quad (\neg q \wedge (\neg p \vee q)) \rightarrow \neg p$ [$\because$ $(a \rightarrow b) \Leftrightarrow (\neg a \vee b)$]

$\Leftrightarrow \quad ((\neg q \wedge \neg p) \vee (\neg q \wedge q)) \rightarrow \neg p$ [By distributive law]

$\Leftrightarrow \quad ((\neg q \wedge \neg p) \vee F) \rightarrow \neg p$

$\Leftrightarrow \quad (\neg q \wedge \neg p) \rightarrow \neg p$

$\Leftrightarrow \quad \neg(\neg q \wedge \neg p) \vee \neg p$

$\Leftrightarrow \quad (q \vee p) \vee \neg p$ [By De Morgan's law and since $\neg(\neg a) \Leftrightarrow a$]

$\Leftrightarrow \quad q \vee (p \vee \neg p)$ [By associative law]

$\Leftrightarrow \quad q \vee T \Leftrightarrow T.$

Hence the given argument is valid.

Notes: (*i*) Modus tollens are Latin words meaning 'method of denying'.

(*ii*) Modus tollens follows from modus ponens by using law of contrapositive.

Example 35: *Represent the argument:*

This integer is not divisible by 7.
If this integer is divisible by 21, then it is divisible by 7.
This integer is not divisible by 21.

symbolically and find whether the argument is valid.

Solution: Let p : This integer is divisible by 21 and q : This integer is divisible by 7.

The given argument may be written symbolically as

$$\begin{array}{c} \neg q \\ p \to q \\ \hline \neg p \end{array}$$

Therefore, by modus tollens the given argument is valid.

We list below some important valid inferences along with the inferences already discussed in the following Table 7.

Table 7. Rules of Inferences

Rule of Inference	*Tautology*	*Name*
$p \wedge q$ $\therefore p$	$p \wedge q \to p$	Simplification
p $\therefore p \vee q$	$p \to p \vee q$	Addition
p $p \to q$ $\therefore q$	$(p \wedge (p \to q)) \to q$	Modus ponens
$\neg q$ $p \to q$ $\therefore \neg p$	$(\neg q \wedge (p \to q)) \to \neg p$	Modus tollens
$p \to q$ $q \to r$ $\therefore p \to r$	$((p \to q) \wedge (q \to r)) \to (p \to r)$	Hypothetical syllogism, or transitive rule
$\neg p$ $p \vee q$ $\therefore q$	$(\neg p \wedge (p \vee q)) \to q$	Disjunctive syllogism
$p \vee q$ $p \to r$ $q \to r$ $\therefore r$	$((p \vee q) \wedge (p \to r) \wedge (q \to r)) \to r$	Dilemma

De Morgan's laws : (*i*) $\neg (p \vee q) \Leftrightarrow \neg p \wedge \neg q$ (*ii*) $\neg (p \wedge q) \Leftrightarrow \neg p \vee \neg q$.

Law of contrapositive : $(p \to q) \Leftrightarrow (\neg q \to \neg p)$.

Example 36: *Show that s is a valid conclusion from the premises* $p \to \neg q$, $q \vee r$, $\neg s \to p$, $\neg r$.

(W.B.U.T. 2012)

Solution: Let us construct an argument to prove that our hypotheses lead to the desired conclusion as follows:

Step	*Reason*
1. $p \to \neg q$	Hypothesis or Premise (given)
2. $q \vee r \Leftrightarrow \neg q \to r$	Hypothesis or Premise (given)
3. $\neg s \to p$	Hypothesis or Premise (given)
4. $\neg r$	Hypothesis or Premise (given)
5. $p \to r$	From steps 1 and 2 using transitive rule
6. $\neg s \to r$	From steps 3 and 5 using transitive rule
$\Leftrightarrow s \vee r \Leftrightarrow r \vee s$	
7. s	From steps 4 and 6 using disjunctive syllogism.

Thus s is a valid conclusion from the given hypotheses.

Example 37: *Show that t is a valid conclusion from the premises* $p \to q$, $q \to r$, $r \to s$, $\neg s$ *and* $p \vee t$.

(W.B.U.T. 2012)

Solution: Let us construct an argument to prove that our hypotheses lead to the desired conclusion as follows:

Step	*Reason*
1. $p \to q$	Hypothesis or Premise (given)
2. $q \to r$	Hypothesis or Premise (given)
3. $p \to r$	From steps 1 and 2 using transitive rule
4. $r \to s$	Hypothesis or Premise (given)
5. $\neg s$	Hypothesis or Premise (given)
6. $p \to s$	From steps 3 and 4 using transitive rule
7. $\neg p$	From steps 5 and 6 using modus tollens
8. $p \vee t$	Hypothesis or Premise (given)
9. t	From steps 7 and 8 using disjunctive syllogism.

Thus t is a valid conclusion from the given hypotheses.

Example 38: *Prove that the hypotheses 'It is not sunny this afternoon and it is colder than yesterday', 'We will go swimming only if it is sunny', 'If we do not go swimming, then we will go to a museum' and 'If we go to a museum, then we will be home by sunset' lead to the conclusion 'We will be home by sunset'.*

Solution: Let p, q, r, s and t denote the following propositions:

p : It is sunny this afternoon.

q : It is colder than yesterday.

r : We will go swimming.

s : We will go to a museum.

t : We will be home by sunset.

Then the given hypotheses become $\neg p \wedge q$, $r \to p$, $\neg r \to s$ and $s \to t$. We have to prove that the conclusion is t.

Let us construct an argument to prove that our hypotheses lead to the desired conclusion as follows:

Step	*Reason*
1. $\neg p \wedge q$	Hypothesis (or Premise)
2. $\neg p$	Simplification using step 1
3. $r \to p$	Hypothesis
4. $\neg r$	From steps 2 and 3 using modus tollens
5. $\neg r \to s$	Hypothesis
6. s	From steps 4 and 5 using modus ponens
7. $s \to t$	Hypothesis
8. t	From steps 6 and 7 using modus ponens

Thus t is a valid conclusion from the given hypotheses.

1.10 PREDICATES

Let us consider two statements:

1. Amal is an engineer.
2. Bimal is an engineer.

Obviously if we express these statements by symbols, we need two different symbols to denote them. Such symbols do not reveal the common features of these two statements because both are statements about two different individuals who are engineers. If we introduce some symbol to denote **is an engineer** and a rule to join it with symbols denoting the names of individuals, then we will have a symbolism to denote statement about any individual who is an engineer. The part **is an engineer** is said to be a **predicate.**

We shall symbolize a predicate by a capital letter and the names of objects in general by small letters where the word **object** is being used in a very general sense to include individuals as well.

Let us denote the predicate **is an engineer** symbolically by E, Amal by a and Bimal by b. Then the statements (1) and (2) can be written as $E(a)$ and $E(b)$ respectively. In general, any statement of the type **p is A** where A is a predicate and p is the object can be denoted by $A(p)$. Here the predicate E in (1) and (2) is a **1-place predicate**. Another example is that **G: is greater than** is a **2-place predicate.** A predicate requiring n (> 0) names is called an **n-place predicate.** In order to extend our definition to $n = 0$, we shall call a statement a **0-place predicate** because no names are associated with a statement.

Let B denotes the predicate **is blue** and let s denotes **this shirt**. Then the statement

3. This shirt is blue

can be symbolized by $B(s)$. Further, the connectives defined earlier can now be used to form **compound statements** such as **Amal is an engineer and this shirt is blue** which can be written as $E(a) \wedge B(s)$. Other connectives can also be used to form compound statements such as:

$$E(b) \to B(s),\ E(a) \vee B(s),\ E(b) \wedge (E(a) \to B(s)) \text{ etc.}$$

Consider, now, statements involving the names of two objects, such as

4. Chandan is taller than Deepa.
5. Canada is to the north of the United States.

The predicates **is taller than** and **is to the north of** are 2-place predicates because names of two objects are required to complete a statement involving these predicates. If the letter H symbolizes **is taller than**, c denotes **Chandan** and d denotes **Deepa** then statement (4) can be translated as $H(c, d)$. Observe that the order in which the names appear in the statement as well as in the predicate is important.

In general, an n-place predicate needs n names of objects to be **inserted in fixed positions** to get a statement. If A denotes an n-place predicate and $a_1, a_2, ..., a_n$ are the names of n objects, then $A(a_1, a_2, ..., a_n)$ is a statement.

Let S be the predicate 'is a mortal', a the name 'Mrs. Aarti', b 'Britain' and c 'A computer'. Then $S(a)$, $S(b)$ and $S(c)$ all represent statements (or propositions). In fact, these propositions have a common form. If $S(x)$ represents 'x is mortal', then $S(a)$, $S(b)$, $S(c)$ and others having the same form can be obtained from $S(x)$ by replacing x by an appropriate name. Note that $S(x)$ does not represent a proposition, but it results in a proposition when x is replaced by the name of an object. The letter x used here has a role of place-holder.

Simple Statement Function: An expression consisting of a predicate symbol and an individual variable is called a **Simple Statement Function** of one variable. Such a statement function becomes a proposition when the variable is replaced by the name of an object. For example, $S(x)$ defined earlier is a simple statement function of one variable x.

Compound Statement Function: A statement function obtained from combining one or more simple statement functions and the logical connectives is called a **compound statement function.** For example, if we let $M(x)$ be 'x is a man' and $S(x)$ be 'x is mortal', then we can form compound statement functions such as $M(x) \wedge S(x)$, $M(x) \rightarrow S(x)$, $\neg M(x)$ etc.

Example 39: *Let $P(x)$: x is a person*
$F(x, y)$: x is the father of y
$M(x, y)$: x is the mother of y

Write the compound statement function which symbolizes 'x is the father of the mother of y'.

Solution: Let us name a person called z as the mother of y. So 'x is the father of the mother of y' is equivalent to 'x is the father of z and z is the mother of y'. It is assumed that such a person z exists. Therefore the required compound statement functions is

$$(\exists z)(P(z) \wedge F(x, z) \wedge M(z, y)).$$

Notes: (*i*) The symbol $\exists$ stands for the words 'there exists'. (*ii*) Here $P(x)$ is a simple statement function of one variable x and $F(x, y)$, $M(x, y)$ are simple statement functions of two variables x, y.

1.11 QUANTIFICATIONS

If all the variables in a statement (or propositional) function are assigned values then the resulting statement (or proposition) has a truth value. However, there is another important way to change statement functions into statements, called **quantification.** Quantification has been broadly classified into two types, namely

1. Universal quantification
2. Existential quantification.

Universe of discourse: Many mathematical statements assert that a property has a truth value for all values of a variable (of the underlying statement function) in a particular domain, called the **universe of discourse.** Such a statement is expressed using a **universal quantification** of the appropriate statement function.

Universal quantification: The **universal quantification** of a statement function $P(x)$ is a statement or proposition only when $P(x)$ has a truth value for all values of x in the universe of discourse. It is denoted by the notation $(x)\ P(x)$ or $\forall\ x\ P(x)$ or $(\forall\ x)\ P(x)$ which is read as 'for all x, $P(x)$' or 'for every x, $P(x)$' or 'for any x, $P(x)$'. The symbol $\forall$ (read as 'for all') is called the **Universal quantifier.** The symbol ( ) is also used to denote the universal quantifier.

Example 40: (*i*) Let $P(x) : (x - 1)^2 = x^2 - 2x + 1$. Then $\forall\ x\ P(x)$ is a proposition and not a propositional function, even though a variable x appears in it. There is no need to replace x by a number for obtaining a statement or proposition. The truth-value of $\forall\ x\ P(x)$ is T, where the universe of discourse consists of the set of all complex numbers.

(*ii*) If $Q(x) : 3x > x$, where the universe consists of all real numbers, then the truth-value $\forall\ x\ Q(x)$ is F.

Example 41: Sometimes it is necessary to use more than one universal quantifier in a statement (or proposition). For example consider the following simple statement function of two variables:

$A(x, y) : x$ is taller than y.

It can be stated as 'for any x and any y, if x is taller than y then y is not taller than x' or 'for any x and any y, if x is taller than y then it is not true that y is taller than x'. So, the universal quantification of $A(x, y)$ is

$$(\forall\ x)(\forall\ y)(A(x, y) \rightarrow \neg A(y, x))$$

which is a proposition with truth value T.

Example 42: *Determine the truth value of* $(\forall\ x)P(x)$, *where* $P(x) : x^2 \leq 16$ *and the universe of discourse consists of the positive integers not exceeding 5.*

Solution: Now, $(\forall\ x)P(x) \Leftrightarrow P(1) \wedge P(2) \wedge P(3) \wedge P(4) \wedge P(5)$

(∵ Universe of discourse consists of 1, 2, 3, 4, 5)

$\Leftrightarrow \quad T \wedge T \wedge T \wedge T \wedge F$ (∵ $5^2 \leq 16$ is false)

$\Leftrightarrow \quad F$

Hence the truth value of $(\forall x)P(x)$ is F.

Existential quantification: The universal quantifier was used to translate expressions such as 'for all', 'for every' and 'for any'. Another quantifier will now be introduced to symbolize expressions such as 'for some', 'there is at least one' or 'there exists some' and is denoted by the symbol $\exists$ which is called **existential quantifier.** Note that 'some' is used in the sense of 'at least one'.

The **existential quantification** of a statement function $P(x)$ is the proposition (or statement) 'there exists at least one x (or an x, or some x) in the universe of discourse such that $P(x)$ has a truth value' and is denoted by the notation $\exists\ x\ P(x)$ or $(\exists\ x)\ P(x)$ or $(\exists\ x)\ (P(x))$.

The proposition $\exists\ x\ P(x)$ is read as 'for some x, $P(x)$'.

Note: If the elements of the universe of discourse is finitely many, say $x_1, x_2, ..., x_n$, then $\exists\ x\ P(x)$ is the same as $P(x_1) \vee P(x_2) \vee .... \vee P(x_n)$. Because the truth values of $\exists\ x\ P(x)$ and $P(x_1) \vee P(x_2) \vee ... \vee P(x_n)$ are same.

Example 43: (*i*) Let $P(x) : x > 7$, then $\exists\ x\ P(x)$ is a statement (or proposition) and not a statement function even though a variable x appears in it. The truth value of $\exists\ x\ P(x)$ is T, where the universe of discourse consists of all real numbers, because $x > 7$ is true for $x = 8$.

(*ii*) Consider the following statements:

1. There exists a woman.
2. Some women are clever.
3. Some real numbers are irrational.

Next, consider the following simple statement functions of a single variable x:

$W(x)$: x is a woman
$C(x)$: x is clever
$R(x)$: x is a real number
$I(x)$: x is irrational.

Then the given statements (or propositions) can be expressed as

1. $\exists\, x\, W(x)$
2. $\exists\, x\, (W(x) \wedge C(x))$
3. $\exists\, x\, (R(x) \wedge I(x))$.

(*iii*) Let $P(x) : x + 2 > 10$ and the universe of discourse is $U = \{1, 2, 3, .., 8\}$. Then $\exists\, x\, P(x)$ is a statement and its truth value is F because $x + 2 \leq 10$, $\forall\, x \in U$.

Example 44: *Determine the truth value of $\exists\, x\, P(x)$, where $P(x) : x^2 > 8$ and the universe of discourse is $U = \{1, 2, 3, 4\}$.*

Solution: Now, $\exists\, x\, P(x) \Leftrightarrow P(1) \vee P(2) \vee P(3) \vee P(4)$ $[\because\ U = \{1, 2, 3, 4\}]$

$\Leftrightarrow$ $F \vee F \vee T \vee T$ ($\because\ 1^2 > 8$ is false etc.)

$\Leftrightarrow$ T.

Hence the truth value of $\exists\, x\, P(x)$ is T when the universe of discourse is $U = \{1, 2, 3, 4\}$.

1.12 NEGATION OF A QUANTIFIED STATEMENT FUNCTION

If $P(x)$ symbolizes 'x has studied mathematical logic', then $\forall\, x\, P(x)$ symbolizes the statement 'every student (in the class) has studied mathematical logic'. The negation of this statement is 'It is not the case that every student in the class has studied mathematical logic' or equivalently 'there is a student in the class who has not studied mathematical logic'. Therefore, $\neg \forall\, x\, P(x) \Leftrightarrow \exists\, x \neg P(x)$. ...(1)

Similarly, $\exists\, x\, P(x)$ means 'there is a student in the class who has studied mathematical logic'. The negation of this statement is 'every student in the class has not studied mathematical logic'.

$\therefore$ $\neg \exists\, x\, P(x) \Leftrightarrow \forall\, x \neg P(x)$ ...(2)

Moreover observe that $\neg \forall\, xP(x)$ is true, when there is an x for which $P(x)$ is false and false when $P(x)$ is true for every x.

$\therefore$ $\neg \forall\, x\, P(x) \Leftrightarrow \exists\, x \neg P(x) \Leftrightarrow \neg P(x_1) \vee \neg P(x_2) \vee \ldots \vee \neg P(x_n)$

where the universe of discourse is $U = \{x_1, x_2, \ldots, x_n\}$

Also, $\neg \exists\, x\, P(x)$ is true when $P(x)$ is false for every x and false when there is an x for which $P(x)$ is true.

$\therefore$ $\neg \exists\, x\, P(x) \Leftrightarrow \forall\, x \neg P(x) \Leftrightarrow \neg P(x_1) \wedge \neg P(x_2) \wedge \ldots \wedge \neg P(x_n)$

where the universe of discourse is $U = \{x_1, x_2, \ldots, x_n\}$.

Example 45: *Negate the statement : For all real numbers x, if $x > 4$ then $x^2 > 16$.*

Solution: Let $P(x) : x > 4$; $Q(x) : x^2 > 16$.

Then the given statement can be written as

$\forall\, x(P(x) \rightarrow Q(x)) \Leftrightarrow \forall\, x(\neg P(x) \vee Q(x))$ $[\because\ (a \rightarrow b) \Leftrightarrow (\neg a \vee b)]$

$\therefore$ $\neg \forall\, x(P(x) \rightarrow Q(x)) \Leftrightarrow \neg \forall\, x(\neg P(x) \vee Q(x))$

$\Leftrightarrow$ $\exists\, x \neg (\neg P(x) \vee Q(x))$ [By (1)]

$\Leftrightarrow$ $\exists\, x\, (P(x) \wedge \neg Q(x))$ [By De Morgan's Law]

Thus the negation of the given statement is 'there exists a real number x such that $x > 4$ and $x^2 \leq 16$'.

Negations of quantified statements with more than one variable

Negation of a quantified statement with more than one variable can be obtained by applying the rule for negating a statement with only one quantifier successively. For example,

$$\neg [\forall x \exists y P(x, y)] \Leftrightarrow \exists x [\neg \exists y P(x, y)] \qquad \text{[By (1)]}$$
$$\Leftrightarrow \quad \exists x \forall y \neg P(x, y) \qquad \text{[By (2)]}$$

Example 46: *Write down the negation of the proposition for every real number x there exists a real number y such that x < y.*

Solution: The given proposition can be written in symbolic form as

$$\forall x \exists y P(x, y), \text{ where } P(x, y) : x < y$$
$$\therefore \quad \neg [\forall x \exists y P(x, y)] \Leftrightarrow \exists x [\neg \exists y P(x, y)] \Leftrightarrow \exists x \forall y \neg P(x, y)$$

Hence the negation of the given proposition is 'there exists a real number x such that for every real number y, $x \geq y$'.

1.13 FREE AND BOUND VARIABLES

Given a formula containing a part of the form $\forall x P(x)$ or $\exists x P(x)$, such a part is called an x-**bound** part of the formula. Any occurrence of x in an x-bound part of a formula is called a **bound occurrence** of x and the variable x is said to be a **bound variable.** Any occurrence of x or of any variable that is not a bound occurrence (*i.e.*, x is not a bound variable) is called a **free occurrence** and the variable x is said to be a **free variable**. Moreover, the statement function $P(x)$ either in $\forall x P(x)$ or in $\exists x P(x)$ is described as the **scope** of the corresponding quantifier. In other words, the scope of a quantifier is the statement function immediately following the quantifier. For examples, consider the following formulas:

$$\forall x P(x, y) \qquad ...(1)$$
$$\forall x (Q(x) \rightarrow R(x)) \qquad ...(2)$$
$$\forall x (Q(x) \rightarrow \exists y S(x, y)) \qquad ...(3)$$
$$\forall x (A(x) \rightarrow Q(x)) \vee \forall x (A(x) \rightarrow R(x)) \qquad ...(4)$$
$$\exists x (Q(x) \wedge R(x)) \qquad ...(5)$$
$$\exists x Q(x) \wedge R(x) \qquad ...(6)$$

In (1): $P(x, y)$ is the scope of the quantifier, occurrence of x is a bound occurrence and occurrence of y is a free occurrence.

In (2): The scope of the universal quantifier is $Q(x) \rightarrow R(x)$ and all occurrences of x are bound occurrences.

In (3): The scope of $\forall x$ is $Q(x) \rightarrow \exists y S(x, y)$ and the scope of $\exists y$ is $S(x, y)$. All occurrences of both x and y are bound occurrences.

In (4): The scope of the first quantifier is $A(x) \rightarrow Q(x)$ and the scope of the second quantifier is $A(x) \rightarrow R(x)$. x is a bound variable.

In (5): The scope of $\exists x$ is $Q(x) \wedge R(x)$ and x is a bound variable.

In (6): The scope of $\exists x$ is $Q(x)$. The first occurrence of x in $Q(x)$ is bound and the last occurrence of x in $R(x)$ is free. Here it is better to write $\exists x Q(x) \wedge R(y)$ instead of $\exists x Q(x) \wedge R(x)$, so as to separate the bound and free variables.

Notes: (*i*) In a statement (or proposition) every variable must be a bound variable, *i.e.*, no variable should have a free occurrence. In case where a free variable occurs in a formula then we have a statement function.

(*ii*) In the bound occurrence of a variable, the letter which is used to represent the variable is not important. In fact, any other letter can be used as a variable without affecting the formula provided the new letter is not used elsewhere in that formula. Thus the formulas $\exists x\, P(x, y)$ and $\exists z\, P(z, y)$ are same.

(*iii*) A bound variable cannot be substituted by a constant, only a free variable can be. As for example, $\exists x\, Q(x) \wedge R(a)$ is a substitution instance of $\exists x\, Q(x) \wedge R(y)$. In fact, $\exists x\, Q(x) \wedge R(a)$ can be expressed as 'there exists x which has the property Q and a has the property R'.

Example 47: *Symbolize the statement:* ***All the world loves a lover.***

Solution: The given statement can be stated as 'everybody loves a lover'. Let us define the following statement functions:

$P(x)$: x is a person

$L(x)$: x is a lover

$R(x, y)$: x loves y

The required symbolic expression is

$$\forall x\, (P(x) \rightarrow \forall y\, (P(y) \wedge L(y) \rightarrow R(x, y)))$$

Example 48: *Symbolize the following statements:*

(*i*) *Some dogs are black but all buffaloes are black.*

(*ii*) *Sum of two positive integers is greater than both of the integers.*

(*iii*) *Every student in an Institution is either good in studies or good in sports.*

Solution: (*i*) $D(x)$: x is a dog.

$B(y)$: y is a buffalo.

$C(x)$: x is black.

The required symbolic expression is

$$(\exists x)\, (\forall y)\, ((D(x) \wedge C(x)) \wedge (B(y) \rightarrow C(y)))$$

(*ii*) $I(x)$: x is a positive integer

$G(x, y)$: x is greater than y

The required symbolic expression is

$$(\forall x)\, (\forall y)\, ((I(x) \wedge I(y)) \rightarrow (G(x + y, x) \wedge G(x + y, y)))$$

(*iii*) $S(x)$: x is a student of the Institution

$E(x)$: x is good in studies.

$A(x)$: x is good in sports.

The required symbolic expression is

$$\forall x\, (S(x) \rightarrow (E(x) \vee A(x))).$$

1.14 RULES OF INFERENCE FOR QUANTIFIED STATEMENTS

Universal specification (US): It is the rule by which it can be concluded that $P(a)$ is true if $\forall x\, P(x)$ is true where a is an arbitrary but fixed member of the universe of discourse.

Existential specification (ES): It is the rule by which it can be concluded that $P(a)$ is true if $\exists x\, P(x)$ is true where a is not an arbitrary member of the universe of discourse, but one for which $P(a)$ is true.

Universal generalization (UG): It is the rule which states that $\forall x\, P(x)$ is true if $P(a)$ is true where a is an arbitrary member of the universe of discourse.

Existential generalization (EG): It is the rule by which it can be concluded that $\exists x\, P(x)$ is true when $P(a)$ is true, where a is a particular member of the universe of discourse.

Example 49: *Prove that* $\exists y \forall x\, P(x, y) \Rightarrow \forall x \exists y\, P(x, y)$

Solution:

	$\exists y \forall x\, P(x, y)$	[Premise (given)]
$\Rightarrow$	$\forall x\, P(x, b)$	[By existential specification]
$\Rightarrow$	$P(x, b)$	[By universal specification]
$\Rightarrow$	$\exists y\, P(x, y)$	[By existential generalization]
$\Rightarrow$	$\forall x \exists y\, P(x, y)$	[By universal generalization]
$\therefore$	$\exists y \forall x\, P(x, y) \Rightarrow \forall x \exists y\, P(x, y)$.	**(Proved)**

Notes: (*i*) The converse is not true.

(*ii*) The following implications can be proved:

$$\forall y \forall x\, P(x, y) \Rightarrow \exists x \forall y\, P(x, y)$$
$$\exists x \forall y\, P(x, y) \Rightarrow \forall y \exists x\, P(x, y)$$
$$\forall x \exists y\, P(x, y) \Rightarrow \exists y \exists x\, P(x, y)$$
$$\forall y \exists x\, P(x, y) \Rightarrow \exists x \exists y\, P(x, y)$$

We summarise these rules of inference for quantified statements in the following Table 8, where U is the universe of discourse.

Table 8. Rules of inference for quantified statements

Rule of inference	
$\forall x\, P(x)$ $\therefore\ P(a)$ if $a \in U$	Universal specification or Universal instantiation
$P(a)$ for an arbitrary $a \in U$ $\therefore\ \forall x\, P(x)$	Universal generalization
$\exists x\, P(x)$ $\therefore\ P(a)$ for some $a \in U$	Existential specification or Existential instantiation
$P(a)$ for some $a \in U$ $\therefore\ \exists x\, P(x)$	Existential generalization

Universal Modus Ponens

The rule of universal specification can be combined with modus ponens to obtain the rule known as **universal modus ponens:**

$$\begin{array}{l} P(a) \text{ for a particular } a \in U \\ \forall x\, (P(x) \to Q(x)) \\ \hline \therefore\ Q(a) \end{array}$$

Example 50: *Show that the hypotheses 'Everyone in this discrete mathematics class has taken a course in probability and statistics' and 'Mini is a student in this class' imply the conclusion 'Mini has taken a course in probability and statistics'.*

Solution: Let $D(x)$ denote 'x is in this discrete mathematics class' and let $P(x)$ denote 'x has taken a course in probability and statistics'. Then the given hypotheses are $\forall x\,(D(x) \to P(x))$ and $D(\text{Mini})$. We have to show that the conclusion is $P(\text{Mini})$.

The following steps can be used to establish the conclusion from the hypotheses.

Step	*Reason*
1. $\forall x(D(x) \to P(x))$	Hypothesis
2. $D(\text{Mini}) \to P(\text{Mini})$	From step 1 using universal specification
3. D (Mini)	Hypothesis
4. $P(\text{Mini})$	From steps 2 and 3 using modus ponens

Thus $P(\text{Mini})$, *i.e.*, 'Mini has taken a course in probability and statistics' is a valid conclusion from the given hypotheses.

Universal Modus Tollens

The rule of universal specification can be combined with modus tollens to obtain the rule called **universal modus tollens:**

$$\begin{array}{l} \neg Q(a) \text{ for a particular } a \in U \\ \quad \forall x\,(P(x) \to Q(x)) \\ \hline \therefore \quad \neg P(a) \end{array}$$

Example 51: *Symbolize the following argument using quantifier, variable and predicates:*

All healthy people eat fruit a day.
Soham does not eat fruit a day.
____
Soham is not a healthy boy.

Find whether the argument is valid.

Solution: Let $P(x)$ denote 'x is a healthy person' and let $F(x)$ denote 'x eats fruit a day'. Then the given hypotheses are $\forall x\,(P(x) \to F(x))$ and $\neg F$ (Soham). We have to check whether the conclusion is $\neg P$ (Soham).

The following steps can be used to establish the conclusion from the hypotheses.

Step	*Reason*
1. $\forall x\,(P(x) \to F(x))$	Hypothesis
2. $P(\text{Soham}) \to F(\text{Soham})$	From step 1 using universal specification
3. $\neg F$ (Soham)	Hypothesis
4. $\neg P(\text{Soham})$	From steps 2 and 3 using modus tollens

Thus $\neg P$ (Soham), *i.e.*, 'Soham is not a healthy boy' is a valid conclusion from the given hypotheses. Hence the given argument is valid.

MISCELLANEOUS EXAMPLES

Example 1: Multiple Choice Questions.

(*i*) If p being any statement, then which one of the following is not a fallacy (or contradiction)?

(*a*) $p \vee F$ (*b*) $p \wedge F$

(*c*) $p \wedge \neg p$ (*d*) none of these.

(*ii*) For any statement p, which one of the following is a tautology?

(*a*) $p \vee F$ (*b*) $p \wedge F$

(*c*) $p \wedge T$ (*d*) $p \vee \neg p$.

(*iii*) Which of the following is not true?

(*a*) $\neg (p \wedge q) \Leftrightarrow (\neg p) \vee (\neg q)$ (*b*) $\neg (p \vee q) \Leftrightarrow (\neg p) \vee (\neg q)$

(*c*) $\neg (p \vee q) \Leftrightarrow (\neg p) \wedge (\neg q)$ (*d*) $p \rightarrow q \Leftrightarrow \neg p \vee q$.

(*iv*) Which of the following is different from the others?

(*a*) $\neg p \vee q$ (*b*) $p \rightarrow q$

(*c*) $\neg q \rightarrow \neg p$ (*d*) none of these.

(*v*) If each of the statements $p \rightarrow \neg q$, $\neg r \rightarrow q$ and p is true, then

(*a*) r is true (*b*) r is false

(*c*) q is true (*d*) none of these.

(*vi*) Which one of the following is true?

(*a*) $p \vee \neg p \Leftrightarrow F$ (*b*) $p \wedge \neg p \Leftrightarrow T$

(*c*) $p \rightarrow q \Leftrightarrow q \rightarrow p$ (*d*) $p \rightarrow q \Leftrightarrow (\neg q) \rightarrow (\neg p)$.

(*vii*) Which one of the following arguments is not valid?

(*a*) $p \rightarrow q, q \vdash p$ (*b*) $p \wedge q \vdash p$

(*c*) $p \wedge q \vdash q$ (*d*) $p \vdash p \vee q$.

(*viii*) If each of the propositions $p \rightarrow q$, $q \rightarrow r$ and $\neg r$ is true, then

(*a*) p is true (*b*) q is true

(*c*) p is false (*d*) none of these.

(*ix*) Let p, q and r denote the following propositions:

p : All mothers are women ;

q : All women are not mothers ;

r : There are some women who are not mothers.

Then which one of the following is not true?

(*a*) $p \rightarrow q$ (*b*) $p \rightarrow r$

(*c*) $p \rightarrow (q \wedge r)$ (*d*) none of these

(*x*) $(p \wedge \neg q) \wedge (\neg p \wedge q)$ is

(*a*) a tautology (*b*) a contradiction

(*c*) both a tautology and a contradiction (*d*) neither a tautology nor a contradiction.

(*xi*) Which of the following is not a proposition?

(*a*) 3 is a prime number (*b*) $\sqrt{2}$ is irrational

(*c*) Mathematics is interesting (*d*) 5 is an even integer.

(*xii*) If $(p \wedge \urcorner r) \rightarrow (\urcorner p \vee q)$ is false, then the truth values of p, q and r are respectively.

(*a*) T, F, F (*b*) F, F, T

(*c*) F, T, T (*d*) T, F, T.

(*xiii*) For any two statements p and q, which one of the following is not true ?

(*a*) $\urcorner [p \vee (\urcorner q)] \Leftrightarrow (\urcorner p) \wedge q$ (*b*) $(p \vee q) \vee (\urcorner q)$ is a tautology

(*c*) $(p \wedge q) \wedge (\urcorner q)$ is a contradiction (*d*) $\urcorner [p \wedge (\urcorner q)]$ is a tautology

(*e*) $\urcorner (p \vee q) \Leftrightarrow (\urcorner p) \vee (\urcorner q)$.

(*xiv*) If p, q, r denote the following propositions:

p : 4 is an even prime number; q : 6 is a divisor of 12 ; r : HCF of 4 and 6 is 2, then which of the following is true?

(*a*) $p \wedge q$ (*b*) $(p \vee q) \wedge (\urcorner r)$

(*c*) $(\urcorner (q \wedge r)) \vee p$ (*d*) $(\urcorner p) \vee (q \wedge r)$.

(*xv*) Which one of the following is the inverse of the proposition 'If a number is prime than it is odd'?

(*a*) If a number is not prime then it is odd

(*b*) If a number is not prime then it is not odd

(*c*) If a number is not odd then it is not prime

(*d*) If a number is odd then it is prime.

Solution: (*i*) We know that $p \vee F \Leftrightarrow p, p \wedge F \Leftrightarrow F$ and $p \wedge \urcorner p \Leftrightarrow F$. So, $p \vee F$ is not a fallacy or contradiction in general.

Ans. (*a*).

(*ii*) We know that $p \vee F \Leftrightarrow p, p \wedge F \Leftrightarrow F, p \wedge T \Leftrightarrow p$ and $p \vee \urcorner p \Leftrightarrow T$. Therefore, $p \vee \urcorner p$ is a tautology.

Ans. (*d*).

(*iii*) Using De Morgan's laws, we get

$$\urcorner (p \wedge q) \Leftrightarrow \urcorner p \vee \urcorner q, \urcorner (p \vee q) \Leftrightarrow \urcorner p \wedge \urcorner q,$$

Also, we know that $p \rightarrow q \Leftrightarrow \urcorner p \vee q$.

Ans. (*b*).

(*iv*) We know that $p \rightarrow q \Leftrightarrow \urcorner p \vee q$,

$$\urcorner q \rightarrow \urcorner p \Leftrightarrow (\urcorner (\urcorner q)) \vee \urcorner p \Leftrightarrow q \vee \urcorner p \Leftrightarrow \urcorner p \vee q.$$

Ans. (*d*).

(*v*)

Step	Reason
1. p	Hypothesis or Premise (given)
2. $p \rightarrow \urcorner q$	Hypothesis or Premise (given)
3. $\urcorner q$	From steps 1 and 2 using modus ponens
4. $\urcorner r \rightarrow q$	Hypothesis or Premise (given)
5. $\urcorner (\urcorner r) \Leftrightarrow r$	From steps 3 and 4 using modus tollens

Thus r is a valid conclusion from the given hypotheses.

Ans. (*a*).

(*vi*) We know that $p \vee \neg p \Leftrightarrow T, p \wedge \neg p \Leftrightarrow F,$

$$p \to q \Leftrightarrow \neg p \vee q, q \to p \Leftrightarrow \neg q \vee p.$$

But $(\neg q) \to (\neg p) \Leftrightarrow (\neg (\neg q)) \vee \neg p \Leftrightarrow q \vee \neg p \Leftrightarrow \neg p \vee q$

$\Leftrightarrow$ $p \to q.$

Ans. (*d*).

(*vii*) **Argument in (b) is**

$(p \wedge q) \to p \Leftrightarrow (\neg (p \wedge q)) \vee p \Leftrightarrow (\neg p \vee \neg q) \vee p$ [By De Morgan's law]

$\Leftrightarrow$ $p \vee (\neg p \vee \neg q)$

$\Leftrightarrow$ $(p \vee \neg p) \vee \neg q \Leftrightarrow T \vee \neg q \Leftrightarrow T.$

Argument in (c) is

Similarly, $(p \wedge q) \to q \Leftrightarrow T.$

Argument in (d) is

$$p \to (p \vee q) \Leftrightarrow \neg p \vee (p \vee q) \Leftrightarrow (\neg p \vee p) \vee q \Leftrightarrow T \vee q \Leftrightarrow T.$$

Argument in (a) is invalid, because $p \to q \Leftrightarrow \neg p \vee q$ and so when q is true, $p \to q$ is also true but p may or may not be true. Hence $\{(p \to q) \wedge q\} \to p$ is not a tautology.

Ans. (*a*)

(*viii*) *Step*	*Reason*
1. $p \to q$	Hypothesis or Premise (given)
2. $q \to r$	Hypothesis or Premise (given)
3. $\neg r$	Hypothesis or Premise (given)
4. $p \to r$	From steps 1 and 2 using transitive rule
5. $\neg p$	From steps 3 and 4 using modus tollens.

Ans. (*c*)

(*ix*) The proposition p is 'all mothers are women' which means that every mother is a woman but there may be some women who are not mothers. Therefore, $p \to q$ and $p \to r$ are true and so $p \to (q \wedge r)$ is also true.

Ans. (*d*)

(*x*) Using associative and commutative laws of conjunction, we get

$$(p \wedge \neg q) \wedge (\neg p \wedge q) \Leftrightarrow (p \wedge \neg p) \wedge (q \wedge \neg q) \Leftrightarrow F \wedge F \Leftrightarrow F.$$

Ans. (*b*)

(*xi*) 'Mathematics is interesting' is not a logical sentence because it may be interesting for some persons and may not be interesting for others. So, it has neither 'true' value nor 'false' value and as such it is not a proposition.

Ans. (*c*)

(*xii*) Given, $(p \wedge \neg r) \to (\neg p \vee q)$ is false. So, by truth table of conditional, $p \wedge \neg r$ is true and $\neg p \vee q$ is false. Therefore, both p and $\neg r$ are true and both $\neg p$ and q are false. This gives, p is true, r is false and also q is false.

Ans. (*a*).

(*xiii*) $\neg (p \vee q) \Leftrightarrow (\neg p) \wedge (\neg q)$ by De Morgan's law.

Ans. (*e*)

Note: Verify that others are true.

(*xiv*) Here p is false, both q and r are true. Therefore, $\neg p$ is true and $q \wedge r$ is true and so $(\neg p) \vee (q \wedge r)$ is true.

Ans. (*d*)

Note: Verify that others are false.

(*xv*) The conditional proposition $\neg p \to \neg q$ is called the inverse of the proposition $p \to q$.

Ans. (*b*)

Example 2: *Write the converse, contrapositive and inverse of the following propositions :*

(*i*) *If ABCD is a rectangle then ABCD is a square.*

(*ii*) *If* $5x - 3 = 12$ *then* $x = 3$.

Solution: (*i*) Let p : *ABCD* is a rectangle and q : *ABCD* is a square.

Given : $p \to q$ ...(1)

The converse proposition of (1) is $q \to p$, *i.e.*, if *ABCD* is a square then *ABCD* is a rectangle.

The contrapositive proposition of (1) is $\neg q \to \neg p$, *i.e.*, if *ABCD* is not a square then *ABCD* is not a rectangle.

The inverse proposition of (1) is $\neg p \to \neg q$, *i.e.*, if *ABCD* is not a rectangle then *ABCD* is not a square.

(*ii*) Let $p : 5x - 3 = 12$ and $q : x = 3$.

The converse proposition of (1) is $q \to p$, *i.e.*, if $x = 3$ then $5x - 3 = 12$.

The contrapositive proposition of (1) is $\neg q \to \neg p$, *i.e.*, if $x \neq 3$ then $5x - 3 \neq 12$.

The inverse proposition of (1) is $\neg p \to \neg q$, *i.e.*, if $5x - 3 \neq 12$ then $x \neq 3$.

Example 3: *Construct the truth table for* $(p \wedge q) \vee (q \wedge r) \vee (r \wedge p)$.

Solution:

p	q	r	$p \wedge q$	$q \wedge r$	$r \wedge p$	$(p \wedge q) \vee (q \wedge r)$	$(p \wedge q) \vee (q \wedge r) \vee (r \wedge p)$
T	T	T	T	T	T	T	T
T	T	F	T	F	F	T	T
T	F	T	F	F	T	F	T
T	F	F	F	F	F	F	F
F	T	T	F	T	F	T	T
F	T	F	F	F	F	F	F
F	F	T	F	F	F	F	F
F	F	F	F	F	F	F	F

Example 4: *Construct the truth table for*

(*i*) $(p \wedge q) \to (p \vee q)$ (*ii*) $(p \to q) \to (p \wedge q)$ (*iii*) $p \to (q \wedge r)$.

Solution: (*i*)

p	q	$p \wedge q$	$p \vee q$	$(p \wedge q) \to (p \vee q)$
T	T	T	T	T
T	F	F	T	T
F	T	F	T	T
F	F	F	F	T

Therefore $(p \wedge q) \to (p \vee q)$ is a tautology.

(*ii*)

p	q	$p \to q$	$p \wedge q$	$(p \to q) \to (p \wedge q)$
T	T	T	T	T
T	F	F	F	T
F	T	T	F	F
F	F	T	F	F

(*iii*)

p	q	r	$q \wedge r$	$p \to (q \wedge r)$
T	T	T	T	T
T	T	F	F	F
T	F	T	F	F
T	F	F	F	F
F	T	T	T	T
F	T	F	F	T
F	F	T	F	T
F	F	F	F	T

Example 5: *Construct truth table for each of the following statement formulas and indicate if it is a tautology or contradiction:*

(*i*) $p \to (p \vee q)$ (*ii*) $(p \to \neg p) \to \neg p$

(*iii*) $(\neg q \wedge p) \wedge q$ (*iv*) $(p \vee q) \wedge (\neg p \wedge \neg q)$.

Solution:

p	q	$\neg p$	$\neg q$	$p \vee q$	$\neg p \wedge \neg q$	$p \to \neg p$	$\neg q \wedge p$	(*i*)	(*ii*)	(*iii*)	(*iv*)
T	T	F	F	T	F	F	F	T	T	F	F
T	F	F	T	T	F	F	T	T	T	F	F
F	T	T	F	T	F	T	F	T	T	F	F
F	F	T	T	F	T	T	F	T	T	F	F

From the table it is clear that $p \to (p \vee q)$, $(p \to \neg p) \to \neg p$ are tautologies and $(\neg q \wedge p) \wedge q$, $(p \vee q) \wedge (\neg p \wedge \neg q)$ are contradictions.

Note: $p \Rightarrow p \vee q$; $(p \to \neg p) \Rightarrow \neg p$.

Example 6: *Show that* $p \to (q \to r) \Leftrightarrow (p \wedge q) \to r$.

Solution: $p \to (q \to r) \Leftrightarrow p \to (\neg q \vee r)$ [By Example 15 (*ii*)]

$\Leftrightarrow \neg p \vee (\neg q \vee r)$

$\Leftrightarrow (\neg p \vee \neg q) \vee r$ [By associative law]

$\Leftrightarrow \neg (p \wedge q) \vee r$ [By De Morgan's law]

$\Leftrightarrow (p \wedge q) \to r$.

Example 7: *Prove that* $\neg (p \vee (\neg p \wedge q)) \Leftrightarrow \neg p \wedge \neg q$.

Solution: $\neg (p \vee (\neg p \wedge q)) \Leftrightarrow \neg p \wedge \neg (\neg p \wedge q)$ [By De Morgan's law]

$\Leftrightarrow \neg p \wedge (\neg (\neg p) \vee \neg q)$ [By De Morgan's law]

$\Leftrightarrow \neg p \wedge (p \vee \neg q)$ [$\because \neg (\neg p) \Leftrightarrow p$]

$\Leftrightarrow (\neg p \wedge p) \vee (\neg p \wedge \neg q)$ [By distributive law]

$\Leftrightarrow F \vee (\neg p \wedge \neg q) \Leftrightarrow \neg p \wedge \neg q$. **(Proved)**

Example 8: *Prove that* $(p \wedge q) \to (p \vee q)$ *is a tautology.*

Solution: $(p \wedge q) \to (p \vee q) \Leftrightarrow (\neg (p \wedge q)) \vee (p \vee q)$ [By Example 15 (*ii*)]

$\Leftrightarrow (\neg p \vee \neg q) \vee (p \vee q)$ [By De Morgan's law]

$\Leftrightarrow (\neg q \vee \neg p) \vee (p \vee q)$ [By commutativity]

$\Leftrightarrow \neg q \vee (\neg p \vee p) \vee q$ [By associativity]

$\Leftrightarrow \neg q \vee (T \vee q)$ [$\because \neg p \vee p \Leftrightarrow T$]

$\Leftrightarrow \neg q \vee T \Leftrightarrow T$

$\therefore$ $(p \wedge q) \to (p \vee q)$ is a tautology. **(Proved)**

Note: $(p \wedge q) \Rightarrow (p \vee q)$.

Example 9: *Prove the following equivalences:*

(*i*) $p \to (q \vee r) \Leftrightarrow (p \to q) \vee (p \to r)$

(*ii*) $(p \to q) \wedge (r \to q) \Leftrightarrow (p \vee r) \to q$

(*iii*) $p \to (q \to p) \Leftrightarrow \neg p \to (p \to q)$

(*iv*) $\neg (p \rightleftarrows q) \Leftrightarrow (p \vee q) \wedge \neg (p \wedge q)$

(*v*) $(\neg p \vee q) \wedge (p \wedge (p \wedge q)) \Leftrightarrow p \wedge q$

(*vi*) $\neg p \to (q \to r) \Leftrightarrow q \to (p \vee r)$.

Solution: We know that for any two propositions a and b,

$$(a \to b) \Leftrightarrow (\neg a \vee b). \quad ...(1)$$

(*i*) $p \to (q \vee r) \Leftrightarrow \neg p \vee (q \vee r)$ [by (1)]

$\Leftrightarrow (\neg p \vee \neg p) \vee (q \vee r)$ [By Idempotent law]

$\Leftrightarrow (\neg p \vee q) \vee (\neg p \vee r)$ [By associativity and commutativity]

$\Leftrightarrow (p \to q) \vee (p \to r)$ [by (1)] **(Proved)**

(*ii*) $(p \to q) \wedge (r \to q) \Leftrightarrow (\neg p \vee q) \wedge (\neg r \vee q)$ [by (1)]

$\Leftrightarrow (\neg p \wedge \neg r) \vee q$ [By distributive law]

$\Leftrightarrow (\neg (p \vee r)) \vee q$ [By De Morgan's law]

$\Leftrightarrow (p \vee r) \to q$ [by (1)] **(Proved)**

(*iii*) $p \to (q \to p) \Leftrightarrow p \to (\neg q \vee p)$ [By (1)]

$\Leftrightarrow \neg p \vee (\neg q \vee p)$ [by (1)]

$\Leftrightarrow (p \vee \neg p) \vee (\neg q)$ [By commutativity and associativity]

$\Leftrightarrow T \vee \neg q \Leftrightarrow T$

$\therefore$ $p \to (q \to p) \Leftrightarrow T$ ...(2)

$\neg p \to (p \to q) \Leftrightarrow \neg (\neg p) \vee (p \to q)$ [by (1)]

$\Leftrightarrow p \vee (\neg p \vee q)$

$\Leftrightarrow (p \vee \neg p) \vee q$ (By associativity)

$$\Leftrightarrow T \vee q \Leftrightarrow T$$

$\therefore \quad \neg p \to (p \to q) \Leftrightarrow T$ ...(3)

From (2) and (3), we get

$$p \to (q \to p) \Leftrightarrow \neg p \to (p \to q) \quad \textbf{(Proved)}$$

(iv) $\neg(p \rightleftarrows q) \Leftrightarrow \neg((p \to q) \wedge (q \to p))$

$\Leftrightarrow \neg((\neg p \vee q) \wedge (\neg q \vee p))$ [by (1)]

$\Leftrightarrow (\neg(\neg p \vee q)) \vee (\neg(\neg q \vee p))$ [By De Morgan's law]

$\Leftrightarrow ((\neg(\neg p)) \wedge \neg q) \vee ((\neg(\neg q)) \wedge \neg p)$ (By De Morgan's law)

$\Leftrightarrow (p \wedge \neg q) \vee (q \wedge \neg p)$ [$\because \neg(\neg p) \Leftrightarrow p$]

$\Leftrightarrow (p \vee (q \wedge \neg p)) \wedge (\neg q \vee (q \wedge \neg p))$ [By distributive law]

$\Leftrightarrow (p \vee q) \wedge (p \vee \neg p) \wedge (\neg q \vee q) \wedge (\neg q \vee \neg p)$ [By distributive law]

$\Leftrightarrow (p \vee q) \wedge T \wedge T \wedge (\neg q \vee \neg p)$

$\Leftrightarrow (p \vee q) \wedge \neg(p \wedge q)$ [By De Morgan's law] **(Proved)**

(v) $(\neg p \vee q) \wedge (p \wedge (p \wedge q)) \Leftrightarrow (\neg p \vee q) \wedge (p \wedge p) \wedge q$ [By associative law]

$\Leftrightarrow (\neg p \vee q) \wedge (p \wedge q)$ [By idempotent law]

$\Leftrightarrow (p \wedge q) \wedge (\neg p \vee q)$ [By commutative law]

$\Leftrightarrow ((p \wedge q) \wedge \neg p) \vee ((p \wedge q) \wedge q)$ [By distributive law]

$\Leftrightarrow ((p \wedge \neg p) \wedge q) \vee (p \wedge (q \wedge q))$ [By commutativity and associativity]

$\Leftrightarrow (F \wedge q) \vee (p \wedge q)$

$\Leftrightarrow F \vee (p \wedge q) \Leftrightarrow p \wedge q.$ **(Proved)**

(vi) $\neg p \to (q \to r) \Leftrightarrow \neg p \to (\neg q \vee r)$ [by (1)]

$\Leftrightarrow (\neg(\neg p)) \vee (\neg q \vee r)$ [by (1)]

$\Leftrightarrow p \vee (\neg q \vee r)$ [$\because \neg(\neg p) \Leftrightarrow p$]

$\Leftrightarrow (p \vee \neg q) \vee r$ [By associativity]

$\Leftrightarrow (\neg q \vee p) \vee r$ [By commutativity]

$\Leftrightarrow \neg q \vee (p \vee r)$ [By associativity]

$\Leftrightarrow q \to (p \vee r).$ [by (1)] **(Proved)**

Example 10: *Prove the following implications (without using truth tables):*

(i) $((p \vee \neg p) \to q) \to ((p \vee \neg p) \to r) \Rightarrow q \to r$

(ii) $(p \vee q) \wedge (p \to r) \wedge (q \to r) \Rightarrow r.$

Solution: We know that for any two propositions a and b,

$$(a \to b) \Leftrightarrow (\neg a \vee b) \quad ...(1)$$

(i) $((p \vee \neg p) \to q) \to ((p \vee \neg p) \to r) \to (q \to r)$

$\Leftrightarrow (T \to q) \to (T \to r) \to (q \to r)$ [$\because p \vee \neg p \Leftrightarrow T$]

$\Leftrightarrow (F \vee q) \to (F \vee r) \to (q \to r)$ [By (1)]

$\Leftrightarrow (q \to r) \to (q \to r)$

$\Leftrightarrow T$

$\therefore \quad ((p \vee \neg p) \to q) \to ((p \vee \neg p) \to r) \Rightarrow q \to r.$ **(Proved)**

(ii) $[(p \vee q) \wedge (p \to r) \wedge (q \to r)] \to r$

$\Leftrightarrow [(p \vee q) \wedge (\neg p \vee r) \wedge (\neg q \vee r)] \to r$ [By (1)]

$\Leftrightarrow [(p \vee q) \wedge ((\neg p \wedge \neg q) \vee r)] \rightarrow r$ (By distributive law)

$\Leftrightarrow [(p \vee q) \wedge ((\neg (p \vee q)) \vee r)] \rightarrow r$ (By De Morgan's law)

$\Leftrightarrow [((p \vee q) \wedge \neg (p \vee q)) \vee ((p \vee q) \wedge r)] \rightarrow r$ (By distributive law)

$\Leftrightarrow [F \vee ((p \vee q) \wedge r)] \rightarrow r$ $[\because \; a \wedge 7a \Leftrightarrow F]$

$\Leftrightarrow ((p \vee q) \wedge r) \rightarrow r$

$\Leftrightarrow (\neg ((p \vee q) \wedge r)) \vee r$ [By (1)]

$\Leftrightarrow (\neg ((p \wedge r) \vee (q \wedge r))) \vee r$ (By distributive law)

$\Leftrightarrow (\neg (p \wedge r) \wedge \neg (q \wedge r)) \vee r$ [By De Morgan's law]

$\Leftrightarrow ((\neg p \vee \neg r) \wedge (\neg q \vee \neg r)) \vee r$ [By De Morgan's law]

$\Leftrightarrow (\neg p \vee \neg r \vee r) \wedge (\neg q \vee \neg r \vee r)$ (By distributive law)

$\Leftrightarrow (\neg p \vee (\neg r \vee r)) \wedge (\neg q \vee (\neg r \vee r))$

$\Leftrightarrow (\neg p \vee T) \wedge (\neg q \vee T)$

$\Leftrightarrow T \wedge T$ $[\because \; a \vee T \Leftrightarrow T]$

$\Leftrightarrow T$

$\therefore \quad (p \vee q) \wedge (p \rightarrow r) \wedge (q \rightarrow r) \Rightarrow r.$ **(Proved)**

Example 11: *Find the disjunctive normal forms (DNF) of the following propositions:*

(i) $(p \wedge \neg (q \vee r)) \vee (((p \wedge q) \vee \neg r) \wedge p)$

(ii) $p \vee (\neg p \rightarrow (q \vee (q \rightarrow \neg r)))$

(iii) $(p \wedge \neg (q \wedge r)) \vee (p \rightarrow q)$

(iv) $\neg (\neg (p \rightleftarrows q) \wedge r).$

Solution:

(i) $(p \wedge \neg (q \vee r)) \vee (((p \wedge q) \vee \neg r) \wedge p)$

$\Leftrightarrow (p \wedge (\neg q \wedge \neg r)) \vee (((p \wedge q) \vee \neg r) \wedge p)$ [By De Morgan's law]

$\Leftrightarrow (p \wedge \neg q \wedge \neg r) \vee (p \wedge q \wedge p) \vee (\neg r \wedge p)$

[By distributive and associative laws]

$\Leftrightarrow (p \wedge \neg q \wedge \neg r) \vee ((p \wedge p) \wedge q) \vee (p \wedge \neg r)$ [By commutativity]

$\Leftrightarrow (p \wedge \neg q \wedge \neg r) \vee (p \wedge q) \vee (p \wedge \neg r)$ $[\because \; p \wedge p \Leftrightarrow p]$

This is the required DNF.

(ii) $p \vee (\neg p \rightarrow (q \vee (q \rightarrow \neg r)))$

$\Leftrightarrow p \vee (\neg p \rightarrow (q \vee (\neg q \vee \neg r)))$ $[\because \; (a \rightarrow b) \Leftrightarrow (\neg a \vee b)]$

$\Leftrightarrow p \vee (\neg (\neg p) \vee (q \vee (\neg q \vee \neg r)))$

$\Leftrightarrow p \vee p \vee q \vee \neg q \vee \neg r$ $[\because \; \neg (\neg p) \Leftrightarrow p$ and By associativity$]$

$\Leftrightarrow (p \vee p) \vee q \vee \neg q \vee \neg r$

$\Leftrightarrow p \vee q \vee \neg q \vee \neg r$ $[\because \; (p \vee p) \Leftrightarrow p]$

This is the required DNF.

Observation: $p \vee q \vee \neg q \vee \neg r \Leftrightarrow p \vee (q \vee \neg q) \vee \neg r \Leftrightarrow p \vee T \vee \neg r \Leftrightarrow T.$

$\therefore \quad p \vee (\neg p \rightarrow (q \vee (q \rightarrow \neg r)))$ is a tautology.

(iii) $(p \wedge \neg (q \wedge r)) \vee (p \rightarrow q)$

$\Leftrightarrow (p \wedge \neg (q \wedge r)) \vee (\neg p \vee q)$ $[\because \; (a \rightarrow b) \Leftrightarrow (\neg a \vee b)]$

$\Leftrightarrow (p \wedge (\neg q \vee \neg r)) \vee (\neg p \vee q)$ [By De Morgan's law]

$\Leftrightarrow ((p \wedge \neg q) \vee (p \wedge \neg r)) \vee (\neg p \vee q)$ (By distributive law)

$\Leftrightarrow (p \wedge \neg q) \vee (p \wedge \neg r) \vee \neg p \vee q$ [By associativity]

This is the required DNF.

(*iv*) $\neg (\neg (p \rightleftarrows q) \wedge r)$

$\Leftrightarrow \neg [\neg ((p \to q) \wedge (q \to p)) \wedge r]$

$\Leftrightarrow \neg [\neg ((\neg p \vee q) \wedge (\neg q \vee p)) \wedge r]$ [$\because$ $(a \to b) \Leftrightarrow (\neg a \vee b)$]

$\Leftrightarrow \neg [(\neg (\neg p \vee q) \vee \neg (\neg q \vee p)) \wedge r)$ [By De Morgan's law]

$\Leftrightarrow \neg [((p \wedge \neg q) \vee (q \wedge \neg p)) \wedge r]$ [By De Morgan's law]

$\Leftrightarrow \neg ((p \wedge \neg q) \vee (q \wedge \neg p)) \vee \neg r$ [By De Morgan's law]

$\Leftrightarrow (\neg (p \wedge \neg q) \wedge \neg (q \wedge \neg p)) \vee \neg r$ [By De Morgan's law]

$\Leftrightarrow ((\neg p \vee q) \wedge (\neg q \vee p)) \vee \neg r$ [By De Morgan's law] ...(1)

$\Leftrightarrow (\neg p \wedge (\neg q \vee p)) \vee (q \wedge (\neg q \vee p)) \vee \neg r$ [By distributive law]

$\Leftrightarrow (\neg p \wedge \neg q) \vee (\neg p \wedge p) \vee (q \wedge \neg q) \vee (q \wedge p) \vee \neg r$

[By distributive and associative laws]

This is the required DNF.

Note: From (1), using distributive law, we get

$\neg (\neg (p \rightleftarrows q) \wedge r) \Leftrightarrow (\neg p \vee q \vee \neg r) \wedge (\neg q \vee p \vee \neg r)$

This is CNF.

Example 12: *Obtain the principal disjunctive normal forms (PDNF) of the following propositions:*

(*i*) $(\neg p \to q) \wedge (q \rightleftarrows p)$

(*ii*) $(p \wedge q) \vee (\neg p \wedge q) \vee (q \wedge r)$

(*iii*) $(q \vee (p \wedge r)) \wedge \neg ((p \vee r) \wedge q)$.

Solution:

(*i*) $(\neg p \to q) \wedge (q \rightleftarrows p)$

$\Leftrightarrow (\neg p \to q) \wedge (q \to p) \wedge (p \to q)$

$\Leftrightarrow (\neg (\neg p) \vee q) \wedge (\neg q \vee p) \wedge (\neg p \vee q)$ [$\because$ $(a \to b) \Leftrightarrow (\neg a \vee b)$]

$\Leftrightarrow (p \vee q) \wedge ((\neg q \wedge (\neg p \vee q)) \vee (p \wedge (\neg p \vee q)))$ [By distributive law]

$\Leftrightarrow (p \vee q) \wedge ((\neg q \wedge \neg p) \vee (\neg q \wedge q) \vee (p \wedge \neg p) \vee (p \wedge q))$

[By distributive and associative laws]

$\Leftrightarrow (p \vee q) \wedge ((\neg q \wedge \neg p) \vee F \vee F \vee (p \wedge q))$ [$\because$ $a \wedge \neg a \Leftrightarrow F$]

$\Leftrightarrow (p \vee q) \wedge ((\neg q \wedge \neg p) \vee (p \wedge q))$ [$\because$ $a \vee F \Leftrightarrow a$]

$\Leftrightarrow (p \wedge ((\neg q \wedge \neg p) \vee (p \wedge q))) \vee (q \wedge ((\neg q \wedge \neg p) \vee (p \wedge q)))$

[By distributive law]

$\Leftrightarrow (p \wedge \neg q \wedge \neg p) \vee (p \wedge p \wedge q) \vee (q \wedge \neg q \wedge \neg p) \vee (q \wedge p \wedge q)$

[By distributive and associative laws]

$\Leftrightarrow ((p \wedge \neg p) \wedge \neg q) \vee ((p \wedge p) \wedge q) \vee ((q \wedge \neg q) \wedge \neg p) \vee (p \wedge (q \wedge q))$

[By commutative and associative laws]

$\Leftrightarrow (F \wedge \neg q) \vee (p \wedge q) \vee (F \wedge \neg p) \vee (p \wedge q)$

$\Leftrightarrow F \vee (p \wedge q) \vee F \vee (p \wedge q)$

$\Leftrightarrow (p \wedge q) \vee (p \wedge q)$

$\Leftrightarrow p \wedge q$ [By idempotent law]

This is the required PDNF.

(*ii*) The given statement formula is in *DNF* but not in *PDNF*.

$(p \wedge q) \vee (\neg p \wedge q) \vee (q \wedge r)$

$\Leftrightarrow ((p \wedge q) \wedge T) \vee ((\neg p \wedge q) \wedge T) \vee ((q \wedge r)) \wedge T)$

$\Leftrightarrow ((p \wedge q) \wedge (r \vee \neg r)) \vee ((\neg p \wedge q) \wedge (r \vee \neg r)) \vee ((q \wedge r) \wedge (p \vee \neg p))$

$\Leftrightarrow (p \wedge q \wedge r) \vee (p \wedge q \wedge \neg r) \vee (\neg p \wedge q \wedge r) \vee (\neg p \wedge q \wedge \neg r) \vee (q \wedge r \wedge p) \vee (q \wedge r \wedge \neg p)$

[By distributive and associative laws]

$\Leftrightarrow ((p \wedge q \wedge r) \vee (p \wedge q \wedge r)) \vee (p \wedge q \wedge \neg r) \vee ((\neg p \wedge q \wedge r) \vee (\neg p \wedge q \wedge r)) \vee (\neg p \wedge q \wedge \neg r)$

[By commutative and associative laws]

$\Leftrightarrow (p \wedge q \wedge r) \vee (p \wedge q \wedge \neg r) \vee (\neg p \wedge q \wedge r) \vee (\neg p \wedge q \wedge \neg r)$ [By idempotent law]

This is the required PDNF.

(*iii*) $(q \vee (p \wedge r)) \wedge \neg ((p \vee r) \wedge q)$

$\Leftrightarrow (q \vee (p \wedge r)) \wedge (\neg (p \vee r) \vee \neg q)$ [By De Morgan's law]

$\Leftrightarrow (q \vee (p \wedge r)) \wedge ((\neg p \wedge \neg r) \vee \neg q)$ [By De Morgan's law]

$\Leftrightarrow (q \wedge \neg p \wedge \neg r) \vee (q \wedge \neg q) \vee (p \wedge r \wedge \neg p \wedge \neg r) \vee (p \wedge r \wedge \neg q)$

[By distributive and associative laws]

$\Leftrightarrow (\neg p \wedge q \wedge \neg r) \vee (q \wedge \neg q) \vee ((p \wedge \neg p) \wedge (r \wedge \neg r)) \vee (p \wedge \neg q \wedge r)$

[By commutative and associative laws]

$\Leftrightarrow (\neg p \wedge q \wedge \neg r) \vee F \vee (F \wedge F) \vee (p \wedge \neg q \wedge r)$ [$\because$ $a \wedge \neg a \Leftrightarrow F$]

$\Leftrightarrow (\neg p \wedge q \wedge \neg r) \vee (p \wedge \neg q \wedge r)$ [$\because$ $F \wedge F \Leftrightarrow F$ and $a \vee F \Leftrightarrow a$]

This is the required PDNF.

Example 13: *Obtain the conjunctive normal forms of the following statement formulas:*

(*i*) $(q \vee (p \wedge q)) \wedge \neg ((p \vee r) \wedge q)$

(*ii*) $(p \wedge \neg (q \wedge r)) \vee (p \rightarrow q)$

(*iii*) $(p \wedge \neg (q \vee r)) \vee (((p \wedge q) \vee \neg r) \wedge p)$

(*iv*) $\neg (p \vee q) \rightleftarrows (p \wedge q)$

(*v*) $p \rightarrow ((p \rightarrow q) \wedge \neg (\neg q \vee \neg p))$.

Solution: (*i*) $(q \vee (p \wedge q)) \wedge \neg ((p \vee r) \wedge q)$

$\Leftrightarrow q \wedge \neg ((p \vee r) \wedge q)$ [By absorption law]

$\Leftrightarrow q \wedge (\neg (p \vee r) \vee \neg q)$ [By De Morgan's law]

$\Leftrightarrow q \wedge ((\neg p \wedge \neg r) \vee \neg q)$ [By De Morgan's law]

$\Leftrightarrow q \wedge (\neg p \vee \neg q) \wedge (\neg r \vee \neg q)$ [By distributive and associative laws]

This is the required CNF.

(*ii*) $(p \wedge \neg (q \wedge r)) \vee (p \rightarrow q)$

$\Leftrightarrow (p \wedge \neg (q \wedge r)) \vee (\neg p \vee q)$ [$\because$ $(a \rightarrow b) \Leftrightarrow (\neg a \vee b)$]

$\Leftrightarrow (p \wedge (\neg q \vee \neg r)) \vee (\neg p \vee q)$ [By De Morgan's law] ...(1)

$\Leftrightarrow (p \vee \neg p \vee q) \wedge (\neg q \vee \neg r \vee \neg p \vee q)$ [By distributive and associative laws]

This is the required CNF.

Note: From (1), using distributive law, we get

$$(p \wedge \urcorner (q \wedge r)) \vee (p \rightarrow q)$$

$\Leftrightarrow (p \wedge \urcorner q) \vee (p \wedge \urcorner r) \vee (\urcorner p \vee q)$

$\Leftrightarrow (p \wedge \urcorner q) \vee (p \wedge \urcorner r) \vee \urcorner (p \wedge \urcorner q)$ [By De Morgan's law]

$\Leftrightarrow ((p \wedge \urcorner q) \vee \urcorner (p \wedge \urcorner q)) \vee (p \wedge \urcorner r)$ [By commutativity and associativity]

$\Leftrightarrow T \vee (p \wedge \urcorner r) \Leftrightarrow T.$

$\therefore$ $(p \wedge \urcorner (q \wedge r)) \vee (p \rightarrow q)$ is a tautology.

(iii) $(p \wedge \urcorner (q \vee r)) \vee (((p \wedge q) \vee \urcorner r) \wedge p)$

$\Leftrightarrow (p \wedge \urcorner q \wedge \urcorner r) \vee ((p \vee \urcorner r) \wedge (q \vee \urcorner r) \wedge p)$

[By De Morgan's law, associativity and distributive law]

$\Leftrightarrow (p \wedge \urcorner q \wedge \urcorner r) \vee ((p \wedge (p \vee \urcorner r)) \wedge (q \vee \urcorner r))$

[By commutativity and associativity]

$\Leftrightarrow (p \wedge \urcorner q \wedge \urcorner r) \vee (p \wedge (q \vee \urcorner r))$ [By absorption law]

$\Leftrightarrow ((p \wedge \urcorner q \wedge \urcorner r) \vee p) \wedge ((p \wedge \urcorner q \wedge \urcorner r) \vee (q \vee \urcorner r))$ [By distributive law]

$\Leftrightarrow p \wedge (((p \wedge \urcorner q \wedge \urcorner r) \vee \urcorner r) \vee q)$

[By absorption law, commutativity and associativity]

$\Leftrightarrow p \wedge (\urcorner r \vee q)$ [By absorption law]

This is the required CNF.

(iv) $\urcorner (p \vee q) \rightleftarrows (p \wedge q)$

$\Leftrightarrow (\urcorner (p \vee q) \rightarrow (p \wedge q)) \wedge ((p \wedge q) \rightarrow \urcorner (p \vee q))$

$\Leftrightarrow (\urcorner (\urcorner (p \vee q)) \vee (p \wedge q)) \wedge (\urcorner (p \wedge q) \vee \urcorner (p \vee q))$ [$\because$ $(a \rightarrow b) \Leftrightarrow \urcorner a \vee b$]

$\Leftrightarrow ((p \vee q) \vee (p \wedge q)) \wedge ((\urcorner p \vee \urcorner q) \vee (\urcorner p \wedge \urcorner q))$ [By De Morgan's law]

$\Leftrightarrow (p \vee q \vee p) \wedge (p \vee q \vee q) \wedge (\urcorner p \vee \urcorner q \vee \urcorner p) \wedge (\urcorner p \vee \urcorner q \vee \urcorner q)$

[By distributive law]

$\Leftrightarrow (p \vee q) \wedge (p \vee q) \wedge (\urcorner p \vee \urcorner q) \wedge (\urcorner p \vee \urcorner q)$

[By commutativity and since $a \vee a \Leftrightarrow a$]

$\Leftrightarrow (p \vee q) \wedge (\urcorner p \vee \urcorner q)$ [$\because$ $a \wedge a \Leftrightarrow a$]

This is the required CNF.

(v) $p \rightarrow ((p \rightarrow q) \wedge \urcorner (\urcorner q \vee \urcorner p))$

$\Leftrightarrow \urcorner p \vee ((p \rightarrow q) \wedge \urcorner (\urcorner q \vee \urcorner p))$ [$\because$ $(a \rightarrow b) \Leftrightarrow (\urcorner a \vee b)$]

$\Leftrightarrow \urcorner p \vee ((\urcorner p \vee q) \wedge \urcorner (\urcorner q \vee \urcorner p))$

$\Leftrightarrow \urcorner p \vee ((\urcorner p \vee q) \wedge \urcorner (\urcorner (q \wedge p))$ [By De Morgan's law]

$\Leftrightarrow \urcorner p \vee ((\urcorner p \vee q) \wedge (q \wedge p))$ [$\because$ $\urcorner (\urcorner a) \Leftrightarrow a$]

$\Leftrightarrow (\urcorner p \vee (\urcorner p \vee q)) \wedge (\urcorner p \vee (q \wedge p))$ [By distributive law]

$\Leftrightarrow ((\urcorner p \vee \urcorner p) \vee q) \wedge (\urcorner p \vee q) \wedge (\urcorner p \vee p)$ [By associative and distributive laws]

$\Leftrightarrow (\urcorner p \vee q) \wedge (\urcorner p \vee q) \wedge T$ [By idempotent law]

$\Leftrightarrow (\urcorner p \vee q)$, this is the required CNF.

Example 14: *Find the principal conjunctive normal forms (PCNF) of the following propositions:*

(i) $(p \vee q) \wedge (r \vee \urcorner p) \wedge (q \vee \urcorner r)$

(ii) $(p \rightarrow (q \wedge r)) \wedge (\urcorner p \rightarrow (\urcorner q \wedge \urcorner r))$

(*iii*) $(p \wedge q \wedge r) \vee (\neg p \wedge q \wedge r) \vee (\neg p \wedge \neg q \wedge \neg r)$

(*iv*) $(p \wedge q) \vee (\neg p \wedge q \wedge r)$

(*v*) $(p \vee \neg (q \vee r)) \vee (((p \wedge q) \wedge \neg r) \wedge p)$.

Solution: (*i*) $(p \vee q) \wedge (r \vee \neg p) \wedge (q \vee \neg r)$

$\Leftrightarrow ((p \vee q) \vee F) \wedge ((r \vee \neg p) \vee F) \wedge ((q \vee \neg r) \vee F)$ $[\because \; a \vee F \Leftrightarrow a]$

$\Leftrightarrow ((p \vee q) \vee (r \wedge \neg r)) \wedge ((r \vee \neg p) \vee (q \wedge \neg q)) \wedge ((q \vee \neg r) \vee (p \wedge \neg p))$

$\Leftrightarrow (p \vee q \vee r) \wedge (p \vee q \vee \neg r) \wedge (r \vee \neg p \vee q) \wedge (r \vee \neg p \vee \neg q) \wedge (q \vee \neg r \vee p) \wedge (q \vee \neg r \vee \neg p)$

[By distributive law]

This is the required PCNF.

(*ii*) $(p \to (q \wedge r)) \wedge (\neg p \to (\neg q \wedge \neg r))$

$\Leftrightarrow (\neg p \vee (q \wedge r)) \wedge (\neg(\neg p) \vee (\neg q \wedge \neg r))$ $[\because \; (a \to b) \Leftrightarrow (\neg a \vee b)]$

$\Leftrightarrow (\neg p \vee q) \wedge (\neg p \vee r) \wedge (p \vee \neg q) \wedge (p \vee \neg r)$

[By distributive law and since $\neg(\neg p) \Leftrightarrow p$]

$\Leftrightarrow ((\neg p \vee q) \vee F) \wedge ((\neg p \vee r) \vee F) \wedge ((p \vee \neg q) \vee F) \wedge ((p \vee \neg r) \vee F)$ $[\because \; a \vee F \Leftrightarrow a]$

$\Leftrightarrow ((\neg p \vee q) \vee (r \wedge \neg r)) \wedge ((\neg p \vee r) \vee (q \wedge \neg q)) \wedge ((p \vee \neg q) \vee (r \wedge \neg r)) \wedge ((p \vee \neg r) \vee (q \wedge \neg q))$

$\Leftrightarrow (\neg p \vee q \vee r) \wedge (\neg p \vee q \vee \neg r) \wedge (\neg p \vee r \vee q) \wedge (\neg p \vee r \vee \neg q) \wedge (p \vee \neg q \vee r)$
$\wedge (p \vee \neg q \vee \neg r) \wedge (p \vee \neg r \vee q) \wedge (p \vee \neg r \vee \neg q)$ [By distributive law]

$\Leftrightarrow (\neg p \vee q \vee r) \wedge (\neg p \vee q \vee r) \wedge (\neg p \vee q \vee \neg r) \wedge (\neg p \vee \neg q \vee r) \wedge (p \vee \neg q \vee r)$
$\wedge (p \vee \neg q \vee \neg r) \wedge (p \vee \neg q \vee \neg r) \wedge (p \vee q \vee \neg r)$

[By commutativity law]

$\Leftrightarrow (\neg p \vee q \vee r) \wedge (\neg p \vee q \vee \neg r) \wedge (\neg p \vee \neg q \vee r) \wedge (p \vee \neg q \vee r) \wedge (p \vee \neg q \vee \neg r) \wedge (p \vee q \vee \neg r)$

[By idempotent law]

This is the required PCNF.

(*iii*) Let $A \Leftrightarrow (p \wedge q \wedge r) \vee (\neg p \wedge q \wedge r) \vee (\neg p \wedge \neg q \wedge \neg r)$

$\therefore \quad \neg A \Leftrightarrow (p \wedge \neg q \wedge r) \vee (p \wedge q \wedge \neg r) \vee (\neg p \wedge \neg q \wedge r) \vee (p \wedge \neg q \wedge \neg r) \vee (\neg p \wedge q \wedge \neg r)$

(Remaining minterms)

So, $A \Leftrightarrow \neg(\neg A)$

$\Leftrightarrow \neg[(p \wedge \neg q \wedge r) \vee (p \wedge q \wedge \neg r) \vee (\neg p \wedge \neg q \wedge r) \vee (p \wedge \neg q \wedge \neg r) \vee (\neg p \wedge q \wedge \neg r)]$

$\Leftrightarrow \neg(p \wedge \neg q \wedge r) \wedge \neg(p \wedge q \wedge \neg r) \wedge \neg(\neg p \wedge \neg q \wedge r) \wedge \neg(p \wedge \neg q \wedge \neg r) \wedge \neg(\neg p \wedge q \wedge \neg r)$

[By De Morgan's law]

$\Leftrightarrow (\neg p \vee q \vee \neg r) \wedge (\neg p \vee \neg q \vee r) \wedge (p \vee q \vee \neg r) \wedge (\neg p \vee q \vee r) \wedge (p \vee \neg q \vee r)$

[By De Morgan's law and since $\neg(\neg a) \Leftrightarrow a$]

This is the required PCNF.

(*iv*) Let $A \Leftrightarrow (p \wedge q) \vee (\neg p \wedge q \wedge r)$

$\Leftrightarrow ((p \wedge q) \wedge T) \vee (\neg p \wedge q \wedge r)$ $[\because \; a \wedge T \Leftrightarrow a]$

$\Leftrightarrow ((p \wedge q) \wedge (r \vee \neg r)) \vee (\neg p \wedge q \wedge r)$

$\Leftrightarrow (p \wedge q \wedge r) \vee (p \wedge q \wedge \neg r) \vee (\neg p \wedge q \wedge r)$

$\therefore \quad \neg A \Leftrightarrow (p \wedge \neg q \wedge r) \vee (\neg p \wedge \neg q \wedge r) \vee (p \wedge \neg q \wedge \neg r) \vee (\neg p \wedge q \wedge \neg r) \vee (\neg p \wedge \neg q \wedge \neg r)$

(Remaining minterms)

So, $\quad A \Leftrightarrow \neg(\neg A)$

$\Leftrightarrow \neg[(p \wedge \neg q \wedge r) \vee (\neg p \wedge \neg q \wedge r) \vee (p \wedge \neg q \wedge \neg r) \vee (\neg p \wedge q \wedge \neg r) \vee (\neg p \wedge \neg q \wedge \neg r)]$

$\Leftrightarrow \neg(p \wedge \neg q \wedge r) \wedge \neg(\neg p \wedge \neg q \wedge r) \wedge \neg(p \wedge \neg q \wedge \neg r) \wedge \neg(\neg p \wedge q \wedge \neg r)$
$\wedge \neg(\neg p \wedge \neg q \wedge \neg r)$ [By De Morgan's law]

$\Leftrightarrow (\neg p \vee q \vee \neg r) \wedge (p \vee q \vee \neg r) \wedge (\neg p \vee q \vee r) \wedge (p \vee \neg q \vee r) \wedge (p \vee q \vee r)$
[by De Morgan's law and since $\neg(\neg a) \Leftrightarrow a$]

This is the required PCNF.

(*v*) Let $A \Leftrightarrow (p \vee \neg(q \vee r)) \vee (((p \wedge q) \wedge \neg r) \wedge p)$

$\Leftrightarrow (p \vee (\neg q \wedge \neg r)) \vee ((p \wedge p) \wedge q \wedge \neg r)$
[By De Morgan's law, commutativity and associativity]

$\Leftrightarrow p \vee (\neg q \wedge \neg r) \vee (p \wedge q \wedge \neg r)$ [By idempotent law]

$\Leftrightarrow (p \wedge T) \vee (\neg q \wedge \neg r) \vee (p \wedge q \wedge \neg r)$ [$\because\ a \wedge T \Leftrightarrow a$]

$\Leftrightarrow (p \wedge (q \vee \neg q)) \vee (\neg q \wedge \neg r) \vee (p \wedge q \wedge \neg r)$ [$\because\ q \vee \neg q \Leftrightarrow T$]

$\Leftrightarrow (p \wedge q) \vee (p \wedge \neg q) \vee (\neg q \wedge \neg r) \vee (p \wedge q \wedge \neg r)$ [By distributive law]

$\Leftrightarrow ((p \wedge q) \wedge T) \vee ((p \wedge \neg q) \wedge T) \vee ((\neg q \wedge \neg r) \wedge T) \vee (p \wedge q \wedge \neg r)$

$\Leftrightarrow ((p \wedge q) \wedge (r \vee \neg r)) \vee ((p \wedge \neg q) \wedge (r \vee \neg r)) \vee ((\neg q \wedge \neg r) \wedge (p \vee \neg p)) \vee (p \wedge q \wedge \neg r)$
[$\because\ a \vee \neg a \Leftrightarrow T$]

$\Leftrightarrow (p \wedge q \wedge r) \vee (p \wedge q \wedge \neg r) \vee (p \wedge \neg q \wedge r) \vee (p \wedge \neg q \wedge \neg r) \vee (\neg q \wedge \neg r \wedge p)$
$\vee (\neg q \wedge \neg r \wedge \neg p) \vee (p \wedge q \wedge \neg r)$ [By distributive law]

$\Leftrightarrow (p \wedge q \wedge r) \vee ((p \wedge q \wedge \neg r) \vee (p \wedge q \wedge \neg r)) \vee (p \wedge \neg q \wedge r) \vee ((p \wedge \neg q \wedge \neg r)$
$\vee (p \wedge \neg q \wedge \neg r)) \vee (\neg p \wedge \neg q \wedge \neg r)$ [By commutativity and associativity]

$\Leftrightarrow (p \wedge q \wedge r) \vee (p \wedge q \wedge \neg r) \vee (p \wedge \neg q \wedge r) \vee (p \wedge \neg q \wedge \neg r) \vee (\neg p \wedge \neg q \wedge \neg r)$
[By idempotent law]

$\therefore \quad \neg A \Leftrightarrow (\neg p \wedge q \wedge r) \vee (\neg p \wedge \neg q \wedge r) \vee (\neg p \wedge q \wedge \neg r)$ [remaining minterms]

So, $\quad A \Leftrightarrow \neg(\neg A)$

$\Leftrightarrow \neg[(\neg p \wedge q \wedge r) \vee (\neg p \wedge \neg q \wedge r) \vee (\neg p \wedge q \wedge \neg r)]$

$\Leftrightarrow \neg(\neg p \wedge q \wedge r) \wedge \neg(\neg p \wedge \neg q \wedge r) \wedge \neg(\neg p \wedge q \wedge \neg r)$ [By De Morgan's law]

$\Leftrightarrow (p \vee \neg q \vee \neg r) \wedge (p \vee q \vee \neg r) \wedge (p \vee \neg q \vee r)$
[By De Morgan's law and since $\neg(\neg a) \Leftrightarrow a$]

This is the required PCNF.

Example 15: *Find the PDNF and PCNF of the following statements using truth tables:*

(*i*) $(\neg p \vee \neg q) \rightarrow (p \rightleftarrows \neg q)$

(*ii*) $p \vee (\neg p \rightarrow (q \vee (\neg q \rightarrow r)))$

(*iii*) $(p \rightarrow (q \wedge r)) \wedge (\neg p \rightarrow (\neg q \wedge \neg r))$.

Solution: (*i*)

Row No.	p	q	$\neg p$	$\neg q$	$\neg p \vee \neg q$	$p \rightleftarrows \neg q$	$(\neg p \vee \neg q) \rightarrow (p \rightleftarrows \neg q)$
1.	*T*	*T*	*F*	*F*	*F*	*F*	*T*
2.	*T*	*F*	*F*	*T*	*T*	*T*	*T*
3.	*F*	*T*	*T*	*F*	*T*	*T*	*T*
4.	*F*	*F*	*T*	*T*	*T*	*F*	*F*

The minterms corresponding to the three T values of the last column are given below:

Row 1: $p \wedge q$. Row 2: $p \wedge \neg q$. Row 3: $\neg p \wedge q$.

$\therefore \quad (\neg p \vee \neg q) \rightarrow (p \rightleftarrows \neg q) \Leftrightarrow (p \wedge q) \vee (p \wedge \neg q) \vee (\neg p \wedge q)$

This is the required PDNF.

The maxterm corresponding to the one F value of the last column is given by $p \vee q$ (row 4).

$\therefore \quad (\neg p \vee \neg q) \rightarrow (p \rightleftarrows \neg q) \Leftrightarrow (p \vee q)$, this is the required *PCNF*.

(*ii*)

Row No.	p	q	r	$\neg p$	$\neg q$	$\neg q \rightarrow r$	$q \vee (\neg q \rightarrow r)$	$\neg p \rightarrow (q \vee (\neg q \rightarrow r))$ $\Leftrightarrow a$	$p \vee a$
1.	T	T	T	F	F	T	T	T	T
2.	T	T	F	F	F	T	T	T	T
3.	T	F	T	F	T	T	T	T	T
4.	T	F	F	F	T	F	F	T	T
5.	F	T	T	T	F	T	T	T	T
6.	F	T	F	T	F	T	T	T	T
7.	F	F	T	T	T	T	T	T	T
8.	F	F	F	T	T	F	F	F	F

The minterms corresponding to the seven T values of the last column are given below:

Row 1: $p \wedge q \wedge r$. Row 2: $p \wedge q \wedge \neg r$. Row 3: $p \wedge \neg q \wedge r$.

Row 4: $p \wedge \neg q \wedge \neg r$. Row 5: $\neg p \wedge q \wedge r$. Row 6: $\neg p \wedge q \wedge \neg r$. Row 7: $\neg p \wedge \neg q \wedge r$.

$$\therefore \quad p \vee (\neg p \rightarrow (q \vee (\neg q \rightarrow r)))$$

$$\Leftrightarrow \quad (p \wedge q \wedge r) \vee (p \wedge q \wedge \neg r) \vee (p \wedge \neg q \wedge r) \vee (p \wedge \neg q \wedge \neg r) \vee (\neg p \wedge q \wedge r)$$
$$\vee (\neg p \wedge q \wedge \neg r) \vee (\neg p \wedge \neg q \wedge r).$$

This is the required PDNF.

The maxterm corresponding to the one F value of the last column is given by $p \vee q \vee r$ (Row 8).

$\therefore \quad p \vee (\neg p \rightarrow (q \vee (\neg q \rightarrow r))) \Leftrightarrow (p \vee q \vee r)$, this is the required *PCNF*.

(*iii*)

Row No.	p	q	r	$\neg p$	$\neg q$	$\neg r$	$q \wedge r$	$p \rightarrow (q \wedge r)$ $\Leftrightarrow a$	$\neg q \wedge \neg r$	$\neg p \rightarrow (\neg q \wedge \neg r)$ $\Leftrightarrow b$	$a \wedge b$
1.	T	T	T	F	F	F	T	T	F	T	T
2.	T	T	F	F	F	T	F	F	F	T	F
3.	T	F	T	F	T	F	F	F	F	T	F
4.	T	F	F	F	T	T	F	F	T	T	F
5.	F	T	T	T	F	F	T	T	F	F	F
6.	F	T	F	T	F	T	F	T	F	F	F
7.	F	F	T	T	T	F	F	T	F	F	F
8.	F	F	F	T	T	T	F	T	T	T	T

The minterms corresponding to the two T values of the last column are given below:

Row 1: $p \wedge q \wedge r$. Row 8: $\neg p \wedge \neg q \wedge \neg r$.

$\therefore \quad (p \to (q \wedge r)) \wedge (\neg p \to (\neg q \wedge \neg r))$

$\Leftrightarrow \quad (p \wedge q \wedge r) \vee (\neg p \wedge \neg q \wedge \neg r)$, this is the required *PDNF*.

The maxterms corresponding to the six F values of the last column are given below:

Row 2: $\neg p \vee \neg q \vee r$. Row 3: $\neg p \vee q \vee \neg r$. Row 4: $\neg p \vee q \vee r$.

Row 5: $p \vee \neg q \vee \neg r$. Row 6: $p \vee \neg q \vee r$. Row 7: $p \vee q \vee \neg r$.

$\therefore \quad (p \to (q \wedge r)) \wedge (\neg p \to (\neg q \wedge \neg r))$

$\Leftrightarrow \quad (\neg p \vee \neg q \vee r) \wedge (\neg p \vee q \vee \neg r) \wedge (\neg p \vee q \vee r) \wedge (p \vee \neg q \vee \neg r)$
$\wedge (p \vee \neg q \vee r) \wedge (p \vee q \vee \neg r)$

This is the required PCNF.

Example 16: *Show that the following argument is valid.*

$$\begin{array}{l} p \\ p \to q \\ \underline{q \to r} \\ r \end{array}$$

Solution: First method:

Step	*Reason*
1. p	Hypothesis
2. $p \to q$	Hypothesis
3. q	From steps 1 and 2 using modus ponens
4. $q \to r$	Hypothesis
5. r	From steps 3 and 4 using modus ponens

Thus the given argument is valid.

Second method:

$(p \wedge (p \to q) \wedge (q \to r)) \to r$

$\Leftrightarrow \quad (p \wedge (\neg p \vee q) \wedge (\neg q \vee r)) \to r$ $\quad [\because \ (a \to b) \Leftrightarrow (\neg a \vee b)]$

$\Leftrightarrow \quad \neg (p \wedge (\neg p \vee q) \wedge (\neg q \vee r)) \vee r$

$\Leftrightarrow \quad (\neg p \vee \neg (\neg p \vee q) \vee \neg (\neg q \vee r)) \vee r$ [By De Morgan's law]

$\Leftrightarrow \quad (\neg p \vee (p \wedge \neg q) \vee (q \wedge \neg r)) \vee r$ [By De Morgan's law and since $\neg (\neg a) \Leftrightarrow a$]

$\Leftrightarrow \quad (\neg p \vee (p \wedge \neg q)) \vee ((q \wedge \neg r) \vee r)$ [By associativity]

$\Leftrightarrow \quad ((\neg p \vee p) \wedge (\neg p \vee \neg q)) \vee ((q \vee r) \wedge (\neg r \vee r))$ [By distributive law]

$\Leftrightarrow \quad (T \wedge (\neg p \vee \neg q)) \vee ((q \vee r) \wedge T)$ $\quad [\because \ \neg a \vee a \Leftrightarrow T]$

$\Leftrightarrow \quad (\neg p \vee \neg q) \vee (q \vee r)$ $\quad [\because \ T \wedge a \Leftrightarrow a \wedge T \Leftrightarrow a]$

$\Leftrightarrow \quad \neg p \vee ((\neg q \vee q) \vee r)$ [By associativity]

$\Leftrightarrow \quad \neg p \vee (T \vee r) \Leftrightarrow \neg p \vee T \Leftrightarrow T$ $\quad [\because \ T \vee a \Leftrightarrow a \vee T \Leftrightarrow T]$

Hence the given argument is valid.

Third method:

p	q	r	$p \to q$	$p \wedge (p \to q)$	$q \to r$	$p \wedge (p \to q) \wedge (q \to r) \Leftrightarrow a$	$a \to r$
T	T	T	T	T	T	T	T
T	T	F	T	T	F	F	T
T	F	T	F	F	T	F	T
T	F	F	F	F	T	F	T
F	T	T	T	F	T	F	T
F	T	F	T	F	F	F	T
F	F	T	T	F	T	F	T
F	F	F	T	F	T	F	T

$\therefore \quad (p \wedge (p \to q) \wedge (q \to r)) \to r \Leftrightarrow T.$

Hence the given argument is valid.

Example 17: *Prove that the following argument is not valid:*

$$\begin{array}{c} p \\ \neg q \vee r \\ \underline{\neg q \to q} \\ r \end{array}$$

Solution: $(p \wedge (\neg q \vee r) \wedge (\neg p \to q)) \to r$

$\Leftrightarrow \quad (p \wedge (\neg q \vee r) \wedge (\neg (\neg p) \vee q)) \to r$ $\quad [\because \; (a \to b) \Leftrightarrow (\neg a \vee b)]$

$\Leftrightarrow \quad ((p \wedge (p \vee q)) \wedge (\neg q \vee r)) \to r$ $\quad$ [By commutativity and associativity]

$\Leftrightarrow \quad (p \wedge (\neg q \vee r)) \to r$ $\quad$ [By absorption law]

$\Leftrightarrow \quad \neg (p \wedge (\neg q \vee r)) \vee r$ $\quad [\because \; (a \to b) \Leftrightarrow (\neg a \vee b)]$

$\Leftrightarrow \quad \neg p \vee \neg (\neg q \vee r) \vee r$ $\quad$ [By De Morgan's law]

$\Leftrightarrow \quad \neg p \vee ((q \wedge \neg r)) \vee r)$ $\quad$ [By De Morgan's law and associativity]

$\Leftrightarrow \quad \neg p \vee ((q \vee r) \wedge (\neg r \vee r))$ $\quad$ [By distributive law]

$\Leftrightarrow \quad \neg p \vee ((q \vee r) \wedge T)$

$\Leftrightarrow \quad \neg p \vee q \vee r.$

This is not a tautology. Because if we assign truth values T, F, F for p, q, r respectively then the truth value of $\neg p \vee q \vee r$ becomes $F \vee F \vee F \Leftrightarrow F$. Hence the given argument is not valid.

Example 18: *Prove that s is a valid conclusion from the hypotheses $p \to q$, $p \to r$, $\neg (q \wedge r)$ and $s \vee p$.*

Solution: Let us construct an argument to prove that our hypotheses lead to the desired conclusion as follows:

Step	*Reason*
1. $p \to q$	Hypothesis
2. $p \to r$	Hypothesis
3. $\neg q \to \neg p$, $\neg r \to \neg p$	From steps 1 and 2 using law of contrapositive
4. $\neg (q \wedge r)$	Hypothesis

5. $\neg q \vee \neg r$	From step 4 using De Morgan's law
6. $\neg p$	From steps 3 and 5 using dilemma
7. $s \vee p$	Hypothesis
8. s	From steps 6 and 7 using disjunctive syllogism

Thus s is a valid conclusion from the given hypotheses. **(Proved)**

Example 19: *Prove that the hypotheses 'If you send me an e-mail message, then I will finish writing the research paper', 'If you do not send me an e-mail message, then I will go to sleep early' and 'If I go to sleep early, then I will wake up feeling refreshed' yield the conclusion 'If I do not finish writing the research paper, then I will wake up feeling refreshed'.*

Solution: Let p, q, r and s denote the following propositions:

p : You send me an e-mail message.

q : I will finish writing the research paper.

r : I will go to sleep early.

s : I will wake up feeling refreshed.

Then the given hypotheses become $p \to q$, $\neg p \to r$ and $r \to s$. We have to prove that the conclusion is $\neg q \to s$.

Let us construct an argument to prove that our hypotheses yield the desired conclusion as follows:

Step	*Reason*
1. $p \to q$	Hypothesis (or premise)
2. $\neg q \to \neg p$	From step 1 using law of contrapositive
3. $\neg p \to r$	Hypothesis
4. $\neg q \to r$	From steps 2 and 3 using hypothetical syllogism, or transitive rule
5. $r \to s$	Hypothesis
6. $\neg q \to s$	From steps 4 and 5 using hypothetical syllogism

Thus $\neg q \to s$ is a valid conclusion from the given hypothesis.

Example 20: *Show that the hypotheses 'If I get the job and work hard, then I will get promotion', 'If I get promotion, then I will be happy' and 'I will not be happy' lead to the conclusion 'I will not get the job or I will not work hard'.*

Solution: Let p, q, r and s denote the following propositions:

p : I get the job; q : I work hard;

r : I will get promotion; s : I will be happy.

Then the given hypotheses become $(p \wedge q) \to r$.

$r \to s$ and $\neg s$. We have to prove that the conclusion is $\neg p \vee \neg q$.

Let us construct an argument to prove that our hypotheses lead to the desired conclusion as follows:

Step	*Reason*
1. $(p \wedge q) \to r$	Hypothesis
2. $r \to s$	Hypothesis
3. $(p \wedge q) \to s$	From steps 1 and 2 using hypothetical syllogism, or transitive rule

4. $\neg s$	Hypothesis
5. $\neg (p \wedge q)$	From steps 3 and 4 using modus tollens
6. $\neg p \vee \neg q$	From steps 5 using De Morgan's law.

Thus $\neg p \vee \neg q$, *i.e.*, 'I will not get the job or I will not work hard' is a valid conclusion from the given hypotheses.

Example 21: *Verify that the existentially quantified statement 'for some real number x, $(1 + x^2)^{-1} > 1$' is false.*

Solution: We need to show that the proposition '$(1 + x^2)^{-1} > 1$, for some real number x' is false is equivalent to prove that the proposition '$(1 + x^2)^{-1} \leq 1$, for all real number x' is true.

Now, for all real number x, $x^2 \geq 0 \Rightarrow 1 + x^2 \geq 1 \Rightarrow (1 + x)^{-1} \leq 1$. Hence the proposition '$(1 + x^2)^{-1} \leq 1$, for all real number x' is true and so the proposition '$(1 + x^2)^{-1} > 1$, for some real number x' is false.

Example 22: *Express each of the following propositions using predicates and quantifiers:*

(*i*) *Every computer science student needs a course in mathematical logic.*

(*ii*) *There is a student in this class who owns a laptop.*

(*iii*) *Every student in this class has taken at least one course in mathematics.*

(*iv*) *There is a student in this class who has taken at least one course in mathematics.*

Solution: (*i*) Let $L(x)$: 'x needs a course in mathematical logic', where the universe of discourse consists of all computer science students. Then the symbolization of the given proposition is

$$\forall x\, L(x)$$

(*ii*) Let $C(x)$: 'x owns a laptop', where the universe of discourse consists of all students in this class. Then the symbolization of the given proposition is $\exists x\, C(x)$.

(*iii*) Let $M(x, y)$: 'x has taken y' where the universe of discourse of x consists of all students in this class and that of y consists of all mathematical courses. Then the symbolization of the given proposition is $\forall x \exists y\, M(x, y)$.

(*iv*) Using the same assumptions as in (*iii*), we have the symbolization of the give proposition as $\exists x \exists y\, M(x, y)$.

Example 23: *Let F(x, y) be the simple statement function representing 'x loves y', where the universe of discourse for both x and y consists of all people in the world. Symbolize the following propositions:*

(*i*) *Everybody loves somebody.*

(*ii*) *There is somebody whom everybody loves.*

(*iii*) *Nobody loves everybody.*

(*iv*) *There is somebody whom no one loves.*

Solution: (*i*) It can be stated as 'for all x, there exists some y such that x loves y.' So, the symbolization of the given statement is

$$(\forall x)\, (\exists y)\, F(x, y).$$

(*ii*) The given statement can be stated as 'there exists some y whom all x love'. So, the symbolization of this statement is

$$(\exists y)\, (\forall x)\, F(x, y).$$

(*iii*) The given statement can be stated as 'there is not one who loves everybody'. So, the symbolization of the given statement is

$$\neg \exists x \forall y F(x, y) \Leftrightarrow \forall x \neg \forall y F(x, y) \quad \text{[By (2), Art. 1.12]}$$

$$\Leftrightarrow \forall x \exists y \neg F(x, y) \quad \text{[By (1), Art. 1.12]}$$

(*iv*) The given proposition means that 'there is somebody whom every one does not love'. Therefore the symbolization of the given statement is

$$\neg \forall x \exists y F(x, y) \Leftrightarrow \exists x \neg \exists y F(x, y) \Leftrightarrow \exists x \forall y \neg F(x, y).$$

Example 24: *Express the negations of the following propositions:*

(*i*) *If the teacher is absent then some students do not keep quiet.*

(*ii*) *The teacher is present and all the students keep quiet.*

(*iii*) *Some of the students do not keep quiet or the teacher is absent.*

(*iv*) *No one has solved every problem in the exercise.*

Solution: Let t : presence of the teacher

$Q(x)$: student x keeps quiet

$D(x, y)$: x has solved problem y.

(*i*) Symbolic expression of the given proposition is

$$\neg t \to \exists x \neg Q(x) \Leftrightarrow \neg t \to \neg \forall x Q(x)$$

$$\Leftrightarrow t \vee \neg \forall x Q(x) \quad [\because (a \to b) \Leftrightarrow (\neg a \vee b)]$$

$\therefore$ Negation of the given proposition is

$$\neg (t \vee \neg \forall x Q(x)) \Leftrightarrow \neg t \wedge \forall x Q(x) \quad \text{[By De Morgan's law]}$$

i.e., the teacher is absent and all the students keep quiet.

(*ii*) Symbolization of the given proposition is

$$t \wedge \forall x Q(x)$$

$\therefore$ Negation of the given proposition is

$$\neg (t \wedge \forall x Q(x)) \Leftrightarrow \neg t \vee \neg \forall x Q(x) \quad \text{[By De Morgan's law]}$$

$$\Leftrightarrow \neg t \vee \exists x \neg Q(x)$$

i.e., the teacher is absent or there exists student who does not keep quiet.

(*iii*) Symbolization of the given proposition is $\exists x \neg Q(x) \vee \neg t$

$\therefore$ Negation of the given proposition is

$$\neg (\exists x \neg Q(x) \vee \neg t) \Leftrightarrow \neg (\neg \forall x Q(x) \vee \neg t)$$

$$\Leftrightarrow \forall x Q(x) \wedge t \quad \text{[By De Morgan's law]}$$

i.e., all the students keep quiet and the teacher is present.

(*iv*) Symbolization of the given proposition is

$$\neg \exists x \forall y D(x, y)$$

$\therefore$ Negation of the given proposition is

$$\neg (\neg \exists x \forall y D(x, y)) \Leftrightarrow \exists x \forall y D(x, y)$$

i.e., some one has solved every problem in the exercise.

Alternative

The given proposition is $\forall x \neg \forall y D(x, y) \Leftrightarrow \forall x \exists y \neg D(x, y)$.

$\therefore$ Negation of the given proposition is

$$\neg(\forall x \exists y \neg D(x, y)) \Leftrightarrow \exists x \neg(\exists y \neg D(x, y))$$
$$\Leftrightarrow \quad \exists x \forall y \neg(\neg D(x, y))$$
$$\Leftrightarrow \quad \exists x \forall y D(x, y).$$

Example 25: *Write in the symbolic form and negate the following propositions:*

(*i*) *Everyone who is healthy can do all types of job.*

(*ii*) *Some people are not admired by every one.*

(*iii*) *Every people should help his neighbours, or his neighbours will not help him.*

Solution: (*i*) Let $P(x)$: x is a healthy person

$J(y)$: y is a type of job.

$R(x, y)$: x can do y.

The symbolization of the given proposition is

$$\forall x \forall y (P(x) \wedge J(y) \rightarrow R(x, y))$$

Its negation is given by

$$\neg[\forall x \forall y (P(x) \wedge J(y) \rightarrow R(x, y))]$$
$$\Leftrightarrow \quad \exists x [\neg \forall y (P(x) \wedge J(y) \rightarrow R(x, y))]$$
$$\Leftrightarrow \quad \exists x \exists y \neg(P(x) \wedge J(y) \rightarrow R(x, y))$$
$$\Leftrightarrow \quad \exists x \exists y \neg[\neg(P(x) \wedge J(y)) \vee R(x, y)] \qquad [\because (a \rightarrow b) \Leftrightarrow (\neg a \vee b)]$$
$$\Leftrightarrow \quad \exists x \exists y (P(x) \wedge J(y) \wedge \neg R(x, y)) \qquad \text{[By De Morgan's law]}$$

i.e., there exists a healthy people x and there exists a type of job y such that x cannot do y.

(*ii*) In the universe of people, let $S(x, y)$: x admires y.

The symbolization of the given proposition is

$$\exists y \exists x (\neg S(x, y))$$

Its negation is $\neg[\exists y \exists x (\neg S(x, y))]$

$$\Leftrightarrow \quad \forall y [\neg \exists x (\neg S(x, y))]$$
$$\Leftrightarrow \quad \forall y \forall x \neg(\neg S(x, y))$$
$$\Leftrightarrow \quad \forall y \forall x S(x, y)$$

i.e., every people is admired by every one.

(*iii*) In the universe of people, let

$N(x, y)$: x and y are neighbours

$H(x, y)$: x helps y.

The symbolization of the given proposition is

$$\forall x \forall y (N(x, y) \rightarrow (H(x, y) \vee \neg H(y, x)))$$

Its negation is given by

$$\neg[\forall x \forall y (N(x, y) \rightarrow (H(x, y) \vee \neg H(y, x)))]$$
$$\Leftrightarrow \quad \exists x [\neg \forall y (N(x, y) \rightarrow (H(x, y) \vee \neg H(y, x)))]$$
$$\Leftrightarrow \quad \exists x \exists y \neg(N(x, y) \rightarrow (H(x, y) \vee \neg H(y, x)))$$
$$\Leftrightarrow \quad \exists x \exists y \neg(\neg N(x, y) \vee H(x, y) \vee \neg H(y, x)) \qquad [\because (a \rightarrow b) \Leftrightarrow (\neg a \vee b)]$$
$$\Leftrightarrow \quad \exists x \exists y (N(x, y) \wedge \neg H(x, y) \wedge H(y, x)) \qquad \text{[By De Morgan's law]}$$

i.e., there are some people who should not help their neighbours but their neighbours will help them.

Example 26: *Symbolize the following 'Famous Socrates Argument' using quantifier, variable and predicates:*

All men are mortal
Socrates is a man
Socrates is a mortal

Find whether the argument is valid.

Solution: Let $P(x)$ denote 'x is a man' and let $M(x)$ denote 'x is a mortal'. Then the given hypotheses are $\forall x\, (P(x) \to M(x))$ and P(Socrates). We have to check whether the conclusion is M (Socrates).

The following steps can be used to establish the conclusion from the hypotheses.

Step	*Reason*
1. $\forall x(P(x) \to M(x))$	Hypothesis
2. P(Socrates) $\to M$(Socrates)	From step 1 using universal specification.
3. P(Socrates)	Hypothesis
4. *M(Socrates)*	From steps 2 and 3 using modus ponens.

Thus M(Socrates), *i.e.*, 'Socrates is a mortal' is a valid conclusion from the given hypotheses. Hence the given argument is valid.

MULTIPLE CHOICE QUESTIONS

1. Which of the following sentences is a proposition?
(*a*) Kolkata is the capital of India. (*b*) How beautiful are you?
(*c*) Wish you a happy new year (*d*) $x + y = z$.

2. Let p : 'Ram reads The Anandabazar', q : 'Ram reads The Statesman' and r : 'Ram reads The Times of India. Then the symbolic form of the proposition 'Ram reads The Anandabazar or The Statesman, but not The Times of India' is
(*a*) $(p \vee q) \vee r$ (*b*) $(p \wedge q) \vee r$ (*c*) $(p \vee q) \wedge r$ (*d*) $(p \vee q) \wedge \neg r$.

3. $p \wedge q$ has the truth value T only when at least one of p and q has the truth value T.
(*a*) True (*b*) False.

4. $p \vee q$ has the truth value F only when both p and q have the truth value F.
(*a*) True (*b*) False.

5. Which of the following statements is tautology?
(*a*) $p \vee (q \to p)$ (*b*) $p \to (p \to q)$ (*c*) $p \vee (p \to q)$ (*d*) $(p \vee q) \to p$.

6. The contrapositive of $p \to q$ is
(*a*) $q \to p$ (*b*) $\neg q \to \neg p$ (*c*) $\neg p \to \neg q$ (*d*) none of these.

7. The proposition $p \wedge (\neg p \vee q)$ is
(*a*) logically equivalent to $(p \wedge q)$ (*b*) a tautology
(*c*) a contradiction (*d*) none of these.

8. Let p, q and r be three atomic propositional assertions. If X denotes $(p \vee q) \to r$ and Y denotes $(p \to r) \vee (q \to r)$ then which one of the following is a tautology?
(*a*) $Y \to X$ (*b*) $\overline{Y} \to X$ (*c*) $X \Leftrightarrow Y$ (*d*) $X \to Y$.
(GATE 2005)

9. $(p \vee p) \wedge (p \to (q \vee q))$ is equivalent to
(*a*) $p \vee q$ (*b*) $p \wedge q$ (*c*) $p \to q$ (*d*) $q \to p$.

10. Given that $(p \vee q) \wedge (\neg p \vee \neg q)$ is false, the truth values of p and q are
 (a) p true and q false (b) p false and q true
 (c) both true or both false (d) none of these.
11. $(p \to q) \wedge (r \to q)$ is equivalent to
 (a) $(p \vee r) \to q$ (b) $p \vee (r \to q)$ (c) $p \vee (r \to p)$ (d) $p \to (q \to r)$.
 [Hint: $(p \to q) \wedge (r \to q) \Leftrightarrow (\neg p \vee q) \wedge (\neg r \vee q) \Leftrightarrow (\neg p \wedge \neg r) \vee q \Leftrightarrow \neg(p \vee r) \vee q \Leftrightarrow (p \vee r) \to q$]
12. $p \rightleftarrows q$ and $(p \to q) \wedge (q \to p)$ are equivalent
 (a) True (b) False.
13. The inverse of $p \to q$ is ——.
 (a) $q \to p$ (b) $\neg p \to \neg q$ (c) $\neg q \to \neg p$ (d) none of these.
14. $p \vee \neg p$ is a tautology.
 (a) True (b) False.
15. $p \wedge \neg p$ is a tautology
 (a) True (b) False.
16. $(p \vee q) \to p$ is
 (a) a tautology (b) logically equivalent to $p \vee q$
 (c) logically equivalent to $\neg p \vee q$ (d) logically equivalent to $\neg q \vee p$.
17. If $p \rightleftarrows q$ is a tautology then $p \Leftrightarrow q$.
 (a) True (b) False.
18. If $p \to q$ is a tautology then we write $p \Rightarrow q$.
 (a) True (b) False.
19. $p \Rightarrow q$ guarantees that q has the truth value T whenever p has the truth value T.
 (a) True (b) False.
20. $(p \to \neg p) \to \neg p$ is a contradiction.
 (a) True (b) False.
21. The proposition $p \wedge (\sim p \vee q)$ is
 (a) a tautology (b) logically equivalent to $p \wedge q$
 (c) logically equivalent to $p \vee q$ (d) a contradiction. *(GATE 1993)*
 Note: $\neg p$ is also denoted as $\sim p$
22. Which of the following is/are tautology?
 (a) $a \vee b \to b \wedge c$ (b) $a \wedge b \to b \vee c$ (c) $a \vee b \to (b \to c)$ (d) $a \to b \to (b \to c)$.
 [GATE 1992]
23. $A \wedge B$ is equivalent to which of the following?
 (a) $\neg A \to \neg B$ (b) $\neg A \to B$ (c) $\neg B \to A$ (d) $\neg(A \to \neg B)$.
 (W.B.U.T. 2012)
24. The binary operation $\square$ is defined as follows:

P	Q	$P \square Q$
T	T	T
T	F	T
F	T	F
F	F	T

 Which of the following is equivalent to $P \vee Q$?
 (a) $\neg Q \square \neg P$ (b) $P \square \neg Q$ (c) $\neg P \square Q$ (d) $\neg P \square \neg Q$.
 (GATE 2009)

25. P and Q are two propositions. Which of the following logical expressions are equivalent?

I. $P \vee \neg Q$ II. $\neg(\neg P \wedge Q)$

III. $(P \wedge Q) \vee (P \wedge \neg Q) \vee (\neg P \wedge \neg Q)$ IV. $(P \wedge Q) \vee (P \wedge \neg Q) \vee (\neg P \wedge Q)$

(*a*) only I and II (*b*) only I, II and III

(*c*) only I, II and IV (*d*) all of I, II, III and IV. (*GATE 2008*)

26. Let Graph (x) be a predicate which denotes that (x) is a graph. Let connected (x) be a predicate which denotes that (x) is connected. Which of the following first order logical sentences does not represent the statement 'Not every graph is connected'?

(*a*) $\neg \forall x$ (Graph $(x) \rightarrow$ connected (x)) (*b*) $\exists x$ (Graph $(x) \wedge \neg$ Connected (x))

(*c*) $\neg \forall x$ ($\neg$ Graph $(x) \vee$ Connected (x)) (*d*) $\forall x$ (Graph $(x) \rightarrow \neg$ Connected (x)).

(*GATE 2007*)

27. Which of the following tuple relational calculus expression(s) is/are equivalent to $\forall t \in r (P(t))$?

I. $\neg \exists t \in r (P(t))$ II. $\exists t \notin r (P(t))$ III. $\neg \exists t \in r (\neg P(t))$ IV. $\exists t \notin r (\neg P(t))$

(*a*) I only (*b*) II only (*c*) III only (*d*) III and IV only.

(*GATE 2009*)

28. Which of the following is the most appropriate logical formula to represent the statement: 'Gold and Silver ornaments are precious'?

The following notations are used:

$G(x)$: x is a gold ornament.

$S(x)$: x is a silver ornament.

$P(x)$: x is precious.

(*a*) $\forall x (P(x) \rightarrow (G(x) \wedge S(x)))$ (*b*) $\forall x ((G(x) \wedge S(x)) \rightarrow P(x))$

(*c*) $\exists x ((G(x) \wedge S(x)) \rightarrow P(x))$ (*d*) $\forall x ((G(x) \vee S(x)) \rightarrow P(x))$. (*GATE 2009*)

29. Consider the following well-formed formulae:

I. $\neg \forall x (P(x))$ II. $\neg \exists x (P(x))$ III. $\neg \exists x (\neg P(x))$ IV. $\exists x (\neg P(x))$.

Which of the above are equivalent?

(*a*) I and III (*b*) I and IV (*c*) II and III (*d*) II and IV.

(*GATE 2009*)

30. Which one of the first order predicate calculus statements given below correctly expresses the following English statement?

Tigers and lions attack if they are hungry or threatened.

(*a*) $\forall x$ [(tiger $(x) \wedge$ lion $(x)) \rightarrow$ {(hungry $(x) \vee$ threatened $(x)) \rightarrow$ attacks (x)}]

(*b*) $\forall x$ [(tiger $(x) \vee$ lion $(x)) \rightarrow$ {(hungry $(x) \vee$ threatened $(x)) \wedge$ attacks (x)}]

(*c*) $\forall x$ [(tiger $(x) \vee$ lion $(x)) \rightarrow$ {attacks $(x) \rightarrow$ (hungry $(x) \vee$ threatened (x))}]

(*d*) $\forall x$ [(tiger $(x) \vee$ lion $(x)) \rightarrow$ {(hungry $(x) \vee$ threatened $(x)) \rightarrow$ attacks (x)}]. (*GATE 2006*)

31. Let P be the statement 'Soham races' and let Q be the statement 'Soham wins', then the verbal translation of $\neg(P \vee (\neg Q))$ is

(*a*) Soham does not race and Soham does not win.

(*b*) It is not true that Soham races and that Soham does not win.

(*c*) Soham does not race or Soham wins.

(*d*) It is not true that Soham races or that Soham does not win.

(*e*) It is not true that Soham does not race and Soham does not win.

32. Which of the following propositions is a tautology?

(*a*) $((\neg q) \wedge p) \wedge q$ (*b*) $(\neg q \wedge p) \wedge (p \wedge \neg p)$

(*c*) $(\neg q \wedge p) \vee (p \vee \neg p)$ (*d*) $(p \wedge q) \wedge (\neg(p \wedge q))$.

ANSWERS

1. (*a*) **2.** (*d*) **3.** (*b*) **4.** (*a*)
5. (*c*) **6.** (*b*) **7.** (*a*)
8. (*d*) $\left[\begin{array}{l} X \Leftrightarrow (p \vee q) \rightarrow r \Leftrightarrow \urcorner (p \vee q) \vee r \Leftrightarrow (\urcorner p \wedge \urcorner q) \vee r \Leftrightarrow (p \rightarrow r) \wedge (q \rightarrow r) \\ \therefore \quad X \rightarrow Y \Leftrightarrow \urcorner X \vee Y \Leftrightarrow \urcorner (p \rightarrow r) \vee \urcorner (q \rightarrow r) \vee (p \rightarrow r) \vee (q \rightarrow r) \Leftrightarrow T \end{array}\right]$
9. (*b*) **10.** (*c*) **11.** (*a*) **12.** (*a*)
13. (*b*) **14.** (*a*) **15.** (*b*) **16.** (*d*)
17. (*a*) **18.** (*a*) **19.** (*a*) **20.** (*b*)
21. (*b*) [**Hint:** $p \wedge (\urcorner p \vee q) \Leftrightarrow (p \wedge \urcorner p) \vee (p \wedge q) \Leftrightarrow F \vee (p \wedge q) \Leftrightarrow p \wedge q$]
22. (*b*) [**Hint:** $a \wedge b \rightarrow b \vee c \Leftrightarrow (\urcorner (a \wedge b)) \vee (b \vee c) \Leftrightarrow \urcorner a \vee \urcorner b \vee b \vee c \Leftrightarrow \urcorner a \vee T \vee b \Leftrightarrow T$]
23. (*d*) **24.** (*b*)
25. (*b*) [**Hint of 25:** II. $\urcorner (\urcorner P \wedge Q) \Leftrightarrow \urcorner (\urcorner P) \vee \urcorner Q \Leftrightarrow P \vee \urcorner Q$
III. $(P \wedge Q) \vee (P \wedge \urcorner Q) \Leftrightarrow P \wedge (Q \vee \urcorner Q) \Leftrightarrow P \wedge T \Leftrightarrow P.$
$\therefore (P \wedge Q) \vee (P \wedge \urcorner Q) \vee (\urcorner P \wedge \urcorner Q)$
$\Leftrightarrow \; P \vee (\urcorner P \wedge \urcorner Q) \Leftrightarrow (P \vee \urcorner P) \wedge (P \vee \urcorner Q) \Leftrightarrow T \wedge (P \vee \urcorner Q) \Leftrightarrow P \vee \urcorner Q$].
26. (*b*) **27.** (*c*) **28.** (*b*) **29.** (*b*)
30. (*a*) **31.** (*d*) **32.** (*c*).

SHORT ANSWER QUESTIONS

1. Define the connectives conjunction and disjunction and write down the truth tables for $p \wedge q$ and $p \vee q$.
2. Define conditional and biconditional propositions and give the truth tables for $p \rightarrow q$ and $p \rightleftarrows q$.
3. Define tautology and contradiction with examples.
4. Define tautological implication and show that $(p \wedge q) \Leftrightarrow (p \rightarrow q)$.
5. When do you say that two compound statements (or propositions) are equivalent? Using truth table, show that $\urcorner (p \rightarrow q) \Leftrightarrow p \wedge \urcorner q$.
6. Define disjunctive and conjunctive normal forms of a statement formula. Find a *DNF* of $p \wedge (p \rightarrow q)$.
7. Define *PDNF* and *PCNF* of a statement formula. Find the *PCNF* of $p \rightleftarrows q$.
8. What do you mean by predicate? Define simple and compound statement functions with examples.
9. Define universal and existential quantifiers. What do you mean by negation of a quantified statement function?
10. Define free and bound variables with examples in predicate logic.
11. Define the rules of specification and generalization in predicate logic.
12. Let p, q and r be the propositions
    p : you have been caught by cold
    q : you miss the class
    r : you pass the final examination.
    Express the following in English language:
    (*i*) $p \rightarrow q$ (*ii*) $\urcorner p \rightleftarrows r$ (*iii*) $q \rightarrow \urcorner r$ (*iv*) $p \vee q \vee r$
    (*v*) $(p \rightarrow \urcorner r) \vee (q \rightarrow \urcorner r)$ (*vi*) $(p \wedge q) \vee (\urcorner q \wedge r)$.
13. Let p be 'Mini reads the Anandabazar, let q be 'Mini reads The Statesman' and let r be 'Mini reads The Times of India'. Write each of the following in symbolic form:
    (*i*) Mini reads The Anandabazar or The Statesman, but not The Times of India.

(*ii*) Mini reads The Anandabazar and The Statesman, or she does not read The Anandabazar and The Times of India.

(*iii*) It is not true that Mini reads The Anandabazar but not The Times of India

(*iv*) It is not true that Mini reads The Times of India or The Statesman but not The Anandabazar.

14. Given that truth values of p and q as T and those of r and s as F, find the truth values of the following:

(*i*) $p \vee (q \wedge r)$ (*ii*) $(p \wedge q \wedge r) \vee \urcorner((\urcorner p \vee q) \wedge (r \vee s))$

(*iii*) $(\urcorner(p \wedge q) \vee \urcorner r) \vee (((\urcorner p \wedge q) \vee \urcorner p) \wedge s)$.

15. Write down the truth table for each of the following compound statement:

(*i*) $(p \wedge q) \rightarrow (p \vee q)$ (*ii*) $(p \rightarrow q) \vee (\urcorner p \rightarrow q)$

(*iii*) $(p \rightarrow q) \wedge (\urcorner p \rightarrow q)$ (*iv*) $(p \rightarrow q) \rightleftarrows (\urcorner q \rightarrow \urcorner p)$

(*v*) $(p \rightleftarrows q) \vee (\urcorner p \rightleftarrows q)$.

16. Determine which of the following propositions are tautologies or contradictions.

(*i*) $p \rightarrow (p \vee q)$ (*ii*) $(\urcorner q \rightarrow p) \wedge q$ (*iii*) $(q \rightarrow p) \wedge \urcorner p \wedge q$ (*iv*) $(p \rightarrow \urcorner p) \rightarrow \urcorner p$.

17. Prove the following equivalences

(*i*) $(p \wedge q) \vee (p \wedge \urcorner q) \Leftrightarrow p$ (*ii*) $(p \vee q) \wedge (p \vee \urcorner q) \Leftrightarrow p$

(*iii*) $\urcorner(p \rightleftarrows q) \Leftrightarrow (p \wedge \urcorner q) \vee (\urcorner p \wedge q)$.

18. Prove the following implications (using truth tables):

(*i*) $q \rightarrow p \Rightarrow \urcorner p \rightarrow \urcorner q$ (*ii*) $p \rightarrow q \Rightarrow p \rightarrow (p \wedge q)$

(*iii*) $(p \rightarrow q) \rightarrow q \Rightarrow p \vee q$.

19. Find a *DNF* or a *CNF* of each of the following statement formulas:

(*i*) $q \wedge (p \vee \urcorner q)$ (*ii*) $\urcorner(p \rightarrow q)$ (*iii*) $\urcorner(p \rightleftarrows q)$.

20. Find the *PDNF* of each of the following statement formulas:

(*i*) $(q \rightarrow p) \wedge \urcorner p \wedge q$ (*ii*) $\urcorner(p \vee q) \rightleftarrows (p \wedge q)$.

21. Find the *PCNF* of each of the following statement formulas:

(*i*) $(\urcorner(p \vee q)) \vee (p \wedge q)$ (*ii*) $(p \vee q) \rightarrow (p \wedge q)$

(*iii*) $(q \rightarrow p) \wedge \urcorner p \wedge q$ (*iv*) $p \rightarrow (p \wedge (q \rightarrow p))$.

22. If $A(x)$: x is an animal, $B(x)$: x is black and $D(x)$: x is a dog, translate the following in words:

(*i*) $\forall x\,(D(x) \rightarrow A(x))$ (*ii*) $\exists x\,(D(x) \wedge B(x))$.

23. Symbolize the propositions 'every one who likes fun will enjoy of these plays'.

24. Remove all negations of the quantifiers of the following symbolic expression:

$$\urcorner\ \forall y\,(\exists x\,A(x, y) \wedge \urcorner\ \forall x\,B(x, y))$$

25. Find the *DNF* for the statement $F(p, q, r)$ given in the following table:

p	q	r	$F(p, q, r)$
T	T	T	T
T	T	F	F
T	F	T	F
T	F	F	T
F	T	T	F
F	T	F	T
F	F	T	F
F	F	F	T

26. Check the validity of the argument:

$$p \rightarrow q,\ r \rightarrow \urcorner q \vdash p \rightarrow \urcorner r$$

ANSWERS

6. $p \wedge q$. **7.** $(\neg p \vee q) \wedge (p \vee \neg q)$

13. (*i*) $(p \vee q) \wedge \neg r$ (*ii*) $(p \wedge q) \vee \neg (p \wedge r)$ (*iii*) $\neg (p \wedge \neg r)$ (*iv*) $\neg ((r \vee q) \wedge \neg p)$

14. (*i*) T (*ii*) T (*iii*) T

15. (*i*) T, T, T, T (*ii*) T, T, T, T (*iii*) T, F, T, F (*iv*) T, T, T, T (*v*) T, T, T, T.
(with respect to the conventional order of truth values of p, q)

16. (*i*) Tautology (*ii*) Contradiction (*iii*) Contradiction (*iv*) Tautology.

19. (*i*) $p \wedge q$ (*ii*) $p \wedge \neg q$ (*iii*) $(p \wedge \neg q) \vee (q \wedge \neg p)$

20. (*i*) *PDNF* cannot be found out because the given statement is a contradiction.
(*ii*) $(p \wedge \neg q) \vee (\neg p \wedge q)$

21. (*i*) $(\neg p \vee q) \wedge (p \vee \neg q)$ (*ii*) $(\neg p \vee q) \wedge (p \vee \neg q)$ (*iii*) $(\neg p \vee \neg q) \wedge (\neg p \vee q) \wedge (p \vee \neg q) \wedge (p \vee q)$
(*iv*) PCNF cannot be found out because the given statement is a tautology.

22. (*i*) All dogs are animals. (*ii*) Some dogs are black.

23. $\forall x \forall y (F(x) \wedge P(y) \rightarrow E(x, y))$,
where $F(x)$: x likes fun, $P(y)$: y is a play and $E(x, y)$: x will enjoy y.

24. $\exists y (\forall x \neg A(x, y) \vee \forall x B(x, y))$

25. $(p \wedge q \wedge r) \vee (p \wedge \neg q \wedge \neg r) \vee (\neg p \wedge q \wedge \neg r) \vee (\neg p \wedge \neg q \wedge \neg r)$

26.

Step	*Reason*
1. $p \rightarrow q$	Hypothesis
2. $r \rightarrow \neg q$	Hypothesis
3. $q \rightarrow \neg r$	From step 2 using law of contrapositive
4. $p \rightarrow \neg r$	From steps 1 and 3 using hypothetical syllogism

Hence the given argument is valid.

PROBLEMS

1. Construct the truth table for each of the following composite propositions:
(*i*) $(\neg p \wedge (\neg q \wedge r)) \vee (q \wedge r) \vee (p \wedge r)$ (*ii*) $(p \rightarrow q) \wedge (q \rightarrow r) \rightarrow (p \rightarrow r)$
(*iii*) $((p \vee q) \wedge ((p \rightarrow r) \wedge (q \rightarrow r))) \rightarrow r$ (*iv*) $(p \rightleftarrows q) \vee (\neg q \rightleftarrows r)$
(*v*) $(p \rightleftarrows q) \rightleftarrows (r \rightleftarrows s)$.

2. Prove that the following statements are contradictions:
(*i*) $p \wedge q \wedge \neg (p \vee q)$ (*ii*) $q \vee (p \wedge \neg q) \vee (\neg p \wedge \neg q)$
(*iii*) $(p \vee q) \wedge (\neg p \vee r) \rightarrow (q \vee r)$ (*iv*) $(p \rightarrow (q \rightarrow r)) \rightarrow ((p \rightarrow q) \rightarrow (p \rightarrow r))$
(*v*) $((p \wedge q) \vee \neg (\neg p \vee (\neg q \wedge \neg r))) \wedge (\neg p \vee \neg q) \wedge (\neg p \vee \neg r)$.

3. Prove the following equivalences (without using truth tables):
(*i*) $(p \vee q) \wedge (\neg p \wedge (\neg p \wedge q)) \Leftrightarrow (\neg p \wedge q)$ (*ii*) $(p \vee q) \rightarrow r \Leftrightarrow (p \rightarrow r) \wedge (q \rightarrow r)$
(*iii*) $(\neg p \rightarrow \neg q) \rightarrow (q \rightarrow p) \Leftrightarrow T$ (*iv*) $\neg (p \rightleftarrows q) \Leftrightarrow (p \vee q) \wedge \neg (p \vee q) \wedge (p \wedge q)$.

4. Prove the following implications:
(*i*) $(p \rightarrow q) \wedge (q \rightarrow r) \Rightarrow (p \rightarrow r)$
(*ii*) $(q \rightarrow (p \wedge \neg p)) \rightarrow (r \rightarrow (p \wedge \neg p)) \Rightarrow (r \rightarrow q)$.

5. Find the disjunctive normal forms (*DNF*) of the following propositions:
(*i*) $\neg (p \rightarrow (q \wedge r))$ (*ii*) $(\neg p \rightarrow r) \wedge (p \rightleftarrows q)$
(*iii*) $(q \vee (p \wedge r)) \wedge \neg ((p \vee r) \wedge q)$.

6. Find the conjunctive normal forms (CNF) of the following propositions:
   (i) $\neg((p \vee \neg q) \wedge \neg r)$ (ii) $q \vee (p \wedge r) \wedge \neg((p \vee r) \wedge q)$.
7. Obtain the principal disjunctive normal forms (PDNF) of the following propositions:
   (i) $(p \wedge q) \vee (\neg p \wedge r) \vee (q \wedge r)$ (ii) $(\neg((p \vee q) \wedge r)) \wedge (p \vee r)$
   (iii) $(p \wedge \neg q) \vee (q \wedge \neg p) \vee (r \wedge p)$ (iv) $p \to ((p \to q) \wedge \neg(\neg q \vee \neg p))$.
8. Find the principal conjunctive normal forms (PCNF) of the following statements:
   (i) $p \vee (\neg p \to (q \vee (\neg q \to r)))$ (ii) $p \vee (\neg p \wedge \neg q \wedge r)$
   (iii) $(\neg p \to r) \wedge (q \rightleftarrows p)$.
9. Prove that $p \wedge (p \to q) \rightleftarrows p \wedge (\neg p \vee q)$ is a tautology.
10. Using truth tables, check the validity of each of the following arguments:
    (i) $p \wedge q, \neg p \to q \vdash \neg q$ (ii) $r, p \to \neg q, q \to r \vdash p$
    (iii) $p \wedge r, (p \to q) \to (r \to s) \vdash s$ (iv) $((p \to q) \to r) \to s, r \vdash s$.
11. Show that
    (i) $p \to (s \vee t)$ is a valid conclusion from the given premises (or hypotheses):
    $p \to (q \vee r), q \to s$ and $r \to t$.
    (ii) $t \to s$ is a valid conclusion from the given premises:
    $(p \wedge q) \vee (r \to s), t \to r, \neg(p \wedge q)$
    (iii) $\neg r$ is a valid conclusion from the given premises:
    $p \vee \neg q, \neg q \to r, q$
    (iv) s is a valid conclusion from the premises:
    $p \to q, q \to r, r \to s \wedge t, p$
    (v) $p \to \neg r$ is a valid conclusion from the premises:
    $p \to q, r \to \neg q$
    (vi) t is a valid conclusion from the premises:
    $\neg p \wedge q, r \to p, \neg r \to s$ and $s \to t$.
12. Let $B(x)$: x can speak Bengali and $S(x)$: x knows the language C++.
    Write down each of the following statements in symbolic form:
    (i) There is a student who can speak Bengali and who knows C++.
    (ii) There is a student who can speak Bengali but does not know C++.
    (iii) Every student either can speak Bengali or knows C++.
    (iv) No student can speak Bengali or knows C++.
13. Symbolize the following argument using quantifier, variable and predicates:

    If an integer is even, then its square is even
    m is a particular integer that is even
    ---
    m^2 is even.

    Find whether the argument is valid.

ANSWERS

1. (i) T, F, T, F, T, F, T, T (ii) A tautology (iii) A tautology (iv) T, T, T, F, F, T, T, T
   (v) $T, F, F, T, F, T, T, F, F, T, T, F, T, F, F, T$.
   (with respect to the conventional order of truth values of p, q)
5. (i) $(p \wedge \neg q) \vee (p \wedge \neg r)$ (ii) $(p \wedge q) \vee (p \wedge q \wedge r) \vee (\neg p \wedge \neg q \wedge r)$
   (iii) $(p \wedge \neg q \wedge r) \vee (\neg p \wedge q \wedge \neg r)$.
6. (i) $(\neg p \vee r) \wedge (q \vee r)$ (ii) $(p \vee q) \wedge (q \vee r) \wedge (\neg p \vee \neg q) \wedge (\neg q \vee \neg r)$

7. (i) $(p \wedge q \wedge r) \vee (p \wedge q \wedge \neg r) \vee (\neg p \wedge q \wedge r) \vee (\neg p \wedge \neg q \wedge r)$

(ii) $(p \wedge q \wedge \neg r) \vee (p \wedge \neg q \wedge \neg r) \vee (\neg p \wedge \neg q \wedge r)$

(iii) $(p \wedge q \wedge r) \vee (p \wedge \neg q \wedge r) \vee (p \wedge \neg q \wedge \neg r) \vee (\neg p \wedge q \wedge r) \vee (\neg p \wedge q \wedge \neg r)$

(iv) $(p \wedge q) \vee (\neg p \wedge q) \vee (\neg p \wedge \neg q)$

8. (i) $(p \vee q \vee r)$

(ii) $(p \vee \neg q \vee \neg r) \wedge (p \vee \neg q \vee r) \wedge (p \vee q \vee r)$

(iii) $(\neg p \vee q \vee \neg r) \wedge (\neg p \vee q \vee r) \wedge (p \vee \neg q \vee \neg r) \wedge (p \vee \neg q \vee r) \wedge (p \vee q \vee r)$

10. (i) $\exists\, x\, (B(x) \wedge S(x))$ (ii) $\exists\, x\, (B(x) \wedge \neg S(x))$

(iii) $\forall\, x\, (B(x) \vee S(x))$ (iv) $\forall\, x\, \neg (B(x) \vee S(x))$

Module II

CHAPTER 2

Theory of Numbers

2.1 INTRODUCTION

The theory of numbers, mainly deals with the properties of **natural numbers** 1, 2, 3, 4, ..., also called the **positive integers** and more generally with those of the **integers** ..., – 3, – 2, – 1, 0, 1, 2, 3, ..., plays an important role in discrete mathematics. The set of integers is denoted by Z (the letter Z comes from the word 'Zahlen' which means number in German) and the set of positive integers or natural numbers by N or by Z^+. The aim of this chapter is to introduce the notions of divisibility, greatest common divisors, prime numbers and the concept of modular (congruence) arithmetic with their applications. The theory of numbers finds its applications in computer arithmetic that includes transmission, coding and manipulation of numerical data. It has also applications in cryptology (the study of secret messages).

2.2 PRINCIPLE OF MATHEMATICAL INDUCTION

Well ordering principle: Every non-empty subset of the set of positive integers contains a least element.

This means that if S be a non-empty subset of N then $\exists m \in S$ such that $m \leq n$, $\forall\, n \in S$.

Note: The symbol $\exists$ stands for the words 'there exists', $\in$ stands for 'belongs to' and $\forall$ stands for 'for all'.

Theorem 1: (Principle of Mathematical Induction)

Let S be a subset of N with the properties:

(i) 1 belongs to S,

(ii) if k belongs to S, then k + 1 belongs to S.

Then S = N.

Proof: Let T be the set of all those positive integers which are not in S. The theorem will be proved if we can show that T is an empty set.

Let us assume that T is a non-empty set and so by the well ordering principle T contains a least element, say m. Since $1 \in S$, $m > 1 \quad \Rightarrow \quad 0 < m - 1 < m$. Again since m is the least element of T, $m - 1 \notin T$ and so $m - 1 \in S \quad \Rightarrow \quad (m - 1) + 1 \in S$, by (*ii*), *i.e.*, $m \in S$, a contradiction.

Therefore our assumption is wrong and so T is empty. This completes the proof.

Theorem 2: *Let P be a proposition (or statement) defined on N such that*

(i) P(1) is true

(ii) P(m + 1) is true whenever P(m) is true.

Then P(n) is true for every integer $n \geq 1$.

Proof: Let S be the set of those positive integers n for which the statement $P(n)$ is true.

$\therefore$ S has the properties (given):

(*i*) $1 \in S$

(*ii*) $m + 1 \in S$ whenever $m \in S$.

Then by principle of mathematical induction $S = N$.

Thus $P(n)$ is true for every integer $n \geq 1$.

Notes: (*i*) To establish a theorem (or a formula) involving positive integers by the principle of induction both the conditions (*i*) and (*ii*) must be established. The condition (*i*) is called the **basis of induction** and the assumption made in the condition (*ii*) is known as the **induction hypothesis.**

(*ii*) Theorem 2 is true if we replace 1 by 0 or by any other positive integer.

Example 1: *Use mathematical induction to prove that $n < 2^n$ for every positive integer n.*

Solution: Let $P(n)$ be the proposition '$n < 2^n$'.

Basis step: $P(1)$ is true since $1 < 2^1 = 2$.

Inductive step: Assume that $P(m)$ is true for the positive integer m.

$$\therefore \quad m < 2^m$$

$$\Rightarrow \quad m + 1 < 2^m + 1 < 2^m + 2^m = 2^{m+1}.$$

$\therefore$ $P(m + 1)$ is true whenever $P(m)$ is true.

Therefore, by the principle of mathematical induction,

$$n < 2^n \text{ for every positive integer } n.$$

Example 2: *Use mathematical induction to prove that $2^n < n\,!$ for every positive integer $n \geq 4$.*

Solution: Let $P(n)$ be the proposition $2^n < n!$

Basis step: $P(4)$ is true since $2^4 = 16 < 4! = 24$.

Inductive step: Assume that $P(m)$ is true for the positive integer m.

$$\therefore \quad 2^m < m\,! \;\Rightarrow\; 2.2^m < m\,!\,2 < m\,!\,(m + 1) \qquad [\because \; m \geq 4]$$

$$\Rightarrow \; 2^{m+1} < (m + 1)\,!$$

$\therefore$ $P(m + 1)$ is true whenever $P(m)$ is true.

Therefore, by the principle of mathematical induction,

$$2^n < n! \text{ for every positive integer } n \geq 4.$$

Example 3: *Use mathematical induction to prove that $n^3 - n$ is divisible by 3 for every positive integer n.*

Solution: Let $P(n)$ be the proposition '$n^3 - n$ is divisible by 3'.

Basis step: $P(1)$ is true since $1^3 - 1 = 0$ is divisible by 3.

Inductive step: Assume that $P(m)$ is true for the positive integer m, *i.e.*, $m^3 - m$ is divisible by 3. We have to show that $P(m + 1)$ is true, *i.e.*, $(m + 1)^3 - (m + 1)$ is divisible by 3.

Now, $$(m + 1)^3 - (m + 1) = (m^3 + 3m^2 + 3m + 1) - (m + 1)$$

$$= (m^3 - m) + 3(m^2 + m)$$

This is divisible by 3 since $(m^3 - m)$ is divisible by 3 (by assumption) and $3(m^2 + m)$ is also divisible by 3.

$\therefore$ $P(m + 1)$ is true whenever $P(m)$ is true.

Hence by the principle of mathematical induction, $n^3 - n$ is divisible by 3 for every positive integer n.

Example 4: *Assuming principle of mathematical induction, prove that every non-empty subset of the set of positive integers contains a least element* (*well ordering principle*).

Solution: Suppose S be any non-empty subset of N. If possible let S has no least element. Let M be the set of all those positive integers which are less than every element of S. Then (*i*) $1 \in M$. Otherwise, $1 \in S$ and 1 be a least element of S which is a contradiction.

Suppose $k \in M$, then k is less than every element of S. Therefore $k + 1 \in M$, otherwise $k + 1$ would be a least element of S which is a contradiction. Hence (*ii*) $k + 1 \in M$ whenever $k \in M$.

From (*i*) and (*ii*), by principle of mathematical induction $M = N$ and so S is empty. This contradicts the hypothesis that S is non-empty. Therefore, the original assumption that S has no least element cannot be true. Hence the result.

Note: Well ordering principle is equivalent to the principle of mathematical induction.

2.3 DIVISIBILITY THEORY

Theorem 3: (Division Algorithm)

Given any two integers a and b with b > 0, there exist unique integers q and r such that

$$a = bq + r, \qquad 0 \le r < b.$$

Proof: Let us consider the sequence of multiples of b:

$$\ldots, -2b, -b, 0, b, 2b, \ldots, qb, \ldots .$$

Clearly $a = qb$ or $qb < a < (q + 1)b$, for some $q \in Z$.

$$\therefore \quad qb \le a < (q + 1)\, b$$

$$\Rightarrow \quad 0 \le a - qb < b$$

$$\Rightarrow \quad 0 \le r < b, \text{ where } r = a - qb$$

$$\Rightarrow \quad a = bq + r, \qquad 0 \le r < b$$

Uniqueness: In order to establish uniqueness of q and r, let us assume that a has two representations:

$$a = bq_1 + r_1, \quad 0 \le r_1 < b$$

$$a = bq_2 + r_2, \quad 0 \le r_2 < b$$

$$\therefore \quad a = bq_1 + r_1 = bq_2 + r_2$$

$$\Rightarrow \quad (q_1 - q_2)b = r_2 - r_1 \qquad \ldots(1)$$

$\Rightarrow$ $(r_2 - r_1)$ is an integral multiple of b. ...(2)

Now, $\quad 0 \le r_1 < b \;\Rightarrow\; b > r_1 \ge 0 \;\Rightarrow\; -b < -r_1 \le 0$ ...(3)

Also, $\quad 0 \le r_2 < b$ ...(4)

From (3) and (4), we get

$$-b < r_2 - r_1 < b \;\Rightarrow\; |r_2 - r_1| < b \qquad \ldots(5)$$

From (2) and (5), we get

$$r_2 - r_1 = 0 \;\Rightarrow\; r_1 = r_2$$

Hence from (1), we get

$$q_1 = q_2 \text{ as } r_1 = r_2 \text{ and } b > 0.$$

Therefore, q and r are unique.

Note: When $b \neq 0$, the result of this theorem can be stated as $a = bq + r$, where $0 \le r < |b|$.

Definition 1: The integer q is called the **quotient** and the integer r is called the **remainder** in the division of a by b.

To illustrate the division algorithm, let us take

$$b = 3, a = -16, 1.$$

$\therefore$ $-16 = 3.(-6) + 2$ gives $q = -6$, $r = 2$,

$1 = 3.0 + 1$ gives $q = 0$, $r = 1$

Let us take $b = -2$, $a = -17, 5, 6$.

$\therefore$ $-17 = (-2).9 + 1$ gives $q = 9$, $r = 1$,

$5 = (-2).(-2) + 1$ gives $q = -2$, $r = 1$,

$6 = (-2).(-3) + 0$ gives $q = -3$, $r = 0$.

When the remainder in the division algorithm turns out to be zero, the case is of special interest.

Definition 2: An integer a is said to be **divisible** by an integer $b \neq 0$ if there exists an integer c such that $a = bc$.

It is denoted by the notation $b \mid a$ and read as 'b divides a'.

When b divides a, b is called a **divisor** or **factor** of a and a is called a **multiple of** b.

If b is a divisor of a, then $-b$ is also a divisor of a. Because $a = bc$ can be rewritten as $a = (-b).(-c)$. So divisors of an integer occur in pairs.

Some of basic properties of divisibility of integers are given in the following theorem.

Theorem 4: Let *a, b and c be integers and assume that a divisor is always a non-zero integer.*

(i) If $a \mid b$ *and* $b \neq 0$, *then* $|a| \le |b|$.

(ii) If $a \mid b$ *and* $b \mid c$, *then* $a \mid c$.

(iii) If $a \mid b$ *and* $b \mid a$, *then* $a = \pm b$.

(iv) If $a \mid b$, *then* $a \mid bc$ *for any* $c \in Z$.

(v) If $a \mid b$ *and* $a \mid c$, *then* $a \mid (bx + cy)$ *for arbitrary* $x, y \in Z$.

Proof: (*i*) If $a \mid b$ and $b \neq 0$, then $b = ma$ for some $m \in Z$, with $m \neq 0$. Then $|m| \ge 1$ and so $|b| = |ma| = |m||a| \ge |a|$, *i.e.*, $|a| \le |b|$.

(*ii*) Since $a \mid b$ and $b \mid c$, there exist integers m, n such that $b = am$ and $c = bn$.

$\therefore$ $c = bn = (am)\,n = a\,(mn) \Rightarrow a \mid c$ as $mn \in Z$.

(*iii*) Since $a \mid b$ and $b \mid a$, there exists integers m, n such that $b = am$ and $a = bn$.

$\therefore$ $b = am = (bn)\,m = b\,(nm) \Rightarrow nm = 1$, as $b \neq 0$.

$\Rightarrow$ $m = n = 1$, or $m = n = -1$ $(\because m, n \in Z)$

$\Rightarrow$ $a = \pm b$.

(*iv*) Since $a \mid b$, there exists $f \in Z$ such that $b = af$.

$\therefore$ $bc = (af)c = a(fc) \Rightarrow a \mid bc$ as $fc \in Z$.

(*v*) Since $a \mid b$ and $a \mid c$, there exist integers m, n such that $b = am$ and $c = an$.

$\therefore$ $bx + cy = (am)\,x + (an)y = a(mx + ny) \Rightarrow a \mid (bx + cy)$, as $mx + ny \in Z$.

Example 5: *Prove that the square of an odd integer is of the form $8k + 1$, $k \in Z$.*

Solution: By division algorithm, every integer upon division by 4 leaves one of the remainders 0, 1, 2, 3. Therefore any integer is one of the forms $4q$, $4q + 1$, $4q + 2$, $4q + 3$ among which $4q + 1$ and $4q + 3$ are odd integers where $q \in Z$.

Now, $(4q + 1)^2 = 8(2q^2 + q) + 1$ is of the form $8k + 1$, where $k = 2q^2 + q \in Z$.

Also, $(4q + 3)^2 = 8(2q^2 + 3q + 1) + 1$ is of the form $8k + 1$, where $k = 2q^2 + 3q + 1 \in Z$. **(Proved)**

Example 6: *Show that square of any integer is of the form $4n$ or $4n + 1$ for some integer n.*

Solution: By division algorithm, every integer upon division by 2 leaves one of the remainders 0, 1. Therefore any integer is one of the forms $2q$, $2q + 1$ among which $2q$ is an even integer and $2q + 1$ is an odd integer where $q \in Z$.

Now, $(2q)^2 = 4q^2$ is of the form $4n$, where $n = q^2 \in Z$.

$(2q + 1)^2 = 4(q^2 + q) + 1$ is of the form $4n + 1$, where $n = q^2 + q \in Z$.

Hence square of any integer is of the form $4n$ or $4n + 1$ for some integer n. **(Proved)**

Definition 3: If a and b are integers then an integer $d(\neq 0)$ is said to be a **common divisor** of a and b if $d \mid a$ as well as $d \mid b$ (*i.e.*, if d divides both a and b).

Notes: (*i*) Since 1 is a divisor of every integer, so for an arbitrary pair of integers a, b there exists always a common divisor.

(*ii*) If both a, b be 0, then each integer is a common divisor of a and b. But if at least one of a and b is non-zero then there is only a finite number of **positive** common divisors. The largest of all positive common divisors is called the **greatest common divisor** of a, b and denoted as gcd (a, b).

Definition 4: If a, b are integers, not both zero, the **greatest common divisor** of a and b, denoted as gcd (a, b), is the **positive integer** d satisfying

(*i*) $d \mid a$ and $d \mid b$

(*ii*) If $c \mid a$ and $c \mid b$, then $c \mid d$.

For example, let $a = 12$, $b = -16$. The positive divisors of 12 are 1, 2, 3, 4, 6, 12 and those of -16 are 1, 2, 4, 8, 16.

Therefore the positive common divisors are 1, 2, 4 and gcd $(12, -16) = 4$.

Theorem 5: *For every pair of integers a, b, not both zero, there exist integers u, v such that*

$$\gcd(a, b) = au + bv.$$

Proof: Consider the set $S = \{ax + by \mid x, y \in Z \text{ and } ax + by > 0\}$.

Since at least one of a, b is non-zero, let $a \neq 0$ and so $|a| > 0$.

$\therefore$ $|a| = a \cdot x + b \cdot 0 \in S$, where we choose $x = 1$ if $a > 0$ and $x = -1$ if $a < 0$. Hence S is a non-empty set of positive integers and so by the well ordering principle S contains a least element d, say.

$$\therefore \quad d = au + bv, \text{ for some integers } u, v.$$

Using division algorithm, $a = dq + r$ where q, r are integers and $0 \leq r < d$.

$$\therefore \quad r = a - dq = a - (au + bv)q = a(1 - uq) + b(-vq)$$

If possible let $r \neq 0$, therefore $0 < r < d$ and so $r \in S$. This is contradictory since d is the least element in S. Hence $r = 0$. Therefore $a = dq$, *i.e.*, d is a divisor of a. Similarly it can be proved that d is a divisor of b. Hence d (> 0) is a common divisor of a and b.

Next let c is a common divisor of a and b. Then $c \mid a$, $c \mid b$ and so $c \mid (au + bv)$, by Theorem 4(*v*). Thus $c \mid d$ and this proves that

$$\gcd(a, b) = d = au + bv, \text{ for some integers } u, v.$$

Notes: *(i)* Integers u, v as stated in this theorem are not unique because $au + bv = a(u - b) + b(v + a)$.

(ii) If $d = \gcd(a, b)$, then for $c \in Z$, $c = au + bv$ if and only if $d \mid c$.

For example, suppose if possible, we want to find $u, v \in Z$ such that $14 = 6u + 12v$.

We have $\gcd(6, 12) = 6 = d$ and $c = 14$. Since 6 is not a divisor of 14, it is not possible to write $14 = 6u + 12v$, where $u, v \in Z$.

(iii) This theorem says that it is always possible to express $\gcd(a, b) = au + bv$, where $u, v \in Z$. But this theorem gives no idea how to express $\gcd(a, b)$ in the desired form $au + bv$, *i.e.*, how to find u and v.

For example, $\gcd(12, -16) = 4$ and $4 = 12 \cdot (-1) + (-16) \cdot (-1)$

$\gcd(-3, 21) = 3$ and $3 = (-3) \cdot (-1) + 21 \cdot 0$

$\gcd(14, -35) = 7$ and $7 = 14 \cdot (-2) + (-35) \cdot (-1)$

$\gcd(15, 33) = 3$ and $3 = 15 \cdot (-2) + 33 \cdot (1)$

Definition 5: Two integers a and b, not both zero, are said to be **prime to each other** (or **relatively prime** or **coprime**) if $\gcd(a, b) = 1$.

For example, 2, – 7 are prime to each other since $\gcd(2, -7) = 1$.

Note: If a and b are prime to each other, then a and b have no common factors except 1 or – 1.

Theorem 6: *Two integers a and b, not both zero, are prime to each other if and only if there exist integers u and v such that $1 = au + bv$.*

Proof: First suppose integers a and b are prime to each other. Then $\gcd(a, b) = 1$. Therefore by Theorem 5, there exist integers u and v such that $1 = au + bv$.

Conversely, let us suppose that there exist integers u and v such that $1 = au + bv$ and let $d = \gcd(a, b)$.

Now, $d = \gcd(a, b) \Rightarrow d \mid a, d \mid b$

$\Rightarrow d \mid (au + bv)$

$\Rightarrow d \mid 1$

$\Rightarrow d = 1$ [$\because$ greatest common divisor is to be positive]

Hence $\gcd(a, b) = 1$ and so a and b are prime to each other.

Example 7: *For any integer n, prove that the integers 8n + 3 and 5n + 2 are relatively prime. Hence find integers x, y such that*

$$(8n + 3)x + (5n + 2)y = \gcd(8n + 3, 5n + 2). \qquad (W.B.U.T.\ 2012)$$

Solution: For any integer n, the integers $8n + 3$ and $5n + 2$ will be relatively prime if there exist integers x, y such that

$$(8n + 3)x + (5n + 2)y = 1 \qquad \text{[By Theorem 6]}$$

$$\Rightarrow (8x + 5y)n + (3x + 2y) = 1$$

$$\Rightarrow 8x + 5y = 0 \text{ and } 3x + 2y = 1 \qquad (\because n \text{ is any integer}),$$

Solving we get, $x = -5$ and $y = 8$.

$$\therefore (8n + 3)(-5) + (5n + 2) \times 8 = 1 = \gcd(8n + 3, 5n + 2).$$

Theorem 7: *If $\gcd(a, b) = d$, then $\frac{a}{d}$ and $\frac{b}{d}$ are integers prime to each other.*

Proof: $\gcd(a, b) = d \Rightarrow d \mid a, d \mid b$

$\Rightarrow$ there exist integers m, n such that $a = md$, $b = nd$

$\Rightarrow \frac{a}{d} = m$ and $\frac{b}{d} = n$ are integers.

Also, $\gcd(a, b) = d \Rightarrow$ there exist integers u, v such that $d = au + bv$

$\Rightarrow \quad 1 = \left(\frac{a}{d}\right)u + \left(\frac{b}{d}\right)v$

$\Rightarrow \quad \frac{a}{d}$ and $\frac{b}{d}$ are prime to each other, by Theorem 6.

Theorem 8: *If $a \mid bc$ and $\gcd(a, b) = 1$, then $a \mid c$.*

Proof: Given : $\gcd(a, b) = 1$

$\Rightarrow \quad \exists\, u, v \in Z$ such that $1 = au + bv$, by Theorem 5

$\Rightarrow \quad c = acu + bcv$ ...(1)

Also, $a \mid ac$ and $a \mid bc$ (given) $\Rightarrow \quad a \mid \{(ac)u + (bc)v\} \quad \Rightarrow \quad a \mid c$ [By (1)]

Theorem 9: *If $a \mid c$ and $b \mid c$ with $\gcd(a, b) = 1$, then $ab \mid c$.*

Proof: Given : $a \mid c, b \mid c \quad \Rightarrow \quad \exists m, n \in Z$ such that $c = am = bn$ ...(1)

Given : $\gcd(a, b) = 1 \quad \Rightarrow \quad \exists u, v \in Z$ such that $1 = au + bv$, by Theorem 5

$\Rightarrow \quad c = auc + bvc = ab\,(un + vm)$ [By (1)]

$\Rightarrow \quad ab \mid c$, as $un + vm \in Z$.

Note: If $\gcd(a, b) \neq 1$, then $a \mid c$ and $b \mid c$ together may not imply $ab \mid c$.

For example, $3 \mid 12$ and $6 \mid 12$ do not imply $3.6 \mid 12$.

Theorem 10: *If $\gcd(a, b) = 1$ and $\gcd(a, c) = 1$, then $\gcd(a, bc) = 1$.*

Proof: Given : $\gcd(a, b) = 1 \quad \Rightarrow \quad \exists\, u_1, v_1 \in Z$ such that $au_1 + bv_1 = 1$

$\Rightarrow \quad bv_1 = 1 - au_1$ ...(1)

Given : $\gcd(a, c) = 1 \quad \Rightarrow \quad \exists u_2, v_2 \in Z$ such that $au_2 + cv_2 = 1$

$\Rightarrow \quad cv_2 = 1 - au_2$ ...(2)

From (1) and (2), we have

$$(bv_1)(cv_2) = (1 - au_1)(1 - au_2)$$

$$\Rightarrow \quad bcv_1v_2 = 1 - a(u_1 + u_2) + a^2u_1u_2$$

$$\Rightarrow \quad a(u_1 + u_2 - au_1u_2) + (bc)(v_1 v_2) = 1$$

$$\Rightarrow \quad au_3 + bcv_3 = 1, \text{ where } u_3 = u_1 + u_2 - au_1 u_2 \in Z,\ v_3 = v_1v_2 \in Z.$$

$$\Rightarrow \quad \gcd(a, bc) = 1, \text{ by Theorem 6.}$$

Notes: (*i*) The converse of Theorem 10 is also true, *i.e.*, $\gcd(a, bc) = 1 \quad \Rightarrow \quad \gcd(a, b) = 1$ and $\gcd(a, c) = 1$. See Miscellaneous Ex. 8.

(*ii*) By applying Theorem 10, it can be proved by mathematical induction that

$$\gcd(a, b) = 1 \quad \Rightarrow \quad \gcd(a, b^n) = 1, \forall n \in N.$$

Example 8: *Show that the product of any three consecutive integers is divisible by 6.*

Solution: By division algorithm, every integer upon division by 3 leaves one of the remainders 0, 1, 2. Therefore any integer n has one of the forms $3q$, $3q + 1$, $3q + 2$ where $q \in Z$.

When $n = 3q$, n is divisible by 3.

When $n = 3q + 1$, $n + 2$ is divisible by 3.

When $n = 3q + 2$, $n + 1$ is divisible by 3.

Therefore, whatever integer n may be, $n\,(n + 1)\,(n + 2)$ is divisible by 3.

Again, the product of two consecutive integers is divisible by 2.

$\therefore$ $2 \mid n(n+1)(n+2)$ and $3 \mid n(n+1)(n+2)$

Since $\gcd(2, 3) = 1$, by Theorem 9, $6 \mid n(n+1)(n+2)$.

2.4 EUCLIDEAN ALGORITHM FOR FINDING gcd (a, b)

Statement

Let a and b are any two integers such that $a > b > 0$. If r_1 is the remainder when a is divided by b, r_2 is the remainder when b is divided by r_1, r_3 is the remainder when r_1 is divided by r_2 and so on, then gcd $(a, b) = r_k$ where r_k is the last non-zero remainder and $r_{k+1} = 0$.

Proof: We have the following relations:

$$a = bq_1 + r_1 \qquad 0 < r_1 < b$$
$$b = r_1 q_2 + r_2 \qquad 0 < r_2 < r_1$$
$$r_1 = r_2 q_3 + r_3 \qquad 0 < r_3 < r_2$$
$$\cdots\cdots\cdots$$
$$r_{k-2} = r_{k-1} q_k + r_k \qquad 0 < r_k < r_{k-1}$$
$$r_{k-1} = r_k q_{k+1} + 0$$

We first prove that $\gcd(a, b) = \gcd(b, r_1)$.

Let $d_1 = \gcd(a, b)$ ...(1)

and $d_2 = \gcd(b, r_1)$ ...(2)

Now, by (2), $d_2 \mid b$ and $d_2 \mid r_1$, so $d_2 \mid (bq_1 + r_1)$, *i.e.*, $d_2 \mid a$.

Thus d_2 is a common divisor of a and b.

$\therefore$ $d_2 \le d_1$, as $d_1 = \gcd(a, b)$ ...(3)

Next, by (1), $d_1 \mid a$ and $d_1 \mid b$, so $d_1 \mid (a - b\,q_1)$, *i.e.*, $d_1 \mid r_1$.

So d_1 is a common divisor of b and r_1.

$\therefore$ $d_1 \le d_2$, as $d_2 = \gcd(b, r_1)$ ...(4)

From (3) and (4), we get $d_1 = d_2$,

i.e., $\gcd(a, b) = \gcd(b, r_1)$, where $a = bq_1 + r_1$ ...(5)

Since $r_1, r_2, r_3, \ldots$ form a decreasing sequence of non-negative integers, there must exist an r_{k+1} equal to zero.

Now by (5):

$$\gcd(a, b) = \gcd(b, r_1) = \gcd(r_1, r_2) = \ldots = \gcd(r_{k-1}, r_k) = \gcd(r_k, 0) = r_k$$

Hence $\gcd(a, b) = r_k$, which is the last non-zero remainder.

Note: Since $\gcd(a, b) = \gcd(|a|, |b|)$, it is enough to assume that a and b are positive integers. Without loss of generality, we assume $a > b > 0$.

Example 9: *Calculate gcd (1761, 189) and express gcd (1761, 189) as 1761u + 189 v.*

Solution: By division algorithm:

$$1761 = 9 \times 189 + 60$$
$$189 = 3 \times 60 + 9$$
$$60 = 6 \times 9 + 6$$
$$9 = 1 \times 6 + 3$$
$$6 = 2 \times 3 + 0$$

Since the last non-zero remainder is 3, gcd (1761, 189) = 3.

Also,
$$\begin{aligned} 3 &= 9 - 1 \times 6 \\ &= 9 - (60 - 6 \times 9) \\ &= 7 \times 9 - 60 \\ &= 7 \times (189 - 3 \times 60) - 60 \\ &= 7 \times 189 - 22 \times 60 \\ &= 7 \times 189 - 22 \times (1761 - 9 \times 189) \\ &= 205 \times 189 - 22 \times 1761 \\ &= 1761u + 189v, \text{ where } u = -22,\ v = 205. \end{aligned}$$

2.5 PRIME NUMBERS

Definition 6: An integer $p > 1$ is called a **prime number**, or simply a **prime** if its only positive divisors are 1 and p.

An integer that is greater than 1 and is not prime is called **composite.**

The prime numbers less than 100 are 2, 3, 5, 7, 11, 13, 17, 19, 23, 29, 31, 37, 41, 43, 47, 53, 59, 61, 67, 71, 73, 79, 83, 89 and 97.

Observe that the only even that is a prime is 2 ; rest all other primes are odd integers.

4, 6, 8, 9, ... are composite numbers. The integer 1 is neither prime nor composite.

Note: A positive integer n is composite if there exist positive integers a and b such that $n = ab$, where $1 < a, b < n$.

Theorem 11: *If p be a prime and a is an integer such that $1 \leq a < p$, then p is prime to a.*

Proof: Let $d = \gcd(a, p)$, then $d \mid a$ and $d \mid p$.

Since p is a prime and $d \mid p$, so either $d = p$ or $d = 1$.

But d cannot be p because $d \mid a$ and $a < p$ (given).

$\therefore$ $d = 1$, *i.e.*, p is prime to a.

Theorem 12: If *p be a prime number and a is an integer greater than p such that p is not a divisor of a, then p is prime to a.*

Proof: Let $d = \gcd(a, p)$, then $d \mid a$ and $d \mid p$.

Since p is a prime and $d \mid p$, so either $d = p$ or $d = 1$. But d cannot be p because p is not a divisor of a.

$\therefore$ $d = 1$, *i.e.*, p is prime to a.

Theorem 13: *If p be a prime and a is an integer greater than p such that $p \mid a$, then* $\gcd(a, p) = p$.

Proof: Let $d = \gcd(a, p)$, then $d \mid a$ and $d \mid p$.

Since p is a prime and $d \mid p$, so either $d = p$ or $d = 1$. But $p \mid a$ and $d = \gcd(a, p)$.

$\therefore$ $d = p$, *i.e.*, $\gcd(a, p) = p$.

Theorem 14: *If p be a prime and $p \mid ab$, then either $p \mid a$* or *$p \mid b$.*

Proof: If $p \mid a$, then the theorem is proved. Let us assume that p is not a divisor of a.

Let $d = \gcd(a, p)$, then $d \mid a$ and $d \mid p$.

Since p is a prime and $d \mid p$, so either $d = p$ or $d = 1$. But $d \neq p$, because p is not a divisor of a. Therefore $d = 1$, *i.e.*, $\gcd(a, p) = 1$. Hence there exist integers u and v such that $1 = au + pv$ and so $b = abu + pbv$.

Since $p \mid ab$ (given) and $p \mid pb$, it follows that $p \mid \{(ab)u + (pb)v\}$ which means $p \mid b$. This completes the proof.

Theorem 15: *If p be a prime and $p \mid a_1 a_2 \ldots a_n$, then $p \mid a_k$ for some k where $1 \leq k \leq n$.*

Proof: If $p \mid a_1$ then the theorem is proved, we need not go further. If p is not a divisor of a_1, then $p \mid a_2 a_3 \ldots a_n$. If p is not a divisor a_2, then $p \mid a_3 a_4 \ldots a_n$. Proceeding in a similar manner, in a finite number of steps we arrive at the desired result.

Theorem 16: *A composite number has at least one prime divisor.*

Proof: Let q be a composite number. Since q is not prime, it has a positive divisor other than 1 and q. Let S be the set of those positive divisors of q which are different from 1 and q. Then S is non-empty and so by the well ordering principle, S contains a least element, say d where $1 < d < q$. Let us prove that d is a prime. If d is not a prime then d has a divisor d_1 other than 1 and d, so $1 < d_1 < d < q$. But $d_1 \mid d$ and $d \mid q \Rightarrow d_1 \mid q$ and this contradicts that d is the least member of S. Hence d is prime and the theorem is proved.

Theorem 17: (*Fundamental Theorem of Arithmetic*)

Any integer greater than 1 is either a prime or can be expressed as a product of primes, the representation being unique except for the order of the prime factors.

Proof: Let $n > 1$ be an integer. If n is a prime, then there is nothing to prove. Suppose n is a composite number then n has a smallest divisor (other than 1) which is a prime (see the proof of Theorem 16). Let this be p_1 and so $n = p_1 n_1$ for some $n_1 \in Z$ such that $n > n_1 > 1$. If n_1 is a prime then the theorem is proved, otherwise n_1 has a smallest divisor (other than 1) which is a prime, say p_2.

Hence $n_1 = p_2 n_2 \Rightarrow n = p_1 p_2 n_2$ such that $n > n_1 > n_2 > 1$ and $p_1 \leq p_2$. The argument is continued. The decreasing sequence $n > n_1 > n_2 > \ldots . > 1$ cannot continue indefinitely. After a finite number of steps n_r becomes a prime, say p_r. This yields the prime factorization: $n = p_1 p_2 \ldots p_r$ where $p_1 \leq p_2 \leq \ldots \leq p_r$.

Uniqueness: Now, if possible, suppose $n = p_1 p_2 \ldots p_r$ and $n = q_1 q_2 \ldots q_s$ are two different prime factorizations (except for the order of the prime factors) with $r < s$. Since $p_1 \mid n = q_1 q_2 \ldots q_s$, there exists at least one integer i such that $p_1 \mid q_i$. But p_1 and q_i are both primes, so $p_1 = q_i$. Without loss of generality we can take $i = 1$ by proper reordering and so $p_1 = q_1$. Cancelling this common factor we obtain $p_2 p_3 \ldots p_r = q_2 q_3 \ldots q_s$.

Repeating this process till all p_k's get exhausted as $r < s$, we are eventually left with $1 = q_{r+1} \cdot \ldots q_s$. But it is not possible as q_k's are primes. Hence r cannot be less than s. Similarly r cannot be greater than s. Therefore, $r = s$. This completes the proof.

Note: Some of the primes which appear in the factorization of n may recur and so the factorization of n takes the form

$$n = p_1^{\alpha_1} p_2^{\alpha_2} \ldots p_m^{\alpha_m} \text{ where } p_1 < p_2 < \ldots < p_m.$$

Theorem 18. *Suppose that the prime factorizations of the integers m and n are*

$$m = p_1^{a_1} p_2^{a_2} \ldots p_k^{a_k} = \Pi\, p_i^{a_i} \quad \text{and} \quad n = p_1^{b_1} p_2^{b_2} \ldots p_k^{b_k} = \Pi\, p_i^{b_i}$$

where each exponent is a non-negative integer. Then gcd $(m, n) = \Pi\, p_i^{\min(a_i, b_i)}$.

where min (x, y) represents the minimum of the two numbers x and y.

Proof: (*i*) Since $\min(a_i, b_i) \leq a_i$, $\Pi\, p_i^{\min(a_i, b_i)}$ divides $\Pi\, p_i^{a_i} = m$. Similarly $\Pi\, p_i^{\min(a_i, b_i)}$ divides $\Pi\, p_i^{b_i} = n$.

(*ii*) Let an integer c divides both the integers m and n. It follows that c is the product of primes not outside $p_1, p_2, ..., p_k$. Let us take

$$c = p_1^{h_1}\, p_2^{h_2} \dots p_k^{h_k} = \Pi\, p_i^{h_i}$$

So, $\Pi\, p_i^{h_i} \mid \Pi\, p_i^{a_i}$ and $\Pi\, p_i^{h_i} \mid \Pi\, p_i^{b_i} \Rightarrow h_i \le a_i$ and $h_i \le b_i$

$\Rightarrow$ $h_i \le \min\,(a_i, b_i)$

$\Rightarrow$ $\Pi\, p_i^{h_i} \mid \Pi\, p_i^{\min(a_i, b_i)}$

From (*i*) and (*ii*), we conclude that

$$\gcd\,(m, n) = \Pi\, p_i^{\min(a_i, b_i)}.$$

Example 10: *Use prime factorization to find the greatest common divisor of 792 and 21296.*

Solution: Prime factorization of 792 and 21296 are $792 = 2^3 \times 3^2 \times 11^1$ and $21296 = 2^4 \times 3^0 \times 11^3$.

Hence $\gcd\,(792, 21296) = 2^{\min(3, 4)} \times 3^{\min(2, 0)} \times 11^{\min(1, 3)}$

$$= 2^3 \times 3^0 \times 11 = 88.$$

Note: All primes occurring in the prime factorization of either m or n are to include in both factorization, with zero exponent if necessary.

Theorem 19: *The number of primes is infinite.*

Proof: We prove the theorem by contradiction. Let us assume that the number of primes be finite and let p be the greatest prime. We write the primes 2, 3, 5, 7, ... in order of magnitude and p be the last in the enumeration.

The product $2 \,.\, 3 \,.\, 5 \dots p$ in which each prime occurs only once is divisible by each prime and therefore the integer $(2 \,.\, 3 \,.\, 5 \dots p) + 1$ is not divisible by any of the primes 2, 3, 5, ..., p.

Hence this integer is either itself a prime or, being a composite, divisible by a prime greater than p because a composite number has a smallest divisor (other than 1) which is a prime (see the proof of Th. 16). In both the cases p fails to be the greatest prime and hence the number of primes is infinite.

Note: Alternative statement

If p be a prime, then there exists a prime q such that $p < q$.

Prime Testing

It is often an important task for computers of deciding whether a positive integer is a prime or composite. Primality testing have become important in the application of number theory to cryptology where large primes are used in some methods for making messages secret. One procedure for checking that an integer is prime is based on the observation of the following theorem.

Theorem 20: *If $n > 1$ is a composite integer, then n has a prime divisor less than or equal to $\sqrt{n}$.*

Proof: If $n > 1$ is composite, then it has a factor a with $1 < a < n$. Hence $n = ab$, where both a and b are integers with $1 < a, b < n$. Observe that $a \le \sqrt{n}$, or $b \le \sqrt{n}$, since otherwise $ab > \sqrt{n} \,.\, \sqrt{n} = n$. Thus n has a positive divisor not exceeding $\sqrt{n}$. This divisor is either a prime or, by the Fundamental Theorem of Arithmetic, has a prime divisor. So, in either case, n has a prime divisor less than or equal to $\sqrt{n}$.

Note: From Theorem 20, it follows that an integer is prime if it is not divisible by any prime less than or equal to its square root. For example, 101 is a prime. Because the only primes not exceeding $\sqrt{101}$ are 2, 3, 5, 7 and 101 is not divisible by 2, 3, 5 or 7.

MISCELLANEOUS EXAMPLES

Example 1: *If gcd (a, b) = d, a | c, b | c,* then show that *ab | cd.*

Solution: Given : $a \mid c,\ b \mid c \Rightarrow c = am = bn$, for some $m, n \in Z$ ...(1)

Given: $\gcd(a, b) = d \Rightarrow d = au + bv$, for some $u, v \in Z$, by Theorem 5.

$\Rightarrow$ $cd = cau + cbv = (bn)au + (am)bv = ab(nu + mv)$, by (1)

$\Rightarrow$ $ab \mid cd$, as $nu + mv \in Z$.

Example 2: *If* $d > 0, d \mid a, d \mid b, \gcd\left(\frac{a}{d}, \frac{b}{d}\right) = 1$, *the prove that* $\gcd(a, b) = d$.

Solution: Given : $d \mid a,\ d \mid b \Rightarrow d \mid \gcd(a, b) = d_1$ (say) ...(1)

Given: $\gcd\left(\frac{a}{d}, \frac{b}{d}\right) = 1 \Rightarrow 1 = \left(\frac{a}{d}\right)u + \left(\frac{b}{d}\right)v$, for some $u, v \in Z$, by Theorem 5.

$\Rightarrow$ $d = au + bv$ ...(2)

Now, $d_1 = \gcd(a, b) \Rightarrow d_1 \mid a$ and $d_1 \mid b$

$\Rightarrow$ $d_1 \mid (au + bv) \Rightarrow d_1 \mid d$, by (2) ...(3)

From (1) and (3), we have $d_1 = \pm d$.

But d_1 and d are both positive, therefore $d_1 = d$.

Hence $\gcd(a, b) = d$. **(proved)**

Note: This example is the converse of Theorem 7.

Example 3: *If gcd (a, b) = 1, then prove that*

(i) $\gcd(a + b, a - b) = 1$ *or 2*

(ii) $\gcd(2a + b, a + 2b) = 1$ *or 3.*

Solution: *(i)* Let $\gcd(a + b, a - b) = d$.

$\therefore$ $d \mid (a + b)$ and $d \mid (a - b)$

$\Rightarrow$ $d \mid \{(a + b) + (a - b)\}$ and $d \mid \{(a + b) - (a - b)\}$

$\Rightarrow$ $d \mid 2a$ and $d \mid 2b$

$\Rightarrow$ d is a common divisor of $2a$ and $2b$. ...(1)

Given : $\gcd(a, b) = 1 \Rightarrow \gcd(2a, 2b) = 2$, by Ex. 2 ...(2)

From (1) and (2), we have d is a common divisor of $2a, 2b$ and 2 is the greatest common divisor of $2a, 2b$. Therefore, by the definition of greatest common divisor we must have $d \mid 2 \Rightarrow d = 1$ or 2.

(Proved)

(ii) Let $\gcd(2a + b, a + 2b) = d$.

$\therefore$ $d \mid (2a + b)$ and $d \mid (a + 2b)$

$\Rightarrow$ $d \mid \{2(2a + b) - (a + 2b)\}$ and $d \mid \{-(2a + b) + 2(a + 2b)\}$

$\Rightarrow$ $d \mid 3a$ and $d \mid 3b$

$\Rightarrow$ d is a common divisor of $3a$ and $3b$. ...(1)

Given: $\gcd(a, b) = 1 \Rightarrow \gcd(3a, 3b) = 3$, by Ex. 2 ...(2)

From (1) and (2), we have d is a common divisor of $3a, 3b$ and 3 is the greatest common divisor of $3a, 3b$. Therefore, by the definition of greatest common divisor we must have $d \mid 3 \Rightarrow d = 1$ or 3.

$\therefore$ $d = \gcd(2a + b, a + 2b) = 1$ or 3. **(Proved)**

Example 4: *If gcd (a, 4) = gcd (b, 4) = 2, then show that gcd (a + b, 4) = 4.*

Solution: Given : gcd $(a, 4)$ = gcd $(b, 4) = 2$

$\Rightarrow$ $a = 2m$ and $b = 2n$, for some odd integers m, n

$\Rightarrow$ $a + b = 2(m + n) = 2 \times 2r = 4r$, for some integer r ($\because$ $m + n$ is even).

$\Rightarrow$ gcd$(a + b, 4)$ = gcd $(4r, 4) = 4$. **(Proved)**

Example 5: *If d = gcd (a, b)* and *d = au + bv, then show that u, v are relatively prime.*

Solution: Given : $d = \gcd(a, b) \Rightarrow d \mid a$ and $d \mid b$

$\Rightarrow$ $a = md$ and $b = nd$, for some $m, n \in Z$ ...(1)

Given : $d = au + bv \Rightarrow d = mdu + ndv = (mu + nv)d$, by (1)

$\Rightarrow$ $1 = mu + nv$, where $m, n \in Z$.

Hence by Theorem 6, u and v are relatively prime.

Example 6: *Prove that $\sqrt{2}$ is irrational.*

Solution: If possible, suppose $\sqrt{2}$ is rational.

$\therefore$ $\sqrt{2} = \dfrac{a}{b}$, where a and b are integers such that gcd $(a, b) = 1$.

$\Rightarrow$ $2 = \dfrac{a^2}{b^2} \Rightarrow a^2 = 2b^2$ ...(1)

$\Rightarrow$ $2 \mid a^2$

$\Rightarrow$ $2 \mid a$, since 2 is a prime (by Theorem 14)

$\Rightarrow$ $a = 2m$, for some integer m.

$\Rightarrow$ $4m^2 = 2b^2$, from (1) putting $a = 2m$

$\Rightarrow$ $b^2 = 2m^2$

$\Rightarrow$ $2 \mid b^2 \Rightarrow 2 \mid b$, since 2 is a prime (by Th. 14)

Thus 2 divides both a and b which contradicts the assumption that gcd $(a, b) = 1$.

Hence $\sqrt{2}$ is irrational.

Example 7: *If n is a positive integer, then prove that $n \geq 1$.*

Solution: If possible, suppose there exists a positive integer less than 1. By the well-ordering principle, there exists a least positive integer m such that

$$0 < m < 1$$

Multiplying this inequality by the positive integer m, we get

$$0 < m^2 < m$$

Therefore, m^2 is a positive integer less than m which is also less than 1. This contradicts m's property of being the least positive integer less than 1. Hence there exists no positive integer less than 1.

Example 8: *If gcd (a, bc) = 1, then prove that gcd (a, b) = 1 and gcd (a, c) = 1.*

Solution: gcd $(a, bc) = 1 \Rightarrow 1 = au + (bc)v$, for some $u, v \in Z$ (By Theorem 5)

$\Rightarrow$ $1 = au + b(cv)$, where $u, cv \in Z$.

$\Rightarrow$ gcd $(a, b) = 1$, by Theorem 6.

Similarly, gcd $(a, bc) = 1 \Rightarrow 1 = au + (bc)v$, for some $u, v \in Z$

$\Rightarrow$ $1 = au + c(bv)$, where $u, bv \in Z$.

$\Rightarrow$ gcd $(a, c) = 1$, by Theorem 6.

$\therefore$ $\gcd(a, bc) = 1 \Rightarrow \gcd(a, b) = 1$ and $\gcd(a, c) = 1$. **(Proved)**

Note: This example is the converse of Theorem 10.

Example 9: *If gcd $(a, b) = 1$, then show that gcd $(a^2, b^2) = 1$.*

Solution: Let gcd $(a^2, b^2) = d$.

If $d = 1$, then gcd $(a^2, b^2) = 1$ and the proof is complete.

If $d > 1$, then d has a prime factor p, say

$\therefore$ $p \mid d$ and $d = \gcd(a^2, b^2) \Rightarrow p \mid d$ and $d \mid a^2, d \mid b^2$

$\Rightarrow$ $p \mid a^2$ and $p \mid b^2$

$\Rightarrow$ $p \mid a$ and $p \mid b$, since p is a prime (by Th. 14)

$\Rightarrow$ $p \mid \gcd(a, b)$

$\Rightarrow$ $p \mid 1$, which is impossible.

Hence we cannot have $d > 1$.

$\therefore$ $d = \gcd(a^2, b^2) = 1$.

Example 10: *If gcd $(a, b) = 1$, then prove that*

$$\gcd(a + b, a^2 - ab + b^2) = 1 \text{ or } 3.$$

Solution: Let gcd $(a + b, a^2 - ab + b^2) = d$.

$\therefore$ $d \mid (a + b), d \mid (a^2 - ab + b^2)$

$\Rightarrow$ $d \mid (a + b)^2, d \mid (a^2 - ab + b^2)$

$\Rightarrow$ $d \mid \{(a + b)^2 - (a^2 - ab + b^2)\}$

$\Rightarrow$ $d \mid 3ab$

Again, $d \mid (a + b), d \mid 3ab \Rightarrow d \mid 3a(a + b), d \mid 3ab$

$\Rightarrow$ $d \mid \{(3a^2 + 3ab) - 3ab\}$

$\Rightarrow$ $d \mid 3a^2$.

Similarly, $d \mid (a + b), d \mid 3ab \Rightarrow d \mid 3b(a + b), d \mid 3ab$

$\Rightarrow$ $d \mid \{(3ab + 3b^2) - 3ab\}$

$\Rightarrow$ $d \mid 3b^2$.

$\therefore$ d is a common divisor of $3a^2$ and $3b^2$.

Given: gcd $(a, b) = 1 \Rightarrow \gcd(a^2, b^2) = 1$, by Ex. 9

$\Rightarrow$ gcd $(3a^2, 3b^2) = 3$, by Ex. 2

Now d is a common divisor of $3a^2$, $3b^2$ and the greatest common divisor of $3a^2$, $3b^2$ is 3. Therefore, by the definition of greatest common divisor we must have $d \mid 3 \Rightarrow d = 1$ or 3. Hence the result.

Example 11: *Let n be a positive integer. If $n^3 + 1$ is a prime, then show that $n = 1$.*

Solution: Let $n^3 + 1$ be a prime and assume, if possible, that $n \neq 1$. So $n \geq 2$. ...(1)

Now, $n^3 + 1 = (n + 1)(n^2 - n + 1)$, where

$n + 1 > 1$ and $n^2 - n + 1 = n(n - 1) + 1 > 1$, by (1).

It follows that $n^3 + 1$ is not a prime, producing a contradiction. Hence the result.

Example 12: *Show that for every positive integer n, the integer $n^2 - 3n$ is even.*

Solution: Assume that the given statement is false. Then $n^2 - 3n$ is odd for some positive integers n and let m be the smallest one among these positive integers. If $n = 1$, then $n^2 - 3n = -2$ which is even and so $m \geq 2$. Therefore we can write $m = k + 1$, where $1 \leq k < m$. Since $1 \leq k < m$, it follows that $k^2 - 3k$ is even (because m is the smallest positive integer for which $n^2 - 3n$ is odd). Hence $k^2 - 3k = 2u$, for some integer u.

Now,
$$\begin{aligned} m^2 - 3m &= (k+1)^2 - 3(k+1) \\ &= (k^2 + 2k + 1) - 3(k+1) \\ &= (k^2 - 3k) + 2k - 2 \\ &= 2u + 2k - 2 = 2(u + k - 1) \end{aligned}$$

Since $u + k - 1$ is an integer, $m^2 - 3m$ is even, producing a contradiction. Hence our assumption is wrong. Therefore, for every positive integer n, the integer $n^2 - 3n$ is even.

Example 13: *If $2^n - 1$ is a prime, prove that n is a prime.*

Solution: Let n is not a prime and let $p = 2^n - 1$. Then there exist positive integers r and s such that $n = rs$, where $1 < r < n$, $1 < s < n$.

$$\therefore \quad p = 2^n - 1 = 2^{rs} - 1 = (2^r)^s - 1 = a^s - 1, \text{ where } a = 2^r.$$

$$\Rightarrow \quad p = (a-1)(a^{s-1} + a^{s-2} + \ldots + a + 1) \quad \ldots(1)$$

Now, $\quad a - 1 > 1$, because $a = 2^r > 2$.

Also, $\quad a - 1 = 2^r - 1 < p$, because $p = (2^r)^s - 1$.

Therefore, $(a - 1)$ is a positive integer such that $1 < a - 1 < p$ and also from (1), $(a - 1)$ is a divisor of p. Hence p is not a prime, producing a contradiction. Therefore n must be a prime.

Notes: (*i*) The converse of the result stated in this example is not true in general. Because $2^{11} - 1 = 2047 = 23 \times 89$ is not a prime but 11 is a prime.

(*ii*) The primes of the form $2^n - 1$ are called **Mersenne primes**, after the French monk Marin Mersenne who studied them in the seventeenth century. The numbers $2^2 - 1 = 3$, $2^3 - 1 = 7$, $2^5 - 1 = 31$ are Mersenne primes while $2^{11} - 1 = 2047 = 23 \times 89$ is not a Mersenne prime.

Example 14: *Using principle of mathematical induction prove that every integer $n \geq 2$ is either a prime or can be expressed as a product of primes.*

Solution: Basis step: Since 2 is prime, the given statement is true for $n = 2$.

Inductive step: For an arbitrary integer $k \geq 2$, let us assume that every integer u (with $2 \leq u \leq k$) is either a prime or can be expressed as a product of primes. We have to show that $k + 1$ is either a prime or can be expressed as a product of primes.

(*i*) If $k + 1$ is a prime, then there is nothing further to prove.

(*ii*) If $k + 1$ is composite, then there exist integers a and b such that $k + 1 = ab$, where $2 \leq a, b \leq k$. So by induction hypothesis, each of a and b is a prime or can be expressed as a product of primes. In both cases, $k + 1 = ab$ is a product of primes.

Hence by the principle of mathematical induction, every integer $n \geq 2$ is either a prime or can be expressed as a product of primes.

Example 15: *Prove that $\log_2 7$ is irrational.*

Solution: If possible, suppose $\log_2 7$ is rational.

$$\therefore \quad \log_2 7 = \frac{a}{b}, \text{ where } a \text{ and } b \text{ are positive integers.}$$

$$\Rightarrow \quad 2^{a/b} = 7 \quad \Rightarrow \quad 2^a = 7^b = n, \text{ say.}$$

This means that the integer $n > 1$ is expressed as a product of primes in two different ways, which contradicts the fundamental theorem of arithmetic.

$\therefore \quad \log_2 7$ is an irrational number.

Example 16: *If gcd (n, 7) = 1, show that $n^{12} - 1$ is divisible by 7.*

Solution: Since 7 is a prime and gcd $(n, 7) = 1$, therefore by Fermat's theorem, we have

$$n^{7-1} \equiv 1 \pmod 7 \Rightarrow n^6 \equiv 1 \pmod 7$$

$\Rightarrow$ $7 \mid (n^6 - 1)$

$\Rightarrow$ $7 \mid (n^6 - 1)(n^6 + 1)$

$\Rightarrow$ $7 \mid (n^{12} - 1)$.

2.6 CONGRUENCES

Definition 7: Let m be a fixed positive integer. Two integers a and b are said to be **congruent modulo** m (or a is said to be **congruent to b modulo m**) if $a - b$ is divisible by m. This is symbolically expressed as

$$a \equiv b \pmod m \qquad ...(1)$$

For example, $37 \equiv 12 \pmod 5$, since $37 - 12 = 25$ is divisible by 5.

If $a - b$ is not divisible by m, then a is said to be **incongruent** to b modulo m, or we say that **a is not congruent to b modulo m** and we write

$$a \not\equiv b \pmod m.$$

For example, $1 \not\equiv -5 \pmod 4$, since $1 - (-5) = 6$ is not divisible by 4.

Notes: (*i*) Expression (1) is called the **congruence,** m is called the **modulus** of congruence and b is called a **residue** of $a \pmod m$.

(*ii*) $a \equiv b \pmod m \Leftrightarrow a - b = mk$, for some integer k.

$\Leftrightarrow a = b + mk$, for some integer k.

(*iii*) When $m = 1$, every two integers are congruent modulo m. This case is not so useful and interesting. Therefore m is usually taken to be a positive integer greater than 1.

(*iv*) Some authors denote by **a (mod m)** the remainder when a is divided by m.

The great German mathematician Karl Friedrich Gauss (1777 – 1855) introduced the concept of congruence at the end of the eighteenth century which laid the foundation of modern theory of numbers.

The following theorem provides a useful way to work with congruences.

Theorem 21: *If a and b are two integers, then $a \equiv b \pmod m$ if and only if a and b have the same remainder when divided by m.*

Proof: Suppose a ≡ b (mod m)

Let r be the remainder when a is divided by m.

$\therefore$ $a = qm + r$, where q is an integer and $0 \le r < m$ ...(1)

Now, $a \equiv b \pmod m$

$\Rightarrow$ $a - b = km$, where k is an integer

$\Rightarrow$ $b = a - km = (qm + r) - km$, by (1)

$\Rightarrow$ $b = (q - k)m + r$

This shows that b leaves the same remainder r when divided by m.

Conversely, let r be the same remainder when a and b are divided by m.

$\therefore$ $a = q_1 m + r, b = q_2 m + r$, where q_1, q_2 are integers and $0 \le r < m$.

$\therefore$ $a - b = (q_1 - q_2) m \Rightarrow m \mid a - b \Rightarrow a \equiv b \pmod m$.

This completes the proof.

To illustrate, let $m = 6$.

Since $31 = 5 \times 6 + 1$ and $-17 = (-3) \times 6 + 1$, so 31 and -17 leave the same remainder upon division by 6.

$\therefore \quad 31 \equiv -17 \pmod{6}$.

Basic Properties

1. $a \equiv a \pmod{m}$ [Reflexive]
2. If $a \equiv b \pmod{m}$, then $b \equiv a \pmod{m}$ [Symmetric]
3. If $a \equiv b \pmod{m}$ and $b = c \pmod{m}$, then $a \equiv c \pmod{m}$ [Transitive]
4. If $a \equiv b \pmod{m}$, then for any integer c :

$$a + c \equiv b + c \pmod{m}$$
$$ac \equiv bc \pmod{m}$$

5. If $a \equiv b \pmod{m}$ and $c \equiv d \pmod{m}$, then

$$a + c \equiv b + d \pmod{m}$$
$$ac \equiv bd \pmod{m}$$

6. If $ac \equiv bc \pmod{m}$ and c is prime to m, then

$$a \equiv b \pmod{m}.$$

Proof:

1. $a - a = 0$ is divisible by m. Hence $a \equiv a \pmod{m}$
2. $a \equiv b \pmod{m} \Rightarrow m \mid (a - b) \Rightarrow m \mid -(a - b) \Rightarrow m \mid (b - a)$.

$\therefore \quad a \equiv b \pmod{m} \Rightarrow b \equiv a \pmod{m}$

3. $a \equiv b \pmod{m}$, $b \equiv c \pmod{m}$

$\Rightarrow m \mid (a - b), m \mid (b - c)$

$\Rightarrow m \mid \{(a - b) + (b - c)\}$

$\Rightarrow m \mid (a - c)$

$\Rightarrow a \equiv c \pmod{m}$

Note: The relation **congruent modulo** m is an **equivalence relation** on Z because it is reflexive, symmetric and transitive.

4. $a \equiv b \pmod{m} \Rightarrow m \mid (a - b) \Rightarrow m \mid \{(a + c) - (b + c)\}, m \mid c(a - b)$, for any $c \in Z$.

$\Rightarrow a + c \equiv b + c \pmod{m}$, $ac \equiv bc \pmod{m}$, for any integer c.

5. $a \equiv b \pmod{m} \Rightarrow a - b = km$, where $k \in Z$.

$c \equiv d \pmod{m} \Rightarrow c - d = lm$, where $l \in Z$.

$\therefore \quad (a + c) - (b + d) = (k + l)\, m$

$\Rightarrow \quad a + c \equiv b + d \pmod{m}$, since $k + l \in Z$.

By property 4:

$$a \equiv b \pmod{m} \Rightarrow ac \equiv bc \pmod{m} \quad \text{...(1)}$$
$$c \equiv d \pmod{m} \Rightarrow bc \equiv bd \pmod{m} \quad \text{...(2)}$$

From (1), (2) and property 3,

$$a \equiv b \pmod{m} \text{ and } c \equiv d \pmod{m} \Rightarrow ac \equiv bd \pmod{m}.$$

6. Given, $ac \equiv bc \pmod{m}$

$\Rightarrow \quad ac - bc = km$, where $k \in Z$.

$\Rightarrow \quad a - b = \dfrac{km}{c}$ ...(1)

$\Rightarrow \quad c \mid km$ $(\because \; a - b$ is an integer)

$\Rightarrow \quad c \mid k$ $(\because \; c$ is prime to m).

$\Rightarrow \quad k = qc$, where $q \in Z$,

$\Rightarrow \quad a - b = qm$, by (1)

$\Rightarrow \quad a \equiv b \pmod{m}$.

Notes: (*i*) Without the condition 'c is prime to m', $ac \equiv bc \pmod{m}$ does not necessarily imply $a \equiv b$ (mod m). For example , $7.2 \equiv 5.2 \pmod{4}$ does not imply $7 \equiv 5 \pmod{4}$. Also, observe that $14.2 \equiv 10.2 \pmod{4}$ implies $14 \equiv 10 \pmod{4}$. Thus, we can cancel the common factor c freely from both sides of the congruence (mod m) provided c is prime to m.

$$(-2) \,.\, 3 \equiv 13 \,.\, 3 \pmod{5}$$

Cancelling the factor 3 which is prime to 5, we get the correct congruence $-2 = 13 \pmod{5}$. However, cancellation is allowed in some restricted sense which is stated in the following Theorem 22.

(*ii*) **Alternative proof of property 6**

Given, $ac \equiv bc \pmod{m}$

$\Rightarrow \quad m \mid (ac - bc)$

$\Rightarrow \quad m \mid (a - b)c$

$\Rightarrow \quad m \mid (a - b)$ $[\because \; c$ is prime to $m]$

$\Rightarrow \quad a \equiv b \pmod{m}$

Theorem 22: *If $ac \equiv bc \pmod{m}$ and $\gcd(c, m) = d$,*

$$\text{then } a \equiv b \left(\text{mod } \frac{m}{d}\right).$$

Proof: Since $\gcd(c, m) = d$, we have $c = dr$ and $m = ds$ where r, s are integers prime to each other.

Given, $ac \equiv bc \pmod{m}$

$\Rightarrow \quad m \mid (ac - bc)$

$\Rightarrow \quad ds \mid (a - b)\, dr$ $[\because \; m = ds$ and $c = dr]$

$\Rightarrow \quad s \mid (a - b)\, r$

$\Rightarrow \quad s \mid (a - b)$ $[\because \; s$ is prime to $r]$

$\Rightarrow \quad a \equiv b \pmod{s} \;\Rightarrow\; a \equiv b\left(\text{mod } \dfrac{m}{d}\right)$ $[\because \; m = ds]$

Corollary: If $ac \equiv bc \pmod{m}$ and $c \mid m$, then

$$a \equiv b\left(\text{mod } \frac{m}{c}\right).$$

For example, $(-2) \,.\, 6 \equiv 4.\, 6 \pmod{9}$. Cancellation of 6 from both sides does not give a correct congruence because 6 is not prime to 9. Since $\gcd(6, 9) = 3$, we get the correct congruence

$$-2 \equiv 4\left(\text{mod } \frac{9}{3}\right).$$

Again, $7.2 \equiv 5.2 \pmod{4}$. Since 2 divides 4, we get the correct congruence $7 \equiv 5 \pmod{2}$ from the corollary.

Example 1: *Find the remainder when the sum 1 ! + 2 ! + 3 ! + 4 ! + ... + 99 ! + 100 ! is divided by 30.*

Solution: Observe that $5! = 120 \equiv 0 \pmod{30}$.

Thus, for $k \geq 5$, we have $k! = 5!\,6.7 \ldots k \equiv 0 \pmod{30}$

Now, $1! + 2! + 3! + 4! = 1 + 2 + 6 + 24 = 33 \equiv 3 \pmod{30}$

Therefore, using property 5, we get

$$1! + 2! + 3! + 4! + 5! + \ldots + 99! + 100 \equiv 3 \pmod{30}.$$

Hence the given sum leaves a remainder 3 when divided by 30.

Theorem 23: *If $a \equiv b \pmod{m}$, then $a^n \equiv b^n \pmod{m}$ for all positive integers n.*

Proof: We use the principle of mathematical induction to prove this theorem.

Basis step: The theorem is true for $n = 1$.

Inductive step : Assume that the theorem is true for some positive integer k.

Then $a^k \equiv b^k \pmod{m}$

Now, $a^k \equiv b^k \pmod{m}$ and $a \equiv b \pmod{m}$

$\Rightarrow$ $a^k.\,a \equiv b^k.\,b \pmod{m}$, by property 5

$\Rightarrow$ $a^{k+1} \equiv b^{k+1} \pmod{m}$.

Therefore the theorem is true for the positive integer $k + 1$ if it is true for k. Hence by the principle of mathematical induction,

$$a^n \equiv b^n \pmod{m},\ \forall\, n \in N$$

Note: The converse of this theorem does not hold good in general, *i.e.*, $a^n \equiv b^n \pmod{m}$ does not necessarily imply $a \equiv b \pmod{m}$.

For example, $5^3 \equiv 8^3 \pmod{9}$, but $5 \not\equiv 8 \pmod{9}$.

$5^3 \equiv 8^3 \pmod{3} \Rightarrow 5 \equiv 8 \pmod{3}$.

Example 2: *Use theory of congruence to prove that for $n \geq 1$, $17 \mid (2^{3n+1} + 3.\,5^{2n+1})$.*

(W.B.U.T. 2012)

Solution: Now, $8 \equiv 25 \pmod{17}$, since $17 \mid (8 - 25)$.

$\therefore$ $8^n \equiv 25^n \pmod{17}$, *i.e.*, $2^{3n} \equiv 5^{2n} \pmod{17}$. [By Theorem 23]

Also, $2 \equiv -15 \pmod{17}$, since $17 \mid \{2 - (-15)\}$.

Using property 5, we have

$$2\,.\,2^{3n} \equiv -15\,.\,5^{2n} \pmod{17} \Rightarrow 2^{3n+1} \equiv -3\,.\,5^{2n+1} \pmod{17}$$

$\Rightarrow$ $17 \mid (2^{3n+1} + 3\,.\,5^{2n+1})$. **(Proved)**

Example 3: *Prove that for all integers $n \geq 1$,*

(*i*) $3^{3n+1} \equiv 3\,.\,4^{2n} \pmod{11}$ (*ii*) $11 \mid (3^{3n+2} + 2\,.\,5^n)$.

Solution: (*i*) Now, $27 \equiv 16 \pmod{11}$, since $11 \mid (27 - 16)$.

$\therefore$ $(27)^n \equiv (16)^n \pmod{11}$, *i.e.*, $3^{3n} \equiv 4^{2n} \pmod{11}$. [By Theorem 23]

$\therefore$ $11 \mid (3^{3n} - 4^{2n}) \Rightarrow 11 \mid 3\,(3^{3n} - 4^{2n}) \Rightarrow 11 \mid (3^{3n+1} - 3\,.\,4^{2n})$

$\therefore$ $3^{3n+1} \equiv 3\,.\,4^{2n} \pmod{11}$. **(Proved)**

(*ii*) Now, $27 \equiv 5 \pmod{11}$, since $11 \mid (27 - 5)$.

$\therefore$ $(27)^n \equiv 5^n \pmod{11}$, *i.e.*, $3^{3n} \equiv 5^n \pmod{11}$. [By Theorem 23]

Also, $9 \equiv -2 \pmod{11}$, since $11 \mid \{9 - (-2)\}$.

Using property 5, we have

$$9 \cdot 3^{3n} \equiv -2 \cdot 5^n \pmod{11} \Rightarrow 3^{3n+2} \equiv -2 \cdot 5^n \pmod{11}$$

$$\Rightarrow 11 \mid (3^{3n+2} + 2.5^n). \quad \textbf{(Proved)}$$

Example 4: *Prove that if $n = a^2 + b^2$, for $a, b \in Z$ (the set of integers), then $n = 4q$ or $n = 4q + 1$ or $n = 4q + 2$ for some $q \in Z$. Deduce that 1235567 can not be written as $a^2 + b^2$, for some $a, b \in Z$.*

Solution: The set of remainders modulo 4 is {0, 1, 2, 3}.

Therefore, we have four possibilities:

$$a \equiv 0 \pmod 4 \Rightarrow a^2 \equiv 0 \pmod 4$$

$$a \equiv 1 \pmod 4 \Rightarrow a^2 \equiv 1 \pmod 4$$

$$a \equiv 2 \pmod 4 \Rightarrow a^2 \equiv 4 \pmod 4 \equiv 0 \pmod 4$$

$$a \equiv 3 \pmod 4 \Rightarrow a^2 \equiv 9 \pmod 4 \equiv 1 \pmod 4.$$

Therefore, given two integers a and b, we have

$$a^2 + b^2 \equiv 0 \text{ or } 1 \text{ or } 2 \pmod 4$$

$$\Rightarrow \quad n = a^2 + b^2 = 4q \quad \text{or} \quad 4q + 1 \quad \text{or} \quad 4q + 2. \quad \textbf{(Proved)}$$

Again, since $1235567 \equiv 3 \pmod 4$ and so if $a^2 + b^2 = 1235567$, then $a^2 + b^2 \equiv 3 \pmod 4$ which is a contradiction (where $a, b \in Z$). Hence the result.

Example 5: *Determine the remainder when 9^{10} is divided by 11.*

Solution: We have $\quad 9^2 \equiv 4 \pmod{11}$, since $11 \mid (81 - 4)$. ...(1)

$\therefore \quad 9^8 \equiv 4^4 \pmod{11}$, *i.e.*, $9^8 \equiv 256 \pmod{11}$ ...(2)

Also, $\quad 256 \equiv 3 \pmod{11} \quad [\because \ 11 \mid (256 - 3)]$ ...(3)

From (2) and (3), by transitive property,

$$9^8 \equiv 3 \pmod{11} \quad \text{...(4)}$$

From (1) and (4) [using property 5], we have

$9^2 \times 9^8 \equiv 4 \times 3 \pmod{11}$, *i.e.*, $9^{10} \equiv 12 \pmod{11}$ ...(5)

Also, $\quad 12 \equiv 1 \pmod{11} \quad [\because \ 11 \mid (12 - 1)]$ ...(6)

From (5) and (6), by transitive property,

$$9^{10} \equiv 1 \pmod{11}.$$

Hence the required remainder is 1.

Theorem 24: *Let $f(x) = a_n x^n + a_{n-1} x^{n-1} + ... + a_1 x + a_0$ be a polynomial having integral coefficients a_i. If $a \equiv b \pmod m$, then $f(a) \equiv f(b) \pmod m$.*

Proof: $a \equiv b \pmod m \Rightarrow a^k \equiv b^k \pmod m, \ \forall \ k \in N$ (By Th. 23)

$\Rightarrow \quad a_k a^k \equiv a_k b^k \pmod m$, by property 4

Adding these congruences, by property 5, for $k = 1, 2, ..., n$, we get

$$a_1 a + a_2 a^2 + ... + a_n a^n \equiv a_1 b + a_2 b^2 + ... + a_n b^n \pmod m$$

$$\Rightarrow \quad a_0 + a_1 a + a_2 a^2 + ... + a_n a^n \equiv a_0 + a_1 b + a_2 b^2 + ... + a_n b^n \pmod m$$

$$[\because \ a_0 \equiv a_0 \pmod m]$$

$$\Rightarrow \quad f(a) \equiv f(b) \pmod m.$$

Theorem 25: (Fermat's Theorem)

If p is a prime and a is an integer such that p is not a divisor of a, then $a^{p-1} \equiv 1 \pmod p$.

Proof: Let us consider the integers $a, 2a, 3a, ..., (p - 1)\,a$.

Observations:

(*i*) None of these is divisible by p. Because if p be a prime and $p \mid ab$ then either $p \mid a$ or $p \mid b$, by Theorem 14.

(*ii*) No two of these are congruent modulo p. Because if $ra \equiv sa \pmod{p}$ for some integers r, s such that $1 \leq r < s \leq p - 1$, then cancelling a (since a is prime to p and using property 6), we have $r \equiv s \pmod{p}$, a contradiction.

So we conclude that the integers $a, 2a, 3a, ..., (p-1)a$ are congruent to $1, 2, 3, ..., (p-1)$ modulo p, taken in some order. Using property 5, we get

$$a \,.\, 2a \,.\, 3a \,...\, (p-1)\, a \equiv 1.2.3. \,...\, (p-1) \pmod{p}$$

$$\Rightarrow \quad (p-1)\,!\, a^{p-1} \equiv (p-1)\,! \pmod{p}$$

$$\Rightarrow \quad a^{p-1} \equiv 1 \pmod{p} \; [\because \;\; \gcd((p-1)\,!, p) = 1 \text{ and by property 6}] \qquad ...(1)$$

Corollary 1. Alternative form of Fermat's Theorem

If p is a prime and a is any integer, then

$$a^p \equiv a \pmod{p}$$

Proof: When p is not a divisor of a.

By (1), $a^{p-1} \equiv 1 \pmod{p}$. Also, $a \equiv a \pmod{p}$.

Using property 5, $a^{p-1} \,.\, a \equiv 1 \,.\, a \pmod{p} \Rightarrow a^p \equiv a \pmod{p}$.

When p is a divisor of a.

$$a^p \equiv 0 \pmod{p}. \text{ Also, } 0 \equiv a \pmod{p}.$$

Using property 3, $a^p \equiv a \pmod{p}$.

Corollary 2. Let $p > 2$. Since p is prime $p - 1$ is an even integer. From (1), we get

$$a^{p-1} - 1 \equiv 0 \pmod{p}$$

$$\Rightarrow \quad \left(a^{\frac{p-1}{2}} - 1\right)\left(a^{\frac{p-1}{2}} + 1\right) \equiv 0 \pmod{p}$$

$$\Rightarrow \quad \text{either } p \,\Big|\, a^{\frac{p-1}{2}} - 1 \quad \text{or} \quad p \,\Big|\, a^{\frac{p-1}{2}} + 1 \qquad \text{[By Theorem 14]}$$

$$\Rightarrow \quad \text{either } a^{\frac{p-1}{2}} = pk + 1 \quad \text{or} \quad a^{\frac{p-1}{2}} = pk - 1, \text{ where } k \in Z.$$

For example, let $p = 11$. Then

(*i*) For any integer a not divisible by 11, either $a^5 = 11k + 1$, or $a^5 = 11k - 1$

(*ii*) When a is divisible by 11, $a^5 = 11k$.

$\therefore$ a^5 is one of the forms : $11k$, $11k + 1$, $11k - 1$, where $k \in Z$.

Note: Theorem 25 is also known as **Fermat's Little Theorem.**

Example 6: *Show that* $2^{41} \equiv 3 \pmod{23}$.

Solution: Now, 23 is a prime number and 2 is prime to 23.

$$\therefore \quad 2^{22} \equiv 1 \pmod{23} \qquad \text{[By Fermat's theorem]}$$

$$\Rightarrow \quad 2^{44} \equiv 1 \pmod{23} \qquad [\because \;\; a \equiv b \pmod{m} \Rightarrow a^n \equiv b^n \pmod{m}]$$

$$\Rightarrow \quad 2^{44} \equiv 24 \pmod{23} \qquad [\because \;\; 1 \equiv 24 \pmod{23} \text{ and by transitive property}]$$

$$\Rightarrow \quad 2^{41} \,.\, 8 \equiv 3 \,.\, 8 \pmod{23}$$

$$\Rightarrow \quad 2^{41} \equiv 3 \pmod{23} \qquad [\because \;\; 8 \text{ is prime to 23 and by property 6}]$$

Alternative form: Find the least positive residue in $2^{41} \pmod{23}$.

Example 7: *If p is a prime and a, b are any two integers, then prove that*

$$(a+b)^p \equiv a^p + b^p \pmod{p}$$

Solution: Since p is a prime, so by Fermat's theorem (see Corollary 1):

$$a^p \equiv a \pmod{p} \quad ...(1)$$

$$b^p \equiv b \pmod{p} \quad ...(2)$$

and

$$(a+b)^p \equiv a + b \pmod{p} \quad ...(3)$$

Adding the congruences (1) and (2), by property 5:

$$a^p + b^p \equiv a + b \pmod{p} \Rightarrow a + b \equiv a^p + b^p \pmod{p} \quad ...(4)$$

(By the symmetric property 2)

From (3) and (4), by the transitive property 3, we have

$$(a+b)^p \equiv a^p + b^p \pmod{p}.$$ **(Proved)**

Example 8: *Prove that $a^{12} - b^{12}$ is divisible by 91 if a and b are both prime to 91.*

Solution: Since a is prime to 91, a is prime to both 13 and 7. Therefore, by Fermat's theorem:

$$a^{12} \equiv 1 \pmod{13} \Rightarrow 13 \mid a^{12} - 1 \quad ...(1)$$

$$a^6 \equiv 1 \pmod{7} \Rightarrow a^{12} \equiv 1 \pmod{7}$$

$[\because \; a \equiv b \pmod{m} \Rightarrow a^n \equiv b^n \pmod{m}]$

$$\Rightarrow \quad 7 \mid a^{12} - 1 \quad ...(2)$$

Since gcd (13, 7) = 1, so by Theorem 9, (1) and (2) together imply $91 \mid a^{12} - 1$. Similarly, $91 \mid b^{12} - 1$.

$$\therefore \quad 91 \mid \{a^{12} - 1) - (b^{12} - 1)\} \Rightarrow 91 \mid (a^{12} - b^{12}).$$ **(Proved)**

Example 9: *Show that $n^5 - n$ is divisible by 30.*

Solution: Since 5 is a prime, therefore by Fermat's theorem (see Corollary 1):

$$n^5 \equiv n \pmod{5} \Rightarrow 5 \mid (n^5 - n) \quad ...(1)$$

Also, $n^5 - n = n(n^4 - 1) = n(n-1)(n+1)(n^2+1) = (n-1)n(n+1)(n^2+1)$.

Now $(n-1)\,n\,(n+1)$ is a product of three consecutive integers and so it is divisible by $2 \times 3 = 6$ (see Ex. 8 of Art. 2.3).

$$\therefore \quad 6 \mid (n^5 - n) \quad ...(2)$$

Since gcd (5, 6) = 1, so by Theorem 9, (1) and (2) together imply $5 \times 6 \mid (n^5 - n)$, *i.e.*, $30 \mid (n^5 - n)$.

Example 10: *Prove that the 8th power of any integer is of the form 17k, or 17k ± 1, where k ∈ Z.*

Solution: Let n be any integer.

Case 1. If 17 is a divisor of n, then $n = 17q$, where $q \in Z$. Hence $n^8 = 17k$, for some $k \in Z$.

Case 2. If 17 is not a divisor of n, then since 17 is a prime we have by Fermat's theorem:

$$n^{17-1} \equiv 1 \pmod{17} \Rightarrow n^{16} \equiv 1 \pmod{17}$$

$\Rightarrow \quad 17 \mid (n^{16} - 1)$

$\Rightarrow \quad 17 \mid (n^8 - 1)(n^8 + 1)$

$\Rightarrow \quad 17 \mid (n^8 - 1)$, or $17 \mid (n^8 + 1)$ [By Theorem 14]

Now, $17 \mid (n^8 - 1) \Rightarrow n^8 - 1$ is a multiple of 17

$\Rightarrow \quad n^8 - 1 = 17k$, for some $k \in Z$.

$\Rightarrow \quad n^8 = 17k + 1.$

Again, $17 \mid (n^8 + 1) \Rightarrow n^8 + 1$ is a multiple of 17

$\Rightarrow \quad n^8 + 1 = 17k$, for some $k \in Z$.

$$\Rightarrow \qquad n^8 = 17k - 1$$

Therefore, from case 1 and case 2 we conclude that if $n \in Z$, then n^8 is one of the forms $17k$, $17k + 1$, $17k - 1$ for some $k \in Z$.

2.7 APPLICATIONS OF CONGRUENCES

Check Digits

We are familiar with the bar code bearing some identification number consisting of 13 numerical digits as a pattern of black and white strips of various thicknesses. They are on groceries, cosmetics, books and so on.

The identification number on book is called **International Standard Book Number (ISBN)** which is a unique numerical commercial book identifier. The ISBN is 13 digits long (if assigned after January 1, 2007) and it consists of five blocks identifying the industry (978 denotes book publishing), the country (81 for India), the publisher, the number assigned to the book by its publishing company and finally a single check digit. The ISBN separates its five blocks with either a hyphen or a space.

The identification number on grocery is called **Universal Product Code (UPC)** which identifies the manufacturer of a product and the product itself. The last digit of the number is a check digit. It is not unusual for a scanning device to misread a bar code and so the use of check digit provides a way to alert the cashier that the number must be read again.

Check Digit Calculation

If the identification number is a 13-digit number, say $x_1 \, x_2 \, x_3 \ldots x_{13}$, the check digit is calculated by the following rule:

(sum of digits in odd positions) + 3 (sum of digits in even positions) $\equiv 0 \pmod{10}$.

In vector notation it can be written as

$$(1, 3, 1, 3, \ldots 3, 1).(x_1, x_2, x_3, x_4, \ldots, x_{12}, x_{13}) \equiv 0 \pmod{10}.$$

Example 11: *The first 12 digits of the book titled 'A Textbook of Engineering Mathematics-III' by Guruprasad Samanta are 978-81-224-3447. What is the check digit of this book?*

Solution: The check digit x_{13} is given by

$$(1, 3, 1, 3, 1, 3, 1, 3, 1, 3, 1, 3, 1) . (9, 7, 8, 8, 1, 2, 2, 4, 3, 4, 4, 7, x_{13}) \equiv 0 \pmod{10}$$

$$\Rightarrow \qquad 9 + 21 + 8 + 24 + 1 + 6 + 2 + 12 + 3 + 12 + 4 + 21 + x_{13} \equiv 0 \pmod{10}$$

$$\Rightarrow \qquad 9 + 1 + 8 + 4 + 1 + 6 + 2 + 2 + 3 + 2 + 4 + 1 + x_{13} \equiv 0 \pmod{10}$$

$$[\because \quad 21 \equiv 1 \pmod{10},\ 24 \equiv 4 \pmod{10},\ 12 \equiv 2 \pmod{10}]$$

$$\Rightarrow \qquad 43 + x_{13} \equiv 0 \pmod{10}$$

$$\therefore \qquad x_{13} = 7 \text{ since } 43 + 7 = 50 \text{ is divisible by } 10.$$

So, the required check digit is 7.

Example 12: *The ISBN of the book titled 'A Textbook of Engineering Mathematics-II' by Guruprasad Samanta is 978-81-224-32C9-0 where C is a digit. Find C.*

Solution: The ISBN of the book is 978-81-224-32C9-0, so it satisfies

$$(1, 3, 1, 3, 1, 3, 1, 3, 1, 3, 1, 3, 1) . (9, 7, 8, 8, 1, 2, 2, 4, 3, 2, C, 9, 0) \equiv 0 \pmod{10}$$

$$\Rightarrow \quad 9 + 21 + 8 + 24 + 1 + 6 + 2 + 12 + 3 + 6 + C + 27 + 0 \equiv 0 \pmod{10}$$

$\Rightarrow\quad 9 + 1 + 8 + 4 + 1 + 6 + 2 + 2 + 3 + 6 + C + 7 \equiv 0 \pmod{10}$

$[\because\ 21 \equiv 1 \pmod{10},\ 24 \equiv 4 \pmod{10},\ 12 \equiv 2 \pmod{10},\ 27 \equiv 7 \pmod{10}]$

$\Rightarrow\quad 49 + C \equiv 0 \pmod{10}$

Hence $C = 1$ since $49 + 1 = 50$ which is divisible by 10.

Divisibility tests

One of the more interesting applications of congruence theory involves to find criteria under which a given integer is divisible by another integer.

Example 13: *Let $N = a_k 10^k + a_{k-1} 10^{k-1} + ... + a_2 10^2 + a_1 10 + a_0$ where a_i are integers and $0 \le a_i \le 9$, be a decimal representation of a positive integer N. Show that 2^m divides N if and only if the number represented by the last m digits is divisible by 2^m.*

Solution: Let us consider the polynomial

$$f(x) = a_k x^k + a_{k-1} x^{k-1} + ... + a_2 x^2 + a_1 x + a_0.$$

We have $10 \equiv 0 \pmod 2$.

$\Rightarrow\quad f(10) \equiv f(0) \pmod 2$, by Theorem 24

$\Rightarrow\quad N \equiv a_0 \pmod 2$

$\Rightarrow\quad 2 \mid (N - a_0)$

$\Rightarrow$ N is divisible by 2 if and only if a_0 is divisible by 2.

Also, $a_k 10^k + a_{k-1} 10^{k-1} + ... + a_2 10^2 \equiv 0 \pmod{2^2}$ and $a_1 10 + a_0 \equiv a_1 10 + a_0 \pmod{2^2}$

$\therefore\quad (a_k 10^k + a_{k-1} 10^{k-1} + ... + a_2 10^2) + (a_1 10 + a_0) \equiv 0 + a_1 10 + a_0 \pmod{2^2}$

$\Rightarrow\quad N \equiv a_1 10 + a_0 \pmod{2^2}$

$\Rightarrow$ N is divisible by 2^2 if and only if $a_1 10 + a_0$ is divisible by 2^2.

Likewise it can be proved that N is divisible by 2^3 if and only if $a_2 10^2 + a_1 10 + a_0$ is divisible by 2^3 and so on. Hence the result.

For illustration, 2313711552 is divisible by 16 because the number represented by the last four digits, namely 1552 is divisible by 16 $(= 2^4)$.

Example 14: *Let $N = a_m 10^m + a_{m-1} 10^{m-1} + ... + a_2 10^2 + a_1 10 + a_0$ where a_i are integers and $0 \le a_i \le 9$ be the decimal representation of a positive integer N. Let $S = a_0 + a_1 + ... + a_m$ and $T = a_0 - a_1 + a_2 - ... + (-1)^m a_m$. Then prove that*

(i) N is divisible by 3 if and only if S is divisible by 3,

(ii) N is divisible by 9 if and only if S is divisible by 9,

(iii) N is divisible by 11 if and only if T is divisible by 11.

Solution: Let us consider the polynomial

$$f(x) = a_m x^m + a_{m-1} x^{m-1} + ... + a_1 x + a_0.$$

(*i*) We have $10 \equiv 1 \pmod 3$

$\Rightarrow\quad f(10) \equiv f(1) \pmod 3$, by Theorem 24

$\Rightarrow\quad N \equiv S \pmod 3$

$\Rightarrow\quad 3 \mid (N - S)$

$\Rightarrow$ N is divisible by 3 if and only if S is divisible by 3.

(*ii*) We have $10 \equiv 1 \pmod 9$

$\Rightarrow\quad f(10) \equiv f(1) \pmod 9$, by Theorem 24

$\Rightarrow\quad N \equiv S \pmod 9$

$\Rightarrow \quad 9 \mid (N - S)$

$\Rightarrow$ N is divisible by 9 if and only if S is divisible by 9.

(*iii*) We have $10 \equiv -1 \pmod{11}$

$\Rightarrow \quad f(10) \equiv f(-1) \pmod{11}$, by Theorem 24

$\Rightarrow \quad N \equiv T \pmod{11}$

$\Rightarrow \quad 11 \mid (N - T)$

$\Rightarrow$ N is divisible by 11 if and only if T is divisible by 11.

For illustration: (*i*) 643422 is divisible by 3 because $S = 2 + 2 + 4 + 3 + 4 + 6 = 21$ is divisible by 3.

(*ii*) 879375231 is divisible by 9 since $S = 1 + 3 + 2 + 5 + 7 + 3 + 9 + 7 + 8 = 45$ is divisible by 9.

(*iii*) 6520437 is divisible by 11 because $T = 7 - 3 + 4 - 0 + 2 - 5 + 6 = 11$ is divisible by 11.

2.8 RESIDUE CLASSES

Definition 8: We know that if m is a fixed positive integer, then **congruent modulo m** is an equivalence relation (since it is reflexive, symmetric and transitive) on the set of integers Z. Consequently it partitions (or decomposes) Z into a collection of mutually disjoint equivalence classes. These equivalence classes are called **residue classes modulo m.**

A **residue class** consists of all those integers with the same remainder when divided by m. Therefore, there are m such residue classes and each residue class contains exactly one of the integers 0, 1, 2, ..., $m - 1$. This set of integers is called the **set of least non-negative residues modulo m.**

The symbol $[x]_m$ or simply $[x]$ or $\bar{x}$ is used to denote the residue class (modulo m) containing an integer x.

$\therefore \quad [x] = \{a \in Z : a \equiv x \pmod{m}\}$.

Accordingly, the residue classes can be denoted by [0], [1], [2], ..., [m – 1], or by $\bar{0}, \bar{1}, \bar{2}, \ldots, \overline{m-1}$.

Definition 9: The **complete set of residue classes modulo m** is denoted and defined as Z_m = {[0], [1], [2], ..., [m – 1]}. It is the set of all residue classes of integers modulo m.

Illustration: Let us consider the equivalence relation given by 'congruent modulo 4' on the set Z of all integers. This equivalence relation generates four residue classes [0], [1], [2], [3]:

$$[0] = \{\ldots, -8, -4, 0, 4, 8, \ldots\}$$
$$[1] = \{\ldots, -7, -3, 1, 5, 9, \ldots\}$$
$$[2] = \{\ldots, -6, -2, 2, 6, 10, \ldots\}$$
$$[3] = \{\ldots, -5, -1, 3, 7, 11, \ldots\}$$

Observations:

(*i*) Any two elements of the same residue class are equivalent under modulo m. For example, $0 \equiv 4 \pmod{m}$, $4 \equiv 8 \pmod{m}$, $-8 \equiv 4 \pmod{m}$ etc. in [0], where $m = 4$.

This implies that any element of a residue class can serve as a representative for that class (*e.g.* [0] = [4] = [– 8] etc.)

(*ii*) No element of Z belongs to more than one residue class.

(*iii*) Any element of Z belongs to one and only one of the four residue classes [0], [1], [2] and [3].

What we have done is **'partitioning the set Z into four residue classes'.** By partition (or decompose) of a non-empty set S we mean breaking the set S into **pairwise disjoint subsets** such that **every element** of S belongs to **one and only one** of these subsets. So, a partition of a non-empty set S is a set P of non-empty subsets of S such that each $x \in S$ is contained to one and only one member of P.

Theorem 26: *Congruent modulo m on the set of integers Z induces a partition of Z into residue classes.*

Proof: For any integer x we form a residue class $[x]$ which is the set of all integers congruent to x modulo m.

(*i*) $x \in [x]$ because $x \equiv x \pmod{m}$.

(*ii*) $y, z \in [x] \Rightarrow y \equiv x \pmod{m}, z \equiv x \pmod{m}$, by definition

$\Rightarrow y \equiv x \pmod{m}, x \equiv z \pmod{m}$, by symmetric property

$\Rightarrow y \equiv z \pmod{m}$, by transitive property

Hence any two integers of the same residue class are congruent modulo m. In particular, if $y \equiv x \pmod{m}$ then $[x] = [y]$, *i.e.*, any integer of a residue class will serve as a representative for that class.

(*iii*) No integer can belong to more than one residue class. Suppose that $x \in [a]$ as well as $x \in [b]$ then by (*ii*), $[x] = [a] = [b]$.

(*iv*) For **any integer** c we may form $[c]$, so we have shown that each integer belongs to one and only one of the residue classes formed under congruent modulo m.

These arguments thus prove that congruent modulo m on Z provides a partition of Z into residue classes.

Definition 10: Let $[a]$, $[b]$ are two residue classes modulo m, their sum is defined as

$$[a] + [b] = [r]$$

where r is the least non-negative remainder when $a + b$ is divided by m.

$$\therefore \qquad [a] + [b] = \begin{cases} [a+b], & \text{if } 0 \le a+b < m \\ [r], & \text{otherwise} \end{cases}$$

where r is the least non-negative remainder when $a + b$ is divided by m.

Illustration: If [0], [1], [2], [3], [4] are residue classes modulo 5, then

$[0] + [2] = [2]$, since $0 < 0 + 2 = 2 < 5$

$[1] + [3] = [4]$, since $0 < 1 + 3 = 4 < 5$

$[2] + [3] = [0]$, since $2 + 3 = 5 = 1 \times 5 + 0$

$[3] + [4] = [2]$, since $3 + 4 > 5$ and $3 + 4 = 7 = 1 \times 5 + 2$.

Definition 11: Let $[a]$, $[b]$ are two residue classes modulo m, their product is defined as

$$[a] \,.\, [b] = [r]$$

where r is the least non-negative remainder when ab is divided by m.

$$\therefore \qquad [a] \,.\, [b] = \begin{cases} [ab], & \text{if } 0 \le ab < m \\ [r], & \text{otherwise} \end{cases}$$

where r is the least non-negative remainder when ab is divided by m.

Illustration: If [– 3], [0], [1], [3], [4] are residue classes modulo 5, then

$[-3] \,.\, [0] = [0]$, since $(-3) \,.\, 0 = 0$

$[-3] \,.\, [3] = [1]$, since $(-3) \,.\, 3 = -9 < 0$ and $-9 = (-2) \,.\, 5 + 1$

$[-3] . [1] = [2]$, since $(-3) . 1 = -3 < 0$ and $-3 = (-1) . 5 + 2$

$[3] . [4] = [2]$, since $3 . 4 = 12 > 5$ and $12 = 2 . 5 + 2$.

2.9 LINEAR CONGRUENCES

Definition 12: A congruence of the form $ax \equiv b \pmod{m}$ in which m is a fixed positive integer, a, b are fixed integers and x is an unknown integer is said to be a **linear congruence.**

Definition 13: If there exists an integer x_0 such that $ax_0 \equiv b \pmod{m}$, then x_0 is said to be a **solution of the linear congruence** $ax \equiv b \pmod{m}$.

Observation

If x_1 is a solution of $ax \equiv b \pmod{m}$, then any other integer $x_2 \equiv x_1 \pmod{m}$ is also a solution. So, the entire residue class $[x_1]$ is counted as a single solution.

For example, a solution of the linear congruence $7x \equiv 5 \pmod{4}$ is the integer 3 because $7 \times 3 \equiv 5 \pmod{4}$. Here all members of the residue class (under modulo 4) $[3] = \{..., -5, -1, 3, 7, 11, ...\}$ are solutions. So, [3] is counted as a single solution of $7x = 5 \pmod{4}$.

Definition 14: Suppose x_1 and x_2 are two solutions of the linear congruence $ax \equiv b \pmod{m}$. If $x_1 \not\equiv x_2 \pmod{m}$, then x_1 and x_2 are said to be **incongruent solutions** of $ax \equiv b \pmod{m}$.

For example, consider the linear congruence $10x \equiv 2 \pmod{4}$. Observe that 1 and 3 are its solutions. Since $1 \not\equiv 3 \pmod{4}$, therefore 1 and 3 are **incongruent solutions** of $10x \equiv 2 \pmod{4}$. On the other hand, 7 is also a solution of $10x \equiv 2 \pmod{4}$, but $7 \equiv 3 \pmod{4}$. Therefore 3 and 7 are **congruent solutions** of the linear congruence $10x \equiv 2 \pmod{4}$.

Theorem 27: (Existence of solutions):

The linear congruence $ax \equiv b \pmod{m}$ has a solution if and only if the greatest common divisor of a and m divides b.

Proof: Let $d = \gcd(a, m)$.

Suppose ax ≡ b (mod m) has a solution x = x_1.

Then $ax_1 \equiv b \pmod{m}$

$\Rightarrow \quad m \mid (ax_1 - b)$

$\Rightarrow \quad ax_1 - b = mk$, for some integer k

$\Rightarrow \quad b = ax_1 - mk$.

Now, $\quad d = \gcd(a, m) \Rightarrow d \mid a$ and $d \mid m$

$\Rightarrow \quad d \mid (ax_1 - mk)$

$\Rightarrow \quad d \mid b$

This shows that if $ax \equiv b \pmod{m}$ has a solution, then $d \mid b$. In other words, **if d does not divide b, then ax = b (mod m) has no solution.**

Conversely, suppose $d \mid b \Rightarrow b = nd$, for some integer n.

Now, $\quad d = \gcd(a, m)$

$\Rightarrow \quad d = au + mv$, for some integers, u, v [By Theorem 5]

$\Rightarrow \quad nd = a(nu) + m(nv)$

$\Rightarrow \quad b = a(nu) + m(nv)$ $[\because \ b = nd]$

$\Rightarrow \quad a(nu) - b = -m(nv)$

$\Rightarrow \quad m \mid \{a(nu) - b\}$

$\Rightarrow \quad a(nu) \equiv b \pmod{m}$

$\Rightarrow \quad x = nu$ is a solution of $ax \equiv b \pmod{m}$

This completes the proof.

Theorem 28: *If gcd $(a, m) = 1$, then the congruence $ax \equiv b \pmod{m}$ has a solution. Further if x_0 is a solution, then the set of all solutions is precisely the residue class $[x_0]$ modulo m, i.e., the congruence $ax \equiv b \pmod{m}$ has a unique incongruent solution modulo m.*

Proof: Existence of solution

$\gcd(a, m) = 1 \quad \Rightarrow \quad 1 = au + mv$, for some integers u, v [By Th. 5]

$\Rightarrow \quad b = bau + bmv$

$\Rightarrow \quad a(bu) - b = -m(bv)$

$\Rightarrow \quad m \mid \{a(bu) - b\}$

$\Rightarrow \quad a(bu) \equiv b \pmod{m}$

$\Rightarrow \quad x = bu$ is a solution of $ax \equiv b \pmod{m}$.

Uniqueness of Solution (modulo m)

Let x_0 be a solution of $ax \equiv b \pmod{m}$

then $\quad ax_0 \equiv b \pmod{m}$ ...(1)

Now, $\quad y \in [x_0] \quad \Rightarrow \quad y \equiv x_0 \pmod{m}$

$\Rightarrow \quad ay \equiv ax_0 \pmod{m}$ [By property 4, Art. 2.6]

$\Rightarrow \quad ay \equiv b \pmod{m}$ [Using (1), by transitive property]

$\Rightarrow$ y is a solution of $ax \equiv b \pmod{m}$.

Conversely, y is a solution of $ax \equiv b \pmod{m}$

$\Rightarrow \quad ay \equiv b \pmod{m}$

$\Rightarrow \quad b \equiv ay \pmod{m}$

$\Rightarrow \quad ax_0 \equiv ay \pmod{m}$ [Using (1), by transitive property]

$\Rightarrow \quad ay \equiv ax_0 \pmod{m}$

$\Rightarrow \quad y \equiv x_0 \pmod{m}$ [$\because$ gcd $(a, m) = 1$, by property 6, Art. 2.6]

$\Rightarrow \quad y \in [x_0]$.

Example 15: *Solve : $7x \equiv 3 \pmod{4}$.*

Solution: The given linear congruence is

$$7x \equiv 3 \pmod{4} \qquad ...(1)$$

Since gcd $(7, 4) = 1$, therefore the congruence (1) has a unique solution modulo 4.

We have $\quad 0 \equiv 4 \pmod{4}$ ...(2)

Adding the congruences (1) and (2) [Using property 5, Art. 2.6]:

$$7x \equiv 7 \pmod{4}$$

$\Rightarrow \quad x \equiv 1 \pmod{4}$, by property 6 (Art. 2.6) since gcd $(7, 4) = 1$.

Hence $x \equiv 1 \pmod{4}$, *i.e.*, $x = 1 + 4n$, where $n \in Z$ is the required solution of the given congruence.

Theorem 29: *If d = gcd (a, m) is a divisor of b, then x_1 is a solution of the linear congruence $ax \equiv b \pmod{m}$ if and only if x_1 is a solution of $a_1x \equiv b_1 \pmod{m_1}$ where $a = a_1d$, $b = b_1d$ and $m = m_1d$.*

Proof: x_1 is a solution of $\quad ax \equiv b \pmod{m}$

$\Leftrightarrow \quad ax_1 \equiv b \pmod{m}$

$\Leftrightarrow \quad a_1dx_1 \equiv b_1d \pmod{m_1d} \qquad [\because \; a = a_1d, b = b_1d, m = m_1d]$

$\Leftrightarrow \quad m_1d \mid (a_1dx_1 - b_1d)$

$\Leftrightarrow \quad m_1 \mid (a_1x_1 - b_1)$

$\Leftrightarrow \quad a_1x_1 \equiv b_1 \pmod{m_1}$

$\Leftrightarrow \quad x_1$ is a solution of $a_1x \equiv b_1 \pmod{m_1}$.

Conversely, x_1 is a solution of $a_1x \equiv b_1 \pmod{m_1}$

$\Leftrightarrow \quad a_1x_1 \equiv b_1 \pmod{m_1}$

$\Leftrightarrow \quad m_1 \mid (a_1x_1 - b_1)$

$\Leftrightarrow \quad m_1d \mid (a_1dx_1 - b_1d)$

$\Leftrightarrow \quad m \mid (ax_1 - b) \qquad [\because \; a = a_1d, b = b_1d, m = m_1d]$

$\Leftrightarrow \quad x_1$ is a solution of $ax \equiv b \pmod{m}$.

Theorem 30: (Number of incongruent solutions)

If $d = \gcd(a, m)$ divides b, then the congruence $ax \equiv b \pmod{m}$ has exactly d incongruent solutions which can be expressed in the form $x_0 + rm_1$ for $r = 0, 1, 2, ..., d-1$, where $m = dm_1$ and x_0 is an arbitrary solution of $ax \equiv b \pmod{m}$.

Proof: Given, $d = \gcd(a, m)$ divides b and so $ax \equiv b \pmod{m}$ possesses a solution (By Theorem 27). Therefore, $a = a_1d$, $b = b_1d$, $m = m_1d$ for some integers a_1, b_1, m_1 and

$$\gcd(a_1, m_1) = 1 \qquad [\because \; d = \gcd(a, m)] \qquad ...(1)$$

Let x_0 be an arbitrary solution of

$$ax \equiv b \pmod{m} \qquad ...(2)$$

and let x_1 be any other solution of (2).

By Theorem 29, x_0 and x_1 are also solutions of

$$a_1x \equiv b_1 \pmod{m_1} \qquad ...(3)$$

Using Theorem 28, (3) has a unique incongruent solution modulo m_1 [by (1)].

$\therefore \quad x_1 \equiv x_0 \pmod{m_1}$

$\Leftrightarrow \quad m_1 \mid (x_1 - x_0)$

$\Leftrightarrow \quad x_1 - x_0 = rm_1$, for some integer r

$\Leftrightarrow \quad x_1 = x_0 + rm_1$

Thus every solution x_1 of (2) can be expressed in the form $x_1 = x_0 + rm_1$, for some integer r. Further for every integer r, we have $x_0 + rm_1 \equiv x_0 \pmod{m_1}$ and so $x_0 + rm_1$ is a solution of (3) and also of (2). Hence **all the solutions of (2) are found among the set of integers $\mathbf{x_0 + rm_1}$.** Let us consider the following d integers in this set:

$$x_0, x_0 + m_1, x_0 + 2m_1, ..., x_0 + (d-1)m_1 \qquad ...(4)$$

Now no two of these d integers are congruent to each other modulo m because if $0 \le i < d$ and $0 \le j < d$, then

$x_0 + im_1 \equiv x_0 + jm_1 \pmod{m}$

$\Leftrightarrow \quad m \mid \{(x_0 + im_1) - (x_0 + jm_1)\}$

$\Leftrightarrow \quad m \mid (i - j)\, m_1$

$\Leftrightarrow \quad m_1d \mid (i - j)\, m_1 \qquad [\because \; m = m_1d]$

$\Leftrightarrow \quad d \mid (i-j)$

$\Leftrightarrow \quad i-j=0 \qquad [\because \; 0 \le i-j < d]$

$\Leftrightarrow \quad i=j \;\Leftrightarrow\; x_0+im_1 = x_0+jm_1$

Hence any two of the d integers given in (4) are incongruent (mod m).

Moreover, if r is any integer, then we shall show that $x_0 + rm_1$ is congruent to one of the d integers given in (4).

Applying division algorithm, we get

$$r = ds + q,\ 0 \le q < d.$$

$\therefore \quad x_0 + rm_1 = x_0 + (ds+q)\,m_1 = x_0 + qm_1 + sdm_1.$

$\Rightarrow \quad (x_0+rm_1)-(x_0+qm_1) = sm \qquad [\because \; m = m_1 d]$

$\Rightarrow \quad m \mid \{(x_0+rm_1)-(x_0+qm_1)\}$

$\Rightarrow \quad x_0+rm_1 \equiv x_0+qm_1 \pmod m$

Since $0 \le q < d$, therefore $x_0 + qm_1$ is one of the integers given in (4).

Therefore, $ax \equiv b \pmod m$ has exactly d incongruent solutions modulo m.

Note: If we are to find the least positive incongruent solutions modulo m of $ax \equiv b \pmod m$, then we should first find the least positive integer, say x_0, satisfying $a_1x \equiv b_1 \pmod{m_1}$ where $d = \gcd(a, m)$, $a = a_1d$, $b = b_1d$, $m = m_1d$. Then $x_0, x_0+m_1, x_0+2m_1, ..., x_0+(d-1)\,m_1$ will give d least positive incongruent solutions modulo m of $ax \equiv b \pmod m$.

Example 16: *Solve the following congruences:*

(i) $234x \equiv 54 \pmod 7$ *(ii) $13x \equiv 9 \pmod{25}$*

(iii) $32x \equiv 16 \pmod{24}$.

Solution: (*i*) The given linear congruence is

$$234x \equiv 54 \pmod 7 \qquad ...(1)$$

Since gcd (234, 7) = 1, so (1) has a unique solution modulo 7.

Here the numbers 234 and 54 are both large, so dividing each by the modulus of congruence 7, we get

$$234 = 7 \times 33 + 3 = 231 + 3, \text{ where } 0 < 3 < 7$$

and $$54 = 7 \times 7 + 5 = 49 + 5, \text{ where } 0 < 5 < 7.$$

$\therefore \quad 231x \equiv 49 \pmod 7 \qquad ...(2)$

From (1) and (2), we get

$$234x - 231x \equiv (54-49) \pmod 7$$

or $$3x \equiv 5 \pmod 7 \qquad ...(3)$$

By trial we see that the least positive integer satisfying the congruence (3) is 4. Note that $3 \times 4 = 12$ and $12 \equiv 5 \pmod 7$.

Thus 4 is the least positive integer satisfying the given congruence and the complete solution consists of the residue class [4] (mod 7) and is given by

$$[4] = \{7k+4 : k \in Z\} = \{..., -17, -10, -3, 4, 11, 18, 25, ...\}.$$

Note: The symbol ':' stands for 'such that'.

(*ii*) The given linear congruence is

$$13x \equiv 9 \pmod{25} \qquad ...(1)$$

Since gcd (13, 25) = 1, so (1) has a unique solution modulo 25. Also, we can find integers m and n such that $1 = 13m + 25n$.

By Euclidean algorithm (see Art. 2.4), we have

$$25 = 1 \times 13 + 12$$
$$13 = 1 \times 12 + 1$$
$$12 = 12 \times 1 + 0$$

Substituting backwards, we have

$$1 = 13 - 1 \times 12 = 13 - 1 \times (25 - 1 \times 13) = 2 \times 13 + (-1) \times 25$$

$\therefore$ $9 = 18 \times 13 + (-9) \times 25.$

So, $13x \equiv 9 \pmod{25}$ is equivalent to

$$13x \equiv \{18 \times 13 + (-9) \times 25\} \pmod{25} \quad \text{...(2)}$$

But $$0 \equiv 9 \times 25 \pmod{25} \quad \text{...(3)}$$

Adding (2) and (3) [using property 5, Art. 2.6], we get

$$13x \equiv 13 \times 18 \pmod{25}$$

showing that $x = 18$ is a solution.

Obviously 18 is the least positive integer in the residue class [18] (mod 25). Therefore,

$$[18] = \{25k + 18 : k \in Z\}$$

is the required solution.

(*iii*) Comparing the given congruence with $ax \equiv b \pmod{m}$, we have $a = 32$, $b = 16$, $m = 24$.

We have gcd (a, m) = gcd (32, 24) = 8 and 8 is a divisor of $b = 16$. Therefore, $32x \equiv 16 \pmod{24}$ has 8 incongruent solutions (mod 24), by Theorem 30.

Now $32x \equiv 16 \pmod{24}$ is equivalent to

$$8 \times 4x \equiv 8 \times 2 \pmod{8 \times 3}$$

or $4x \equiv 2 \pmod{3}$ [by Theorem 29]

showing that $x = 2$ is a solution.

By Theorem 30, the 8 incongruent solutions are given by $x \equiv 2 + 3t \pmod{24}$, where $t = 0, 1, 2, 3, 4, 5, 6, 7$, *i.e.*, $x \equiv 2, 5, 8, 11, 14, 17, 20, 23 \pmod{24}$.

MULTIPLE CHOICE QUESTIONS

1. Subtraction of integers is

(*a*) commutative (*b*) associative
(*c*) neither commutative nor associative (*d*) none of these.

2. If $a, x, y \in Z$, $a \mid (x - y)$ and $a \mid (x + y)$, then

(*a*) $a \mid 2x, a \mid y$ (*b*) $a \mid x, a \mid 2y$ (*c*) $a \mid x, a \mid y$ (*d*) none of these

3. If x and y are odd integers, then $x^2 + y^2$ is

(*a*) a prime (*b*) an even integer (*c*) an odd integer (*d*) none of these.

4. If $n \in N$, then $\forall k \in N$,

(*a*) $(n + 1) \mid (n^{2k} - 1)$ (*b*) $(n + 1) \mid (n^{2k} + 1)$ (*c*) $(n + 1) \mid (n^{3k} - 1)$ (*d*) none of these.

5. The last digit in the decimal representation of 6^{2999} is

(*a*) 2 (*b*) 4 (*c*) 6 (*d*) 8.

6. If n is any even natural number, then $2^{2n} - 1$ is a multiple of

(*a*) 5 (*b*) 15 (*c*) 7 (*d*) 25.

7. For all $n \in N$, $3^{2n} - 1$ is divisible by

(*a*) 8 (*b*) 6 (*c*) 4 (*d*) 3.

8. If $n \in Z^+$, then $n(n^2 - 1)(n^2 - 4)$ is divisible by

(*a*) 4, 5, 6 (*b*) 5, 6, 7 (*c*) 5, 7, 11 (*d*) none of these.

9. Which one of the following statements is not true in respect of a prime number?

(*a*) 2 is the smallest prime

(*b*) The smallest positive divisor (> 1) of an integer (> 1) is a prime number

(*c*) If p be a prime and $p \mid ab$, then either $p \mid a$ or $p \mid b$.

(*d*) Number of primes is finite.

10. The number of positive divisors of 252 is

(*a*) 9 (*b*) 5 (*c*) 18 (*d*) 10.

11. If a and b are positive integers such that $a^2 - b^2$ is a prime, then $a^2 - b^2$ =

(*a*) ab (*b*) $a + b$ (*c*) $a - b$ (*d*) 1.

12. $p \equiv q \pmod{m}$ iff

(*a*) $m \mid (p - q)$ (*b*) $(p - q) \mid m$ (*c*) $m \mid p$ (*d*) $p \mid m$.

13. Which one of the following statements is false?

(*a*) $48 \equiv -2 \pmod{10}$ (*b*) $157 \equiv 7 \pmod{15}$

(*c*) $83 \equiv 8 \pmod{6}$ (*d*) $120 \equiv 3 \pmod{13}$.

14. If $68 \equiv 4 \pmod{n}$, then n can be

(*a*) 12 (*b*) 17 (*c*) 13 (*d*) 16.

15. When $64 \times 65 \times 66$ is divided by 67, the remainder is

(*a*) 61 (*b*) 62 (*c*) 63 (*d*) 64.

16. The units digit in the decimal representation of 11^{132} is

(*a*) 6 (*b*) 4 (*c*) 2 (*d*) 1.

17. If $100 \equiv x \pmod{8}$, then least positive value of x is

(*a*) 3 (*b*) 4 (*c*) 5 (*d*) 6.

18. The remainder obtained when 2^{100} is divided by 11 is

(*a*) 3 (*b*) 2 (*c*) 1 (*d*) 0.

19. The least positive integer to which 97×155 congruent modulo 19, is

(*a*) 6 (*b*) 4 (*c*) 8 (*d*) 2.

20. The last digit in 7^{291} is

(*a*) 9 (*b*) 7 (*c*) 5 (*d*) 3.

21. If $7x \equiv 3 \pmod{5}$, then x can take value

(*a*) 17 (*b*) 19 (*c*) 21 (*d*) 22.

22. The remainder obtained when 5^{124} is divided by 124 is

(*a*) 5 (*b*) 0 (*c*) 2 (*d*) 1.

23. What is the result of $(-3) \times_8 5 +_8 (-3) \times_8 (-5)$ in $[Z_8, +_8, \times_8]$, where Z_8 is the set of integers modulo 8, $+_8$ is the modulo 8 addition operation and $\times_8$ is the modulo 8 multiplication operation?

(*a*) 0 (*b*) 7 (*c*) 8 (*d*) 2. (*WBUT 2012*)

24. Every integer n (> 1) can be written uniquely as a product of

(*a*) perfect squares (*b*) primes (*c*) positive integers (*d*) none of these.

25. Division algorithm states as, let a and b ($\neq 0$) be integers, then there exists integers q and r ($0 \le r < b$) such that

(*a*) $a = qr + b$ (*b*) $a = bq - r$ (*c*) $a - bq = r$ (*d*) all of these.

26. If a, b, c are integers, which of the following are true?

(*a*) if $a \mid b$ and $b \mid c$, then $a \mid c$

(*b*) if $a \mid b$ and $b \mid a$, then $|a| = |b|$

(*c*) if $a \mid b$ and $a \mid c$, then $a \mid (b + c)$ and $a \mid (b - c)$

(*d*) all of these.

27. A positive integer $d = \gcd(a, b)$ and $d = au + bv$, which one of the following is not true?

(*a*) if $x > 0$, then $\gcd(ax, bx) = x + \gcd(a, b)$

(*b*) $\gcd(a, b) = \gcd(b, a)$

(*c*) if $d = \gcd(a, b)$, then $\gcd\left(\frac{a}{d}, \frac{b}{d}\right) = 1$

(*d*) for any integer x, $\gcd(a, b) = \gcd(a, b + ax)$.

28. $\gcd(540, 168) =$

(*a*) 168 (*b*) 12 (*c*) 34 (*d*) none of these.

29. Two integers a, b are said to be relatively prime (or coprime) if

(*a*) $\gcd(a, b) = a$ (*b*) $\gcd(a, b) = ab$ (*c*) $\gcd(a, b) = 1$ (*d*) all of these.

ANSWERS

1. (*c*) [For example, $8 - 3 \neq 3 - 8$ and $(8 - 3) - 4 \neq 8 - (3 - 4)$]

2. (*d*) [**Hint.** $a \mid (x - y)$ and $a \mid (x + y) \Rightarrow a \mid (x - y + x + y)$ and $a \mid \{(x + y) - (x - y)\}$
$\Rightarrow a \mid 2x$ and $a \mid 2y$.
So, we cannot conclude any of (*a*), (*b*) and (*c*)]

3. (*b*) [**Hint.** Let $x = 2p + 1$, $y = 2q + 1$, where $p, q \in Z$.
$\therefore\ x^2 + y^2 = 4(p^2 + q^2 + p + q) + 2$]

4. (*a*) [**Hint.** $(m - 1) \mid (m^k - 1), \forall m \in N, \forall k \in N$
$\Rightarrow (n^2 - 1) \mid \{(n^2)^k - 1\} \Rightarrow (n - 1)(n + 1) \mid (n^{2k} - 1) \Rightarrow (n + 1) \mid (n^{2k} - 1)$]

5. (*c*) [**Hint.** $6^2 = 36$, $6^3 = 216$, $6^4 = 1296$, ...
So, all powers of 6 terminate in 6 on account of 6×6 terminating in 6.]

6. (*b*) [**Hint.** If $n = 2m$, $m \in Z^+$, then $2^{2n} - 1 = 16^m - 1 = (16 - 1)(16^{m-1} + 16^{m-2} + ... + 16 + 1)$]

7. (*a*) [**Hint.** $3^{2n} - 1 = 9^n - 1 = (1 + 8)^n - 1 = {}^nC_0 + {}^nC_1 8 + {}^nC_2 8^2 + ... + {}^nC_n 8^n - 1$
$= 8\{{}^nC_1 + {}^nC_2 8 + ... + {}^nC_n 8^{n-1}\}$]

8. (*a*) [**Hint.** $n(n^2 - 1)(n^2 - 4) = (n - 2)(n - 1)n(n + 1)(n + 2)$, which is the product of 5 consecutive natural numbers (for $n \geq 3$) and hence it is divisible by $5! = 120$. Observe that 120 is a multiple of 4, 5 and 6 but not of 7 and 11. For $n = 1, 2$, the given number = 0 which is divisible by all positive numbers.]

9. (*d*)

10. (*c*) **Theorem.** Let $n > 1$, $n \in N$ and let $n = p_1^{\alpha_1} p_2^{\alpha_2} ... p_r^{\alpha_r}$ be prime power factorisation of n, then number of positive integral divisors of n is $(\alpha_1 + 1)(\alpha_2 + 1) ... (\alpha_r + 1)$.
Here $252 = 2^2 \times 3^2 \times 7^1$. So, number of positive divisors of 252 is $(2 + 1)(2 + 1)(1 + 1) = 18$.

11. (*b*) [**Hint.** Given, $a^2 - b^2 = (a - b)(a + b)$ is prime, so either $a + b$ or $a - b$ must be 1 or – 1. Also, $a^2 - b^2$ is positive and $a + b$ is certainly a positive integer ≥ 2. Therefore, $a - b = 1$ and hence $a^2 - b^2 = a + b$].

12. (*a*) **13.** (*c*) **14.** (*d*)

15. (*a*) [**Hint.** $64 \equiv (-3) \pmod{67}$, $65 \equiv (-2) \pmod{67}$ and $66 \equiv (-1) \pmod{67}$.
$\therefore\ 64 \times 65 \times 66 \equiv (-3)(-2)(-1) \pmod{67} \Rightarrow 64 \times 65 \times 66 \equiv (-6) \pmod{67}$.
Also, $-6 \equiv 61 \pmod{67}$. By transitivity, $64 \times 65 \times 66 \equiv 61 \pmod{67}$].

16. (*d*) [**Hint.** $11^2 = 121 = 12 \times 10 + 1 \Rightarrow 11^2 \equiv 1 \pmod{10} \Rightarrow (11^2)^{66} \equiv 1^{66} \pmod{10}$]
17. (*b*) [**Hint.** $100 = 12 \times 8 + 4$, so $x = 4$]
18. (*c*) [**Hint.** $2^5 = 32 = 3 \times 11 + (-1) \Rightarrow 2^5 \equiv (-1) \pmod{11}$
$\Rightarrow (2^5)^{20} \equiv (-1)^{20} \pmod{11} \Rightarrow 2^{100} \equiv 1 \pmod{11}$]
19. (*a*) [**Hint.** $97 \equiv 2 \pmod{19}$, $155 \equiv 3 \pmod{19} \Rightarrow 97 \times 155 \equiv 2 \times 3 \pmod{19}$.]
20. (*d*) [**Hint.** $7^2 = 5 \times 10 + (-1) \Rightarrow 7^2 \equiv (-1) \pmod{10} \Rightarrow (7^2)^{145} \equiv (-1)^{145} \pmod{10}$
$\therefore 7^{290} \equiv (-1) \pmod{10} \Rightarrow 7^{291} \equiv (-7) \pmod{10}$. Also, $-7 \equiv 3 \pmod{10}$.
$\therefore 7^{291} \equiv 3 \pmod{10}$.]
21. (*b*) [**Hint.** $7x \equiv 3 \pmod 5 \Rightarrow 7x - 3 = 5n, n \in Z \Rightarrow x = \dfrac{5n+3}{7}$. $x = 19$ when $n = 26$]

22. (*a*) 23. (*a*) 24. (*b*) 25. (*c*)
26. (*d*) 27. (*a*) 28. (*b*) 29. (*c*)

PROBLEMS

1. In each of the following cases, find integers q and r such that $a = bq + r$, $0 \le r < b$:
(*i*) $a = 3, b = 10$ (*ii*) $a = 221, b = 17$
(*iii*) $a = -111, b = 11$ (*iv*) $a = -278, b = 12$.
2. Calculate the greatest common divisor of each pair of integers (use Euclidean Algorithm):
(*i*) 12, 18 (*ii*) 60, 90
(*iii*) 414, 662 (*iv*) 7469, 2464.
3. In each of the following cases, calculate gcd (a, b) and express it as $au + bv$:
(*i*) $a = 45, b = 126$ (*ii*) $a = 1819, b = 3587$
(*iii*) $a = 12345, b = 54321$.
4. Show that gcd (a, b) = gcd $(a, b + na)$, where n is an integer.
5. If a, b, c are any three integers such that gcd $(a, c) = 1$ and gcd $(b, c) = 1$, then prove that gcd $(ab, c) = 1$.
6. If $a, b \in Z$ and gcd $(a, b) = 1$, prove that gcd $(a, b^n) = 1$, $\forall n \in N$.
[**Hint.** The result is true for $n = 1$ and let it is true for $n = k \in N$. Then gcd $(a, b^k) = 1$. Also, gcd $(a, b) = 1$.
$\therefore$ gcd $(a, b^{k+1}) = 1$, by Th. 10, Art. 2.3.]
7. If gcd $(a, b) = 1$, prove that gcd $(a + b, a^2 + b^2) = 1$ or 2.
8. Use Fermat's theorem to compute the values of
(*i*) $3^{302} \pmod 5$ (*ii*) $3^{302} \pmod 7$ (*iii*) $3^{302} \pmod{11}$.
[**Hint:** (*i*) $3^4 \equiv 1 \pmod 5 \Rightarrow (3^4)^{75} = 3^{300} \equiv 1 \pmod 5$
$\Rightarrow 3^{302} \equiv 9 \pmod 5$
$\Rightarrow 3^{302} \equiv 4 \pmod 5$]
9. Find the remainder when the sum
$$1! + 2! + 3! + \ldots + 99! + 100!$$
is divided by 12.
10. When n is a positive integer, show that $3^{3n+1} \equiv 3 \times 5^n \pmod{11}$. Obtain a similar result for 2^{4n+3} and hence show that $11 \mid (3^{3n+1} + 2^{4n+3})$.
11. Show that
(*i*) $77 \mid (36^{36} + 41^{41})$ (*ii*) $7 \mid (3^{2n+1} + 2^{n+2}), \forall n \in N$
(*iii*) $12 \mid (5n^3 + 7n^5), \forall n \in Z$.

12. Prove that

(*i*) there is no integer $n > 1$ for which $n \mid (2^n - 1)$

(*ii*) the product of two integers of the form $6n + 1$ is of the same form, where n is an integer.

13. Solve the following linear congruences:

(*i*) $6x \equiv 7 \pmod 9$ (*ii*) $59x \equiv 31 \pmod{67}$

(*iii*) $18x \equiv 30 \pmod{42}$.

ANSWERS

1. (*i*) $q = 0, r = 3$ (*ii*) $q = 13, r = 0$ (*iii*) $q = -11, r = 10$ (*iv*) $q = -24, r = 10$.

2. (*i*) 6 (*ii*) 30 (*iii*) 2 (*iv*) 77

3. (*i*) gcd = 9, $u = 3, v = -1$ (*ii*) gcd = 17, $u = 71, v = -36$ (*iii*) gcd = 3, $u = 3617, v = -822$

8. (*i*) 4 (*ii*) 2 (*iii*) 9

9. 9

13. (*i*) No solution (*ii*) $x \equiv 38 \pmod{67}$ (*iii*) $x \equiv 4, 11, 18, 25, 32, 39 \pmod{42}$.

Module II

CHAPTER 3

Posets and Lattices

3.1 PARTIAL ORDERING RELATIONS

Definition 1: A **binary relation** R on a non-empty set A is a subset of the cartesian product $A \times A$. The subset consists of pairs (a, b) where $a, b \in A$ and a, b are connected by certain given condition. Symbolically, it is denoted as $R \subseteq A \times A$.

If $(a, b) \in R$, we write $a\,R\,b$ which is read as '*a* is **related** to *b* by the relation *R*'. If $(a, b) \notin R$, we say that '*a* is **not related** to *b* by the relation *R*'.

Example 1: Let $A = \{0, 1, 2, 3\}$ and take $a\,R\,b$ iff $b = a - 1$. Then the binary relation R is given by

$$R = \{(1, 0), (2, 1), (3, 2)\}.$$

Notes: (*i*) Relation R is called **binary relation** since it connects two elements.

(*ii*) Let A be a non-empty set having n elements and so $A \times A$ has $n \times n = n^2$ elements. Hence the total number of subsets of $A \times A$ is 2^{n^2} and so the total number of possible binary relations on a non-empty finite set A is 2^{n^2}, where n is the number of elements contained in A.

Definition 2: A binary relation R defined on a non-empty set S is said to be a **partial ordering relation** (or **partial order relation**) if R is **reflexive, antisymmetric** and **transitive.**

There is also an alternative notation for specifying the partial ordering relation—'for each ordered pair $(a, b) \in R$ we write $a \leq b$ instead of $(a, b) \in R$ or $a\,R\,b$'. The hypotheses for a partial ordering relation may thus be written as:

P1 (Reflexivity): $x \leq x, \quad \forall\, x \in S$

P2 (Antisymmetry): $x \leq y$ and $y \leq x \;\Rightarrow\; x = y$, for $x, y \in S$.

P3 (Transitivity): $x \leq y$ and $y \leq z \;\Rightarrow\; x \leq z$, for $x, y, z \in S$.

If $x \leq y$ we say that '*x* is less than or equal to *y*', or that '*y* is greater than or equal to *x*', also written in this case as $y \geq x$. Some other terminology frequently adopted are that '*x* precedes or is smaller than *y*' and that '*y* follows or dominates or is larger than *x*'.

Notes: (*i*) The symbol '$\leq$' does not necessarily mean 'less than or equal to' as is used for real numbers.

(*ii*) A partial order (ordering) is a total order (ordering) (or linear order (ordering)) iff either $x \leq y$ or $y \leq x$, $\forall\, x, y \in S$.

(*iii*) When neither $x \leq y$ nor $y \leq x$, we say that x and y are **incomparable.**

(*iv*) The **strict order (ordering),** $<$, associated with $\leq$ is the binary relation defined by $x < y$ iff $x \leq y$ and $x \neq y$.

(*v*) 'Partial orders' and 'Equivalence relations' are two main kinds of relations that play a very important role in mathematics and computer science.

Remark: Observe that if $<$ is the strict order associated with a partial order, $\leq$, then $<$ is antireflexive and transitive, which means that

1. (Antireflexivity): $x \nless x, \forall x \in S$.

2. (Transitivity): $x < y$ and $y < z \Rightarrow x < z$, for $x, y, z \in S$.

Conversely, let $<$ be a relation on S and assume that $<$ is antireflexive and transitive. Then we can define the relation $\leq$ so that $x \leq y$ iff $x = y$ or $x < y$.

It is easy to check that $\leq$ is a partial order and that the strict order associated with $\leq$ is our original relation $<$.

Example 2: *Show that the relation $\leq$ (less than or equal to) is a partial ordering on the set of integers Z.*

Solution: 1. $a \leq a, \forall a \in Z$, since $a = a$. So, $\leq$ is **reflexive.**

2. $a \leq b$ and $b \leq a \Rightarrow a = b$, for $a, b \in Z$. Thus $\leq$ is **antisymmetric.**

3. $a \leq b$ and $b \leq c \Rightarrow a \leq c$, for $a, b, c \in Z$. Hence $\leq$ is **transitive.**

So, the usual 'less than or equal to ($\leq$)' relation is a partial ordering relation on Z. Observe that here $\leq$ is a total ordering on Z because either $a \leq b$ or $b \leq a, \forall a, b \in Z$.

Note: Similarly, the relation $\geq$ is a partial ordering on the set of integers Z. It is also a total ordering on Z.

Example 3: *Let P(S) denotes the power set of a non-empty set S, i.e., the collection of all subsets of a given non-empty set S. Show that the usual set theoretic inclusion relation $\subseteq$ on pairs A, B of subsets of S defined as '$A \subseteq B$ iff $a \in A \Rightarrow a \in B$' is a partial ordering relation on P(S).*

Solution: 1. $A \subseteq A, \forall A \in P(S)$. So $\subseteq$ is **reflexive.**

2. $A \subseteq B$ and $B \subseteq A \Rightarrow A = B$, for $A, B \in P(S)$. Thus $\subseteq$ is **antisymmetric.**

3. $A \subseteq B$ and $B \subseteq C \Rightarrow A \subseteq C$, for $A, B, C \in P(S)$. Hence $\subseteq$ is **transitive.**

It follows that $\subseteq$ is a partial ordering relation on $P(S)$.

Note: Observe that a natural example of a partial ordering which is not total ordering is provided by the subset ordering ($\subseteq$).

For example, if $S = \{a, b, c\}$, we have $\{a\} \subseteq \{a, b\}$. However, neither $\{a\}$ is a subset of $\{b, c\}$ nor $\{b, c\}$ is a subset of $\{a\}$. The subsets $\{a\}$ and $\{b, c\}$ are said to be **incomparable.**

Example 4: *Let Z^+ be the set of all positive integers and let $\leq$ be a binary relation defined on Z^+ such that $a \leq b$ iff $a \mid b$ (i.e., a divides b). Show that $\leq$ be a partial ordering relation on Z^+.*

Solution: 1. $a \leq a, \forall a \in Z^+$, since $a \mid a$. So, $\leq$ is **reflexive.**

2. Now, $a \mid b$ and $b \mid a$, for $a, b \in Z^+$

$\Rightarrow \quad \exists\, m, n \in Z^+$, s.t. $b = ma$ and $a = nb$

$\Rightarrow \quad ba = m\,n\,a\,b \Rightarrow mn = 1 \Rightarrow m = n = 1 \Rightarrow a = b.$

$\therefore \quad a \leq b$ and $b \leq a \Rightarrow a = b$, for $a, b \in Z^+$. Thus $\leq$ is **antisymmetric.**

3. Now, $a \mid b$ and $b \mid c$, for $a, b, c \in Z^+$

$\Rightarrow \quad \exists\, m, n \in Z^+$, s.t. $b = ma$ and $c = nb$

$\Rightarrow \quad c = n\,(ma) = nma \quad \Rightarrow a \mid c$, since $nm \in Z^+$ as $n, m \in Z^+$

$\therefore \quad a \le b$ and $b \le c \quad \Rightarrow \quad a \le c$, for $a, b, c \in Z^+$. Hence $\le$ is **transitive.**

It follows that $\le$ is a partial ordering relation on Z^+.

Note: Observe that here $\le$ is not a total ordering.

3.2 PARTIALLY ORDERED SET (POSET)

Definition 3: A non-empty set S together with a partial ordering relation $\le$ on S is called a **partially ordered set** or for short, a **poset** and is usually denoted by $(S, \le)$.

In other words: A **partially ordered set (poset**, for short) is a structure $(S, \le)$ where S is a non-empty set and $\le$ is a partial ordering relation on S, *i.e.*, a binary relation on S such that the following conditions hold:

P1 (Reflexivity): $x \le x, \quad \forall\, x \in S$.

P2 (Antisymmetry): $x \le y$ and $y \le x \quad \Rightarrow \quad x = y$, for $x, y \in S$.

P3 (Transitivity): $x \le y$ and $y \le z \quad \Rightarrow \quad x \le z$, for $x, y, z \in S$.

Notes: (*i*) The structures $(Z, \le)$, $(Z, \ge)$, $(P(S), \subseteq)$, $(Z^+, \le)$ given in Examples 2, 3, 4 are posets.

(*ii*) Let $(S, \le)$ be a poset and A be a non-empty subset of S. Then $(A, \le)$ is also a poset.

Principle of Duality

Definition 4: Let $(S, \le)$ be a poset and let R be a binary relation on S such that for $a, b \in S$, $a\,R\,b$ iff $b \le a$. Then the relation R is called the **converse of the partial ordering relation $\le$ and is denoted by $\ge$.**

Theorem 1: *If $(S, \le)$ is a poset, then $(S, \ge)$ is also a poset.*

Proof: Given $(S, \le)$ is a poset. We have to prove that $(S, \ge)$ is also a poset.

1. **(Reflexivity):** $(a \le a, \forall\, a \in S) \quad \Leftrightarrow \quad (a \ge a, \forall\, a \in S)$.

2. **(Antisymmetry):** $(a \le b$ and $b \le a \quad \Rightarrow \quad a = b)$

$\Leftrightarrow \quad (b \ge a$ and $a \ge b \quad \Rightarrow \quad a = b)$

$\Leftrightarrow \quad (a \ge b$ and $b \ge a \quad \Rightarrow \quad a = b)$, for $a, b \in S$.

3. **(Transitivity):** $(a \le b$ and $b \le c \quad \Rightarrow \quad a \le c)$

$\Leftrightarrow \quad (b \ge a$ and $c \ge b \quad \Rightarrow \quad c \ge a)$

$\Leftrightarrow \quad (c \ge b$ and $b \ge a \quad \Rightarrow \quad c \ge a)$, for $a, b, c \in S$.

It follows that $(S, \ge)$ is a poset.

Therefore, we can replace the relation $\le$ in any theorem about posets by the relation $\ge$ throughout without effecting its truth. This is known as the **principle of duality.**

Definition 5: Let $(S, \le)$ be a poset. A non-empty subset A of S is called a **chain** if every two elements of A are related (*i.e.*, comparable). We frequently refer to the number of elements in a chain as the **length of the chain.**

Definition 6: Let $(S, \le)$ be a poset. A non-empty subset B of S is called an **anti-chain** if no two distinct elements in B are related (*i.e.*, comparable).

Example 5: Let Z^+ be the set of all positive integers and let $\le$ be a binary relation defined on Z^+ such that $a \le b$ iff $a \mid b$, where a, b $\in Z^+$. Then $(Z^+, \le)$ is a poset (see Ex. 4).

Observe that $A = \{2, 4, 8, 16, 32\} \subseteq Z^+$ and any two elements of A are related with respect to $\le$. So, A is a chain having length 5.

Now, $B = \{2, 3, 5, 7, 11, 13, 17\} \subseteq Z^+$ and no two distinct elements of B are related and so B is an anti-chain.

Example 6: Let $S = \{a, b, c, d, e\}$ and let a relation $\le$ be defined on S as given by the following table:

	a	b	c	d	e
a	✓	✓	✓	✓	✓
b		✓	✓		✓
c			✓		✓
d				✓	✓
e					✓

Then it is easy to check that $(S, \le)$ is a poset. In this poset the subsets $\{a, b, c, e\}$, $\{a, b, e\}$ etc. are chains and the subsets $\{b, d\}$, $\{c, d\}$ are anti-chains.

Definition 7: A poset $(S, \le)$ is called a **totally ordered** (or **linearly ordered**) **set** or simply **ordered set** if S is a chain. In this case the binary relation $\le$ is called a **total ordering** (or **linear ordering**) **relation.**

Example 7: Under the usual relation of inequality $\le$, the set of all real numbers forms a poset which is a totally ordered set.

Example 8: The poset $(Z^+, \le)$, given in Ex. 5, is not totally ordered.

Definition 8: Let $(S, \le)$ be a poset. An element (if exists) $0 \in S$ is said to be the **least** (or **minimum, first, zero**) **element** or **universal lower bound** of $(S, \le)$ iff $0 \le a$, $\forall\, a \in S$.

Definition 9: Let $(S, \le)$ be a poset. An element (if exists) $1 \in S$ is said to be the **greatest** (or **maximum, last, unit**) **element** or **universal upper bound** of $(S, \le)$ iff $a \le 1$, $\forall\, a \in S$.

Note: Posets need not have any universal bounds. As for example, under the usual relation of inequality the set of all real numbers form a poset which has no universal bounds.

Theorem 2: *A given poset $(S, \le)$ can have at most one least element (universal lower bound) and at most one greatest element (universal upper bound).*

Proof: If possible, let $0 \in S$ and $0^* \in S$ both be the least elements of the poset $(S, \le)$.

$0 \in S$ is the least element $\Rightarrow$ $0 \le 0^*$ (by definition as $0^* \in S$)

$0^* \in S$ is the least element $\Rightarrow$ $0^* \le 0$ (by definition as $0 \in S$)

Now, $0 \le 0^*$ and $0^* \le 0$ $\Rightarrow$ $0 = 0^*$ (by P2).

Similarly if possible, let $1 \in S$ and $1^* \in S$ both be the greatest elements of the poset $(S, \le)$.

$1 \in S$ is the greatest element $\Rightarrow$ $1^* \le 1$ (by definition as $1^* \in S$)

$1^* \in S$ is the greatest element $\Rightarrow$ $1 \le 1^*$ (by definition as $1 \in S$)

Now, $1 \le 1^*$ and $1^* \le 1$ $\Rightarrow$ $1 = 1^*$ (by P2).

Definition 10: A partially ordered set with a minimum (universal lower bound) and a maximum (universal upper bound) element (also called **bounded poset**) is a structure $(S, \le, 0, 1)$, where $\le$ is a partial ordering relation on non-empty set S such that 0 and 1 are the **least (minimum)** and the **greatest (maximum)** element with respect to $\le$, respectively. In other words, $0 \le x \le 1$, $\forall\, x \in S$.

Example 9: Consider the poset $(S, \leq)$ where $S = \{2, 4, 8, 16, 32, 64\}$ and $\leq$ be a partial ordering relation defined on S such that $a \leq b$ iff $a \mid b$ (*i.e.*, a divides b), $a, b \in S$. It is a bounded poset having least element $2 \in S$ and greatest element $64 \in S$.

Definition 11: In a poset $(S, \leq)$ we say that element $a \in S$ **covers** another element $b \in S$ if

$$b < a \ (i.e., b \leq a \quad \text{and} \quad b \neq a)$$

and there is no element $c \in S$ such that $b < c < a$.

Hasse Diagram

A convenient way of displaying the ordering relation among the elements of a finite poset $(S, \leq)$ is by means of a graph whose vertices represent the elements of the set S. Thus we define a graph whose vertices are the different elements $a, b, c, \ldots\ldots$ of S in which a and b are joined by a segment if and only if a covers b or b covers a. If the graph is so drawn that whenever a covers b, the vertex a is at higher level than the vertex b, then the graph is called the **Hasse diagram** of the finite poset $(S, \leq)$.

Note: Hasse diagram is also known as covering diagram.

Example 10: *Draw the Hasse diagram of the poset* $(P(S), \subseteq)$ *where* $P(S)$ *is the power set of* $S = \{a, b\}$ *and* $\subseteq$ *denotes the usual set theoretic inclusion relation.*

Solution: Here $P(S) = \{\varphi, \{a\}, \{b\}, \{a, b\}\}$. The Hasse diagram of the poset $(P(S), \subseteq)$ is shown below (Fig. 3.1):

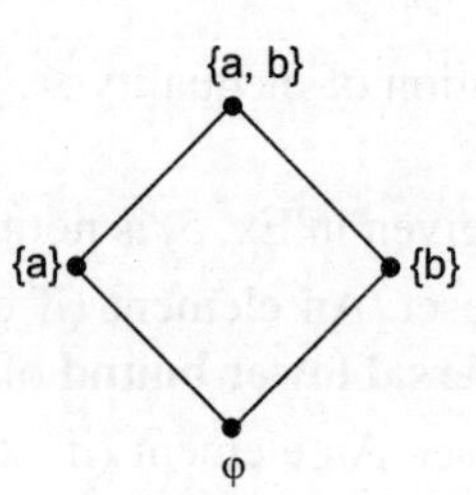

Fig. 3.1

Example 11: *Draw Hasse diagrams of* $(P(S), \subseteq)$ *for*

(*i*) $S = \{a\}$ (*ii*) $S = \{a, b, c\}$.

Solution: (*i*) Here $P(S) = \{\varphi, \{a\}\}$. The Hasse diagram of the poset $(P(S), \subseteq)$ is shown in Fig. 3.2.

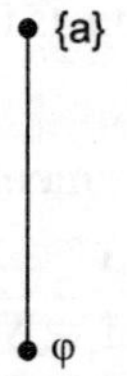

Fig. 3.2

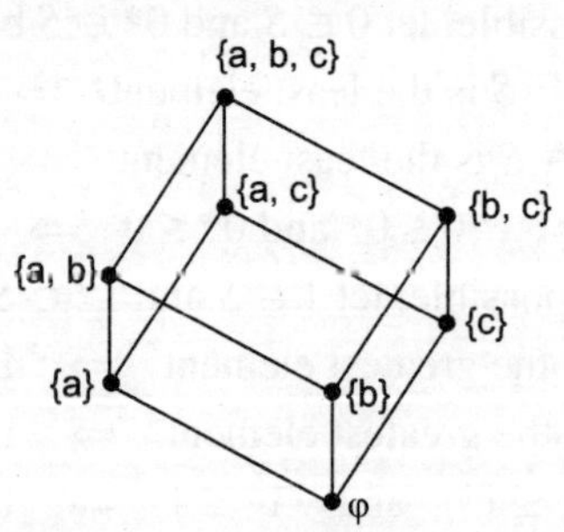

Fig. 3.3

(*ii*) Here $P(S) = \{\varphi, \{a\}, \{b\}, \{c\}, \{a, b\}, \{a, c\}, \{b, c\}, \{a, b, c\}\}$. The Hasse diagram of the poset $(P(S), \subseteq)$ is shown in Fig. 3.3.

Example 12: *Draw Hasse diagrams of the posets (S, ≤) for*

(i) S = {1, 2, 3, 4, 6, 12} *(ii) S = {2, 3, 6, 12, 24, 36}*

(iii) S = {1, 3, 9, 27, 81} *(iv) S = {2, 4, 5, 6, 10, 12, 20, 25}*

where ≤ be a partial ordering relation defined on S such that a ≤ b iff a | b.

Solution: (*i*) The Hasse diagram of the poset (*S*, ≤) is shown in Fig. 3.4.

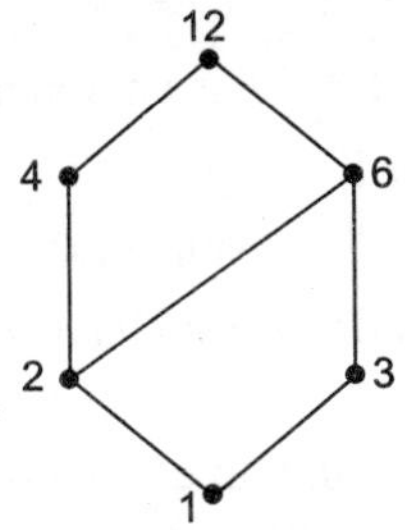

Fig. 3.4

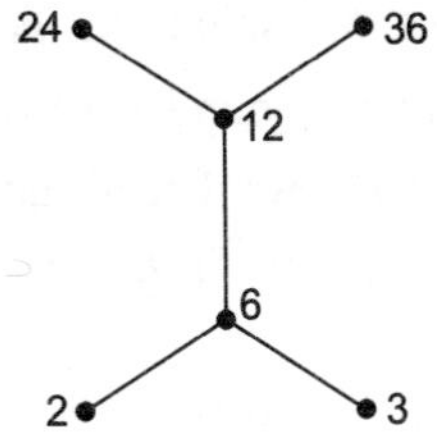

Fig. 3.5

(*ii*) The Hasse diagram of the poset (*S*, ≤) is shown in Fig. 3.5.

(*iii*) The Hasse diagram of the poset (*S*, ≤) is shown in Fig. 3.6. This is a chain having length 5.

Fig. 3.6

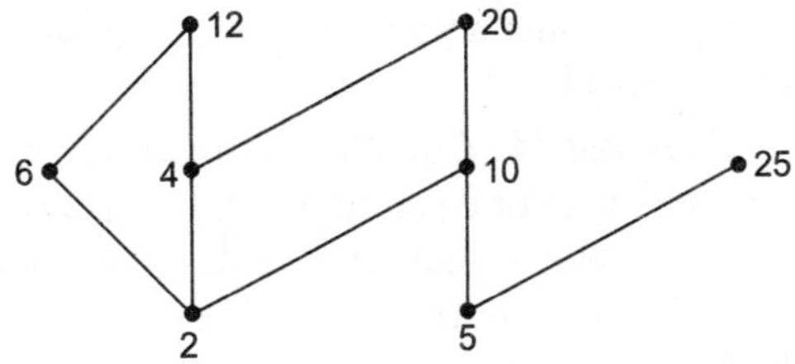

Fig. 3.7

(*iv*) The Hasse diagram of the poset (*S*, ≤) is shown in Fig. 3.7.

Maximal and Minimal Elements

Definition 12: An element *m* of a poset (*S*, ≤) is **minimal** when $x < m$ for no *x* in *S*. It is maximal when $x > m$ for no *x* in *S*.

Example 13: Let us consider the poset whose Hasse diagram is given in Fig. 3.8. Minimal elements are *a*, *b*, *e* and maximal elements are *j*, *k*. Here *f* covers *b*, *c* but *f* does not cover *a*.

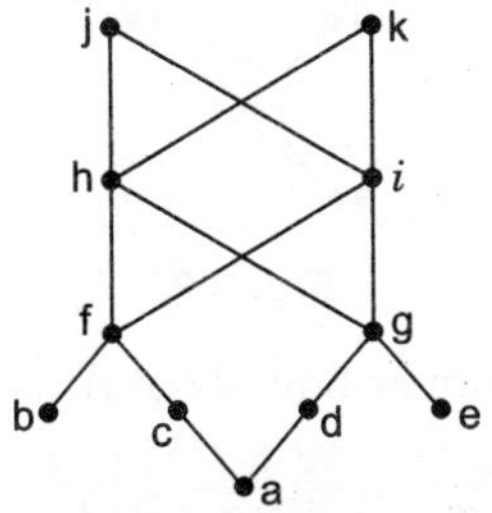

Fig. 3.8

Note: The maximal and minimal elements of a poset need not be unique.

Greatest Lower Bound and Least Upper Bound

Definition 13: Let $(S, \leq)$ be a poset and A be a subset of S. An element $a \in S$ is said to be a **lower bound** of A if $a \leq x$, $\forall x \in A$. An element $b \in S$ is said to be an **upper bound** of A if $x \leq b$, $\forall x \in A$.

Definition 14: Let $(S, \leq)$ be a poset and A be a subset of S. An element $a \in S$ is said to be the **greatest lower bound (g.l.b.)** of A if a is a lower bound of A and $\bar{a} \leq a$ for any other lower bound $\bar{a}$ of A and we write a = g.l.b. (A) or $a = \inf (A)$.

An element $b \in S$ is said to be the **least upper bound** (l.u.b) of A if b is an upper bound of A and $b \leq \bar{b}$ for any other upper bound $\bar{b}$ of A and we write b = lub (A) or $b = \sup (A)$.

Note: The greatest lower bound and the least upper bound of A may not belong to A.

Theorem 3: *A given finite poset $(S, \leq)$ has at least one maximal element and at least one minimal element.*

Proof: Let $a \in S$. If a is not a maximal element, then we can find an element $a_1 \in S$ such that $a < a_1$. If a_1 is not a maximal element of $(S, \leq)$, then we can find another element $a_2 \in S$ such that $a_1 < a_2$. Since $(S, \leq)$ is finite there exists a positive integer r such that we get a chain:

$$a < a_1 < a_2 < a_3 < \ldots < a_{r-1} < a_r$$

which cannot be extended further and so for any $b \in S$ we cannot have $a_r < b$. Hence a_r is a maximal element of $(S, \leq)$.

By the same argument, the dual poset $(S, \geq)$ has a maximal element and hence $(S, \leq)$ has a minimal element.

Example 14: Consider the poset $(R, \leq)$ where R is the set of all real numbers and $\leq$ be the relation 'less than or equal to' on R. Let $A = (0, 1) = \{x \in R : 0 < x < 1\}$, then 0 and all negative real numbers are lower bounds of A while all real numbers greater than or equal to 1 are upper bounds of A. Here 0 is the greatest lower bound [*i.e.*, 0 = g.l.b. $(A) = \inf (A)$] and 1 is the least upper bound [*i.e.*, 1 = lub $(A) = \sup (A)$] of A.

Example 15: Consider the poset $(S, \leq)$ where $S = (0, 1] = \{x \in R : 0 < x \leq 1\}$ and $\leq$ stands for usual less than or equal to relation. Let $A = \left\{\frac{1}{n} : n \in N\right\}$, then A has no g.l.b. but 1 is the l.u.b.

Example 16: Let $P = \{a, b, c, d, e, f, g, h\}$ and consider the poset $(P, \leq)$ whose Hasse diagram is shown below (Fig. 3.9):

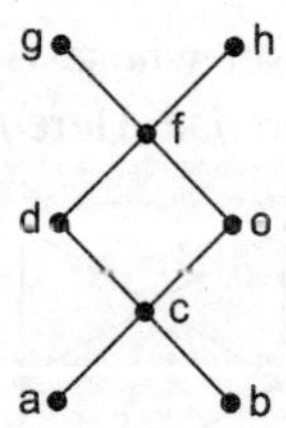

Fig. 3.9

If $A = \{d, e, f\}$, then f, g, h are upper bounds of A while a, b, c are lower bounds of A. Here f = lub (A) and c = glb (A).

Definition 15: *Let* $(S, \leq)$ be any poset and let $a, b \in S$, then an element $d \in S$ is called the **greatest lower bound or meet of a and b** denoted by $d = a \wedge b$ when

(*i*) $d \leq a$, $d \leq b$ and (*ii*) $x \leq a, x \leq b \Rightarrow x \leq d$

Definition 16: Let $(S, \leq)$ be any poset and let $a, b \in S$, then an element $s \in S$ is called the **least upper bound or join of a and b** denoted by $s = a \vee b$ when

(*i*) $a \leq s$, $b \leq s$ and (*ii*) $a \leq x, b \leq x \Rightarrow s \leq x$

Example 17: *Consider the poset $(S, \leq)$ where $S = \{1, 2, 3, 4, 6, 9, 12, 18, 36\}$ and $\leq$ is a partial ordering relation defined on S such that $a \leq b$ iff $a \mid b$. Find the greatest lower bound and the least upper bound of the sets $\{3, 9\}$ and $\{4, 6, 12\}$.*

Solution: An integer belongs to S is a lower bound of $\{3, 9\}$ if 3 and 9 are divisible by that integer. Only such integers are 1 and 3. Since 1/3, 3 is the g.l.b. of $\{3, 9\}$, *i.e.*, glb $\{3, 9\} = 3 \wedge 9 = 3$.

An integer belongs to S is an upper bound of $\{3, 9\}$ if it is divisible by 3 and 9. So, 18 and 36 are upper bounds of $\{3, 9\}$. Since $18 \mid 36$, 18 is the l.u.b. of $\{3, 9\}$, *i.e.*, lub $\{3, 9\} = 3 \vee 9 = 18$.

The lower bounds of $\{4, 6, 12\}$ are 1 and 2. Since $1 \mid 2$, 2 is the g.l.b. of $\{4, 6, 12\}$, *i.e.*, glb $\{4, 6, 12\} = 2$.

The upper bound of $\{4, 6, 12\}$ are 12 and 36. Since $12 \mid 36$, 12 is the l.u.b. of $\{4, 6, 12\}$, *i.e.*, lub $\{4, 6, 12\} = 12$.

Example 18: *Draw the Hasse diagram of the poset $(S, \leq)$ where $S = \{2, 3, 6, 18, 36, 54\}$ and $a \leq b$ iff $a \mid b$. Find*

(*i*) *the l.u.b. and the g.l.b. of $A = \{2, 3, 6\}$*

(*ii*) *the l.u.b. and the g.l.b. of $B = \{2, 3\}$*

(*iii*) *the l.u.b. and the g.l.b. of $C = \{6, 18\}$.*

Solution: The Hasse diagram of the poset $(S, \leq)$ is shown in Fig. 3.10. From the Hasse diagram it is clear that

(*i*) lub $(A) = 6$ and glb (A) does not exist.

(*ii*) lub $(B) = 2 \vee 3 = 6$ but glb $(B) = 2 \wedge 3$ does not exist

(*iii*) lub $(C) = 6 \vee 18 = 18$ and glb $(C) = 6 \wedge 18 = 6$.

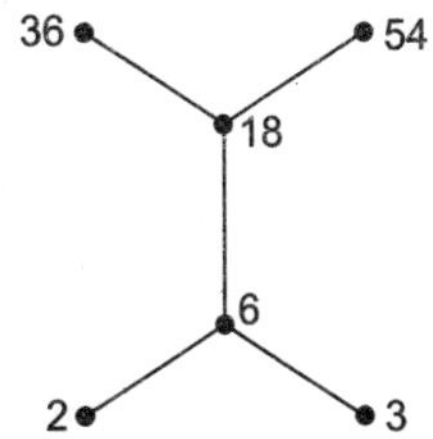

Fig. 3.10

Example 19: *Draw the Hasse diagram of the poset $(S, \leq)$ where $S = \{1, 2, 3, 4, 6, 8, 12, 24\}$ and $a \leq b$ iff $a \mid b$. Hence*

(*i*) *find all lower bounds of 8 and 12.*

(*ii*) *find all upper bounds of 8 and 12.*

(*iii*) *find the g.l.b. and l.u.b. of 8 and 12.*

Solution: The Hasse diagram of the poset $(S, \leq)$ is shown in Fig. 3.11. From the Hasse diagram it is evident that

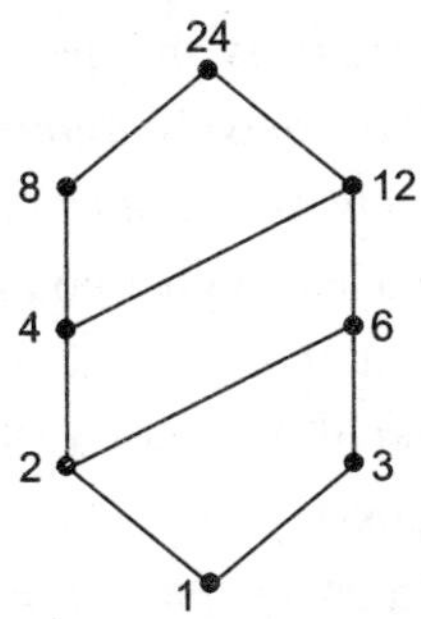

Fig. 3.11

(*i*) lower bounds of 8 and 12 are 1, 2, 4

(*ii*) upper bounds of 8 and 12 is 24.

(*iii*) glb $\{8, 12\} = 8 \wedge 12 = 4$,

lub $\{8, 12\} = 8 \vee 12 = 24$.

Example 20: *Draw the Hasse diagram of the poset* $(S, \leq)$ *where* $S = \{1, 2, 4, 5, 10, 20, 25, 50, 100\}$ *and* $a \leq b$ *iff* $a \mid b$. *Hence*

(*i*) *find the g.l.b. and the l.u.b. of* $\{10, 20\}$

(*ii*) *find the g.l.b. and the l.u.b. of* $\{5, 10, 20, 25\}$

Solution: The Hasse diagram of the poset $(S, \leq)$ is as shown in Fig. 3.12. From the Hasse diagram it is evident that

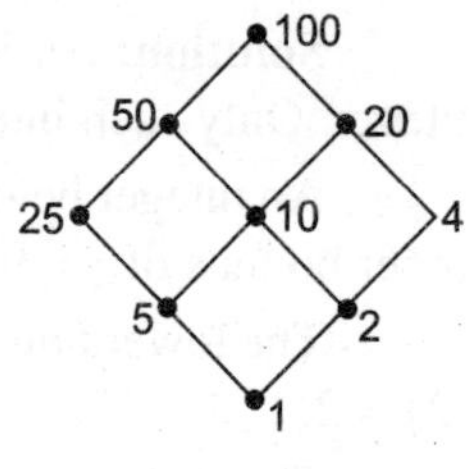

Fig. 3.12

(*i*) glb $\{10, 20\} = 10 \wedge 20 = 10$ and lub $\{10, 20\} = 10 \vee 20 = 20$

(*ii*) glb $\{5, 10, 20, 25\} = 5$ and lub $\{5, 10, 20, 25\} = 100$

Note: Here S denotes the set of all positive integers which are divisors of 100.

Example 21: *Consider the poset* $S = \{2, 4, 6, 9, 12, 18, 27, 36, 48, 60, 72\}$ *under the relation '|' (i.e., 'divides'). Find the following: Maximal element, Minimal element, Greatest element, Least element, lub (2, 9), glb (60, 72).* [*WBUT 2012*]

Solution: The Hasse diagram of the given poset is as shown in Fig. 3.13. From the Hasse diagram it is evident that

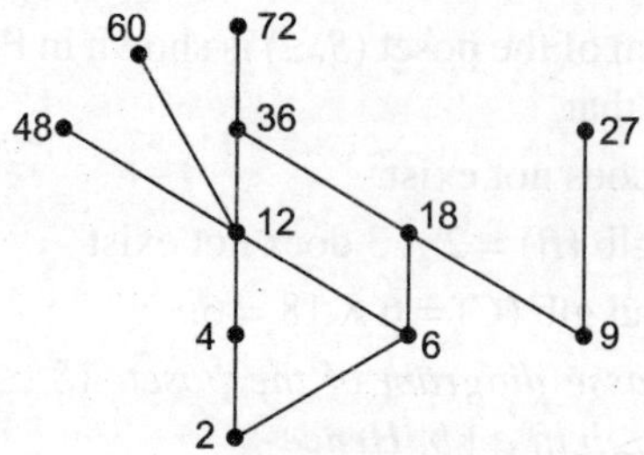

Fig. 3.13

(*i*) maximal elements are 27, 48, 60, 72 and minimal elements are 2, 9.

(*ii*) the given poset has no greatest and least elements.

(*iii*) lub (2, 9) = 18 and glb (60, 72) = 12.

Theorem 4: *Let* $(S, \leq)$ *be a poset and A be a subset of S. Then*

(*i*) *the least upper bound of A, if it exists, is unique*

(*ii*) *the greatest lower bound of A, if it exists, is unique*

i.e., A can have at most one least upper bound and at most one greatest lower bound.

Proof: (*i*) If possible let there be two least upper bounds for A, say x_1 and x_2.

x_1 is the lub and x_2 is an upper bound of $A \Rightarrow x_1 \leq x_2$

x_2 is the lub and x_1 is an upper bound of $A \Rightarrow x_2 \leq x_1$

Since $x_1, x_2 \in S$ and $(S, \leq)$ is a poset,

by P_2: $x_1 \leq x_2$ and $x_2 \leq x_1 \Rightarrow x_1 = x_2$

Hence least upper bound of A is unique.

(*ii*) Left as an exercise.

Example 22: Let us consider the poset whose Hasse diagram is given in Fig. 3.14. The subset $A = \{e, d, g\}$ has four elements a, b, c, d as upper bounds, d is the l.u.b. *of* A. But the subset $B = \{a, b, c\}$ has no upper bound.

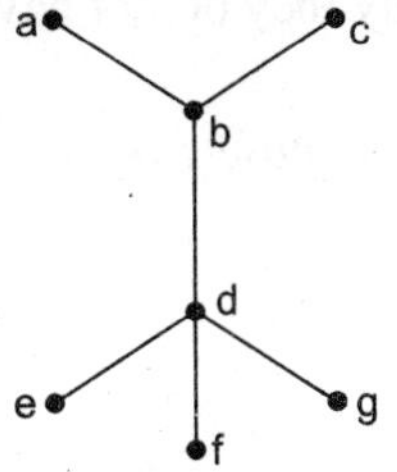

Fig. 3.14

Example 23: In the poset $(N, \leq)$ where $\leq$ is defined on N, the set of positive integers such that $a \leq b$ iff $a \mid b$, for $a, b \in N$.

We see that a common multiple of a and b is an upper bound of a and b and the least common multiple is the l.u.b. of a and b. Similarly, a common divisor of a and b is a lower bound of a and b and the greatest common divisor is the g.l.b. of a and b.

Theorem 5: *If $(S_1, \leq_1)$ and $(S_2, \leq_2)$ be two posets, then $(S_1 \times S_2, \leq)$ is a poset with partial ordering relation $\leq$ defined by $(a, b) \leq (a', b')$ if $a \leq_1 a'$ in S_1 and $b \leq_2 b'$ in S_2.*

Proof:

P1 (Reflexivity): Since $a \leq_1 a$, $\forall\, a \in S_1$ and $b \leq_2 b$, $\forall\, b \in S_2$

$\therefore$ $(a, b) \leq (a, b)$, $\forall\, (a, b) \in S_1 \times S_2$.

P2 (Antisymmetry): Since $a \leq_1 a'$ and $a' \leq_1 a \Rightarrow a = a'$, for $a, a' \in S_1$,

also $b \leq_2 b'$ and $b' \leq_2 b \Rightarrow b = b'$, for $b, b' \in S_2$,

We have, $(a, b) \leq (a', b')$ and $(a', b') \leq (a, b) \Rightarrow (a, b) = (a', b')$,

for $(a, b), (a', b') \in S_1 \times S_2$.

P3 (Transitivity): Since $a \leq_1 a'$ and $a' \leq_1 a'' \Rightarrow a \leq_1 a''$, for $a, a', a'' \in S_1$,

also $b \leq_2 b'$ and $b' \leq_2 b'' \Rightarrow b \leq_2 b''$, for $b, b', b'' \in S_2$,

We have $(a, b) \leq (a', b')$ and $(a', b') \leq (a'', b'')$

$\Rightarrow$ $(a, b) \leq (a'', b'')$, for $(a, b), (a', b'), (a'', b'') \in S_1 \times S_2$.

Consequently, $\leq$ is a partial ordering relation on $S_1 \times S_2$ and hence $(S_1 \times S_2, \leq)$ is a poset.

Note: The partial ordering relation $\leq$ on $S_1 \times S_2$ defined as in this theorem is called the **product partial ordering relation.**

3.3 LATTICES

Definition 17: A **lattice** is a poset $(L, \leq)$ in which any two elements $a, b \in L$ have a greatest lower bound (g.l.b.) called the meet (or intersection), $a \wedge b$ and a least upper bound (l.u.b.) called the join (or union), $a \vee b$.

Note: Every finite subset of a lattice has a g.l.b. and a l.u.b.

Example 24: Consider the poset $(N, \leq)$ where N is the set of all positive integers (*i.e.*, natural numbers) and for $a, b \in N$, $a \leq b$ iff $a \mid b$. Then $(N, \leq)$ is a lattice in which the meet and join of any two elements $a, b \in N$ are their greatest common divisor and least common multiple respectively, *i.e.*, $a \wedge b = gcd\,(a, b)$ and $a \vee b = lcm\,(a, b)$.

Example 25: *Show that the poset* $(S, \leq)$, *where* $S = \{1, 2, 3, 4, 5\}$ *and* $a \leq b$ *iff* $a \mid b$, *is not a lattice.*

Solution: Observe that 2 and 3 have no upper bound because the set S does not contain any common multiple of 2 and 3. Consequently they do not have a least upper bound and hence the given poset is not a lattice.

Example 26: *Determine whether the posets represented by the following Hasse diagrams are lattices:*

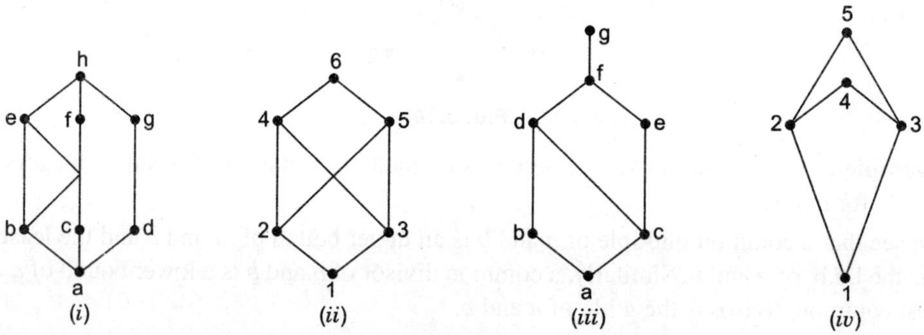

Fig. 3.15

Solution: (*i*) The poset represented by the given Hasse diagram is a lattice since every pair of elements of it has a least upper bound as well as a greatest lower bound.

(*ii*) The poset represented by the given Hasse diagram is not a lattice since the elements 2 and 3 have no least upper bound. The elements 4, 5 and 6 are the upper bounds of the pair 2 and 3 but none of these three elements precedes the other two with respect to the partial ordering relation of the given poset.

(*iii*) The poset represented by the given Hasse diagram is a lattice because every pair of elements of it has a least upper bound as well as a greatest lower bound.

(*iv*) The poset represented by the given Hasse diagram is not a lattice because the elements 2 and 3 have no least upper bound. The elements 4 and 5 are the upper bounds of the pair 2 and 3 but none of these (4 and 5) precedes the other w.r.t. the partial ordering relation of the given poset.

Example 27: *Show that the poset* $(S, \leq)$, *where* $S = \{1, 3, 6, 12, 24, 48\}$ *and* $a \leq b$ *iff* $a \mid b$, *is a lattice.*

Solution: Observe that any two elements of the given poset have a greatest lower bound as well as a least upper bound which are smaller and larger elements respectively. Hence the given poset is a lattice.

Note: The posets whose Hasse diagrams are as shown in Figs. 3.1 – 3.6 are lattices but the posets whose Hasse diagrams are as shown in Figs. 3.7, 3.8 are not lattices.

Example 28: Consider the poset $(R, \leq)$ where R is the set of all real numbers and $\leq$ is the usual 'less than or equal to' relation. $(R, \leq)$ is a lattice since glb $(x, y) = x \wedge y$ is smaller of x and y, lub $(x, y) = x \vee y$ is larger of x and y.

Example 29: Let S be any set and $P(S)$ be the power set of S, then the poset $(P(S), \subseteq)$ is a lattice in which the g.l.b. (meet) and l.u.b. (join) of $A, B \in P(S)$ are $A \cap B$ and $A \cup B$ respectively.

Example 30: *Show that the following posets are lattices and obtain the Hasse diagrams of each of these posets:*

(*i*) $(L_6, \leq)$ (*ii*) $(L_8, \leq)$

(*iii*) $(L_{24}, \leq)$ (*iv*) $(L_{30}, \leq)$

where L_n *is the set of all positive divisors of n and* $a \leq b$ *iff* $a \mid b$.

Solution: (*i*) We have $L_6 = \{1, 2, 3, 6\}$. If we take any two elements of L_6 then their greatest lower bound and least upper bound with respect to the given partial ordering relation will also be in L_6. Therefore, $(L_6, \leq)$ is a lattice.

Here,
$$L_8 = \{1, 2, 4, 8\}$$
$$L_{24} = \{1, 2, 3, 4, 6, 8, 12, 24\}$$
$$L_{30} = \{1, 2, 3, 5, 6, 10, 15, 30\}$$

It can be shown easily that these are also lattices under the given partial ordering relation.

The Hasse diagrams of the given lattices are shown below:

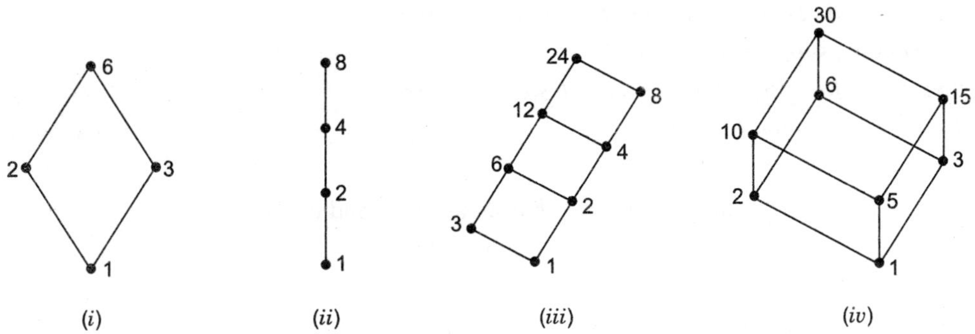

Fig. 3.16

Theorem 6: *If* $(L, \leq)$ *be any lattice, then for any* $x, y, z \in L$, *the following identities hold:*

L1. $x \wedge x = x$ *and* $x \vee x = x$ (*Idempotency*)

L2. $x \wedge y = y \wedge x$ *and* $x \vee y = y \vee x$ (*Commutativity*)

L3. $(x \wedge y) \wedge z = x \wedge (y \wedge z)$ *and* $(x \vee y) \vee z = x \vee (y \vee z)$ (*Associativity*)

L4. $x \wedge (x \vee y) = x$ *and* $x \vee (x \wedge y) = x$ (*Absorption*)

Proof. By the principle of duality which interchanges $\wedge$ and $\vee$, it suffices to prove one of the two identities in each of *L*1 – *L*4.

L1. Since $x \wedge x$ is the g.l.b. of x and x, we have

$$x \wedge x \leq x \qquad ...(1)$$

Also, $x \leq x$, $x \leq x$ and so x is a lower bound of x and x. But as $x \wedge x$ is the g.l.b. of x and x, we have

$$x \leq x \wedge x \qquad ...(2)$$

From (1) and (2), applying P2, we get $x = x \wedge x$.

L2. Since the meaning of glb $\{x, y\}$ does not alter by interchanging x and y, it follows that

$$x \wedge y = y \wedge x.$$

L3. Since both of $(x \wedge y) \wedge z$ and $x \wedge (y \wedge z)$ represent glb $\{x, y, z\}$, the result follows.

L4. Since $x \wedge (x \vee y)$ represents a lower bound of x and $x \vee y$, we have

$$x \wedge (x \vee y) \leq x \qquad ...(3)$$

Now, $x \leq x$ and $x \leq x \vee y$, therefore x is a lower bound of x and $x \vee y$. Since $x \wedge (x \vee y)$ is the glb of x and $x \vee y$, we have

$$x \leq x \wedge (x \vee y) \qquad ...(4)$$

Combining (3) and (4) and applying P2, we get

$$x \wedge (x \vee y) = x.$$

Theorem 7: *If $(L, \leq)$ be a lattice, then for any $a, b \in L$,*

(*i*) $a \leq b \iff a \wedge b = a$ (*ii*) $a \leq b \iff a \vee b = b$

Proof: (*i*) $a \in L \Rightarrow a \leq a$ (by reflexivity).

Let $a \leq b$.

Now, $a \leq b$ and $a \leq a \Rightarrow a$ is a lower bound of $\{a, b\}$

$\Rightarrow a \leq \text{glb}\ \{a, b\}$

$\Rightarrow a \leq a \wedge b$ ...(1)

Since $a \wedge b = \text{glb}\{a, b\}$, we have $a \wedge b \leq a$ ...(2)

From (1) and (2), by antisymmetry, we have $a \wedge b = a$.

Conversely, suppose $a \wedge b = a$.

Now, $a \wedge b = \text{glb}\{a, b\} \Rightarrow a \wedge b \leq b \Rightarrow a \leq b$ $(\because\ a \wedge b = a)$

(*ii*) $b \in L \Rightarrow b \leq b$ (By reflexivity)

Let $a \leq b$.

Now, $a \leq b$ and $b \leq b \Rightarrow b$ is an upper bound of $\{a, b\}$

$\Rightarrow \text{lub}\ \{a, b\} \leq b$

$\Rightarrow a \vee b \leq b$ ...(3)

But $a \vee b = \text{lub}\ \{a, b\} \Rightarrow b \leq a \vee b$ ...(4)

From (3) and (4), by antisymmetry, we have $a \vee b = b$

Conversely, suppose $a \vee b = b$.

Now, $a \vee b = \text{lub}\ \{a, b\} \Rightarrow a \leq a \vee b \Rightarrow a \leq b$ $(\because\ a \vee b = b)$

Theorem 8: *If $(L, \leq)$ be a lattice, then for any $a, b, c, d \in L$,*

(*i*) $a \leq b, c \leq d \Rightarrow a \wedge c \leq b \wedge d$ (*ii*) $a \leq b, c \leq d \Rightarrow a \vee c \leq b \vee d$

Proof: (*i*) Let $a \leq b, c \leq d$

Now, $a \wedge c \leq a$ and $a \wedge c \leq c$

By transitive property,

$a \wedge c \leq a$ and $a \leq b \Rightarrow a \wedge c \leq b$

and $a \wedge c \leq c$ and $c \leq d \Rightarrow a \wedge c \leq d$

Therefore, $a \wedge c$ is a lower bound of b and d and hence $a \wedge c \leq b \wedge d$.

(*ii*) Let $a \leq b, c \leq d$.

Now, $b \leq b \vee d$ and $d \leq b \vee d$.

By transitive property,

$a \leq b$ and $b \leq b \vee d \Rightarrow a \leq b \vee d$

and $c \leq d$ and $d \leq b \vee d \Rightarrow c \leq b \vee d$

Therefore, $b \vee d$ is an upper bound of a and c and so $a \vee c \leq b \vee d$.

Theorem 9: *If $(L, \leq)$ be a lattice, then for any $a, b, c \in L$,*

(i) $a \wedge (b \vee c) \geq (a \wedge b) \vee (a \wedge c)$

(ii) $a \vee (b \wedge c) \leq (a \vee b) \wedge (a \vee c)$

Proof: (*i*) We have $a \wedge b \leq a$ and $a \wedge b \leq b \leq b \vee c$.

Therefore, $a \wedge b$ is a lower bound of a and $b \vee c$ and so

$$a \wedge b \leq a \wedge (b \vee c) \qquad ...(1)$$

Now, $a \wedge c \leq a$ and $a \wedge c \leq c \leq b \vee c$.

Therefore, $a \wedge c$ is a lower bound of a and $b \vee c$ and hence

$$a \wedge c \leq a \wedge (b \vee c) \qquad ...(2)$$

From (1) and (2), we get $a \wedge (b \vee c)$ is an upper bound of $a \wedge b$ and $a \wedge c$ and hence

$$(a \wedge b) \vee (a \wedge c) \leq a \wedge (b \vee c) \iff a \wedge (b \vee c) \geq (a \wedge b) \vee (a \wedge c).$$

(*ii*) We have $a \leq a \vee b$, $b \wedge c \leq b \leq a \vee b$ and so $a \vee b$ is an upper bound of a and $b \wedge c$.

$$\therefore \quad a \vee (b \wedge c) \leq a \vee b \qquad ...(3)$$

Now, $a \leq a \vee c$, $b \wedge c \leq c \leq a \vee c$ and so $a \vee c$ is an upper bound of a and $b \wedge c$.

$$\therefore \quad a \vee (b \wedge c) \leq a \vee c \qquad ...(4)$$

From (3) and (4), we get $a \vee (b \wedge c)$ is a lower bound of $a \vee b$ and $a \vee c$.

$$\therefore \quad a \vee (b \wedge c) \leq (a \vee b) \wedge (a \vee c).$$

Theorem 10: *Dual of a lattice is a lattice.*

Proof: Let $(L, \leq)$ be a given lattice and let $(L, \geq)$ be its dual where $a \geq b$ iff $b \leq a$, for $a, b \in L$. Then $(L, \geq)$ is a poset by Theorem 1, Art. 3.2.

Let $a, b \in L$, then $a \vee b = \text{lub } \{a, b\}$ exists in $(L, \leq)$ since $(L, \leq)$ is a lattice.

Now, $a \leq a \vee b$ and $b \leq a \vee b \Rightarrow a \vee b \geq a$ and $a \vee b \geq b$

$\Rightarrow$ $a \vee b$ is a lower bound of a and b in $(L, \geq)$

Let c be any lower bound of a, b in $(L, \geq)$.

$$\begin{aligned} \therefore \quad c \geq a \text{ and } c \geq b &\Rightarrow a \leq c \text{ and } b \leq c \\ &\Rightarrow c \text{ is an upper bound of } a, b \text{ in } (L, \leq) \\ &\Rightarrow a \vee b \leq c \Rightarrow c \geq a \vee b \\ &\Rightarrow a \vee b = \text{glb } \{a, b\} \text{ in } (L, \geq) \end{aligned}$$

Similarly it can be proved that $a \wedge b = \text{lub } \{a, b\}$ in $(L, \geq)$.

Therefore $(L, \geq)$ is a lattice.

3.4 SUBLATTICES

Definition 18: Let $(L, \leq)$ be a lattice and let L_1 be a non-empty subset of L. Then $(L_1, \leq)$ is said to be a **sublattice** of $(L, \leq)$ if $a, b \in L_1 \Rightarrow a \wedge b, a \vee b \in L_1$.

Example 31: Let $S = \{a, b, c, d\}$ and let $P(S)$ be the power set of S, then the poset $(P(S), \subseteq)$ is a lattice in which the g.l.b. (meet) and the l.u.b. (join) of $A, B \in P(S)$ are $A \cap B$ and $A \cup B$ respectively. Let

$$L_1 = \{\varphi, \{a\}, \{a, c\}, \{c\}, \{a, b, c\}\} \subseteq P(S), L_2 = \{\varphi, \{a\}, \{b\}, \{c\}, \{b, c\}\} \subseteq P(S).$$

Here $(L_1, \subseteq)$ is a sublattice of $(P(S), \subseteq)$ but $(L_2, \subseteq)$ is not a sublattice of $(P(S), \subseteq)$ because $\{a\} \cup \{c\} = \{a, c\} \notin L_2$.

Example 32: *Show with an example that the union of two sublattices may not be a sublattice.*

Solution: Consider the lattice $(L_{24}, \leq)$ where $L_{24} = \{1, 2, 3, 4, 6, 8, 12, 24\}$ and $a \leq b$ iff $a \mid b$ (see Ex. 30). Let $S = \{1, 2\}$ and $T = \{1, 3\}$. Then $(S, \leq)$ and $(T, \leq)$ are sublattices of $(L_{24}, \leq)$. But $S \cup T = \{1, 2, 3\}$ is not a sublattice because $2, 3 \in S \cup T$ but $2 \vee 3 = 6 \notin S \cup T$.

3.5 SOME SPECIAL LATTICES

In a lattice every pair of elements has a least upper bound (l.u.b.) and a greatest lower bound (g.l.b.). As a consequence of this fact, it can be shown by using the principle of mathematical induction that every finite subset of a lattice possesses l.u.b. and g.l.b. For example, let $(L, \leq)$ be a lattice, so

$$a_1, a_2, a_3 \in L \;\Rightarrow\; a_1 \vee a_2, a_3 \in L \;\Rightarrow\; (a_1 \vee a_2) \vee a_3 \in L.$$

Also, $$a_1, a_2, a_3 \leq c \;\Rightarrow\; a_1 \vee a_2, a_3 \leq c \;\Rightarrow\; (a_1 \vee a_2) \vee a_3 \leq c$$
$$\Rightarrow\; (a_1 \vee a_2) \vee a_3 = \text{lub}\,\{a_1, a_2, a_3\}$$

By associativity and commutativity of the operation $\vee$, we have

$$\text{lub}\,\{a_1, a_2, a_3\} = a_1 \vee a_2 \vee a_3.$$

Similarly, $\text{glb}\,\{a_1, a_2, a_3\} = a_1 \wedge a_2 \wedge a_3.$

In general, let $(L, \leq)$ be a lattice and $S = \{a_1, a_2, ..., a_n\}$ be a finite subset of L, then

$$\text{lub}\,(S) = a_1 \vee a_2 \vee ... \vee a_n \text{ and } \text{glb}\,(S) = a_1 \wedge a_2 \wedge ... \wedge a_n.$$

Note: There may not be l.u.b. and g.l.b. for an infinite subset of a lattice. For example, consider the lattice $(N, \leq)$ where N is the set of all positive integers (*i.e.*, natural numbers, and $\leq$ represents usual 'less than or equal to' relation. The subset consisting of even positive integers has no least upper bound.

Definition 19: A lattice is said to be **complete** if each of its non-empty subsets has a least upper bound and a greatest lower bound.

Notes: (*i*) Every finite lattice is complete since every subset here is finite.

(*ii*) Every complete lattice must have a **least element** denoted by 0 (*i.e.*, $0 \leq a$, $\forall\, a \in L$) and a **greatest element** denoted by 1 (*i.e.*, $a \leq 1$, $\forall\, a \in L$). The least and greatest elements 0 and 1 are called **bounds** (or **universal bounds**), of the lattice (see Definitions 8–10, Art. 3.2.).

Definition 20: A lattice is said to be **bounded** if it has a least element 0 and a greatest element 1.

Example 33: Let S be any set and $P(S)$ be the power set of S, then $(P(S), \subseteq)$ is a lattice (see Ex. 29). It is a bounded lattice having least element φ (empty set) and greatest element S.

Example 34: Consider the lattice $(N, \leq)$ where N is the set of all positive integers and $a \leq b$ iff $a \mid b$. It is not a bounded lattice because greatest element does not exist though it has a least element 1.

Example 35: Consider the lattice $(Z, \leq)$ where Z is the set of all integers and $\leq$ represents the usual 'less than or equal to' relation. It is not a bounded lattice since it has neither a least element nor a greatest element.

Theorem 11: *Every complete lattice is a bounded lattice.*

Proof: Let $(L, \leq)$ be a complete lattice and so every non-empty subset of L has least upper bound and greatest lower bound. Therefore L itself has least upper bound and greatest lower bound. Hence $(L, \leq)$ is a bounded lattice.

Theorem 12: *Every finite lattice is complete.*

Proof: Let $(L, \leq)$ be any finite lattice and let S be any non-empty subset of L. Therefore, S is a finite set and let $S = \{a_1, a_2, ..., a_n\}$.

$\therefore$ $\text{glb}\,(S) = a_1 \wedge a_2 \wedge ... \wedge a_n$ and $\text{lub}\,(S) = a_1 \vee a_2 \vee ... \vee a_n$ in $(L, \leq)$. Hence $(L, \leq)$ is complete.

Observations: It is evident that in a bounded lattice $(L, \leq)$, the bounds 0 and 1 satisfy:

$$0 \wedge a = a \wedge 0 = 0, \quad \forall\, a \in L$$
$$0 \vee a = a \vee 0 = a, \quad \forall\, a \in L$$
$$1 \wedge a = a \wedge 1 = a, \quad \forall\, a \in L$$
$$1 \vee a = a \vee 1 = 1, \quad \forall\, a \in L$$

A bounded lattice is also denoted as $(L, \wedge, \vee, 0, 1)$. In a bounded lattice, a complement of an element is defined in the following manner:

Definition 21: In a bounded lattice $(L, \wedge, \vee, 0, 1)$, an element $a' \in L$ is called a **complement** of an element $a \in L$ if $a \wedge a' = 0$ and $a \vee a' = 1$, where 0 and 1 are least and greatest elements of L.

Notes: *(i)* a' is a complement of a, then a is also a complement of a'.

(ii) Any element $a \in L$ may or may not have a complement.

(iii) An element $a \in L$ may have more than one complement in L and so complements are not unique.

(iv) In any bounded lattice, the bounds 0 and 1 are unique complements of each other since $0 \vee 1 = 1$, $0 \wedge 1 = 0$ and by Theorem 4.

Definition 22: A lattice $(L, \leq)$ is called **complemented lattice** if it is bounded and if every element in L has a complement.

Example 36: *Let S be any set and P(S) be the power set of S. Show that $(P(S), \subseteq)$ is a complemented lattice.*

Solution: We know that the poset $(P(S), \subseteq)$ is a lattice in which the g.l.b. (meet) and l.u.b. (join) of $A, B \in P(S)$ are $A \cap B$ and $A \cup B$ respectively. It is a bounded lattice having least element φ (null set) and greatest element S.

We know that the complement of any subset X of S is given by $X' = S - X$.

Now, $\quad X \wedge (S - X) = 0$ and $X \vee (S - X) = 1$

as $\quad X \cap (S - X) = \varphi$ and $X \cup (S - X) = S$

So, $\quad X \wedge X' = 0$ and $X \vee X' = 1, \quad \forall\, X \in P(S)$

This shows that X has its complement in $P(S)$, $\forall\, X \in S$.

Hence $(P(S), \subseteq)$ is a complemented lattice.

Example 37: The Hasse diagrams of two lattices are shown in Figs. 3.17(*a*), (*b*). In Fig. 3.17(*a*), complement of x_1 is x_2 since $x_1 \wedge x_2 = 0$ and $x_1 \vee x_2 = 1$. In Fig. 3.17 (*b*), both x_1 and x_2 are complements of x_3 because $x_1 \wedge x_3 = 0, x_2 \wedge x_3 = 0$ and $x_1 \vee x_3 = 1, x_2 \vee x_3 = 1$ and so here complements are not unique.

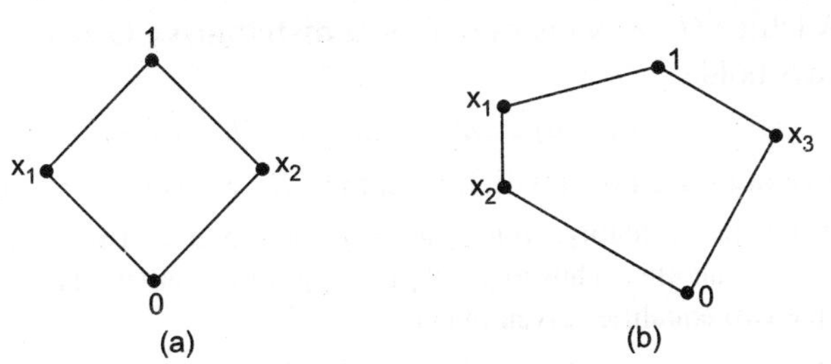

Fig. 3.17

Example 38: *Show that the lattice $(L^3, \leq_3)$ of tuples of 0 and 1 is complemented where $(a_1, a_2, a_3) \leq_3 (b_1, b_2, b_3)$ iff $a_i \leq b_i$, $i = 1, 2, 3$; $(a_1, a_2, a_3), (b_1, b_2, b_3) \in L^3$.*

Solution: The Hasse diagram of the lattice $(L^3, \leq_3)$ is shown below (Fig. 3.18):

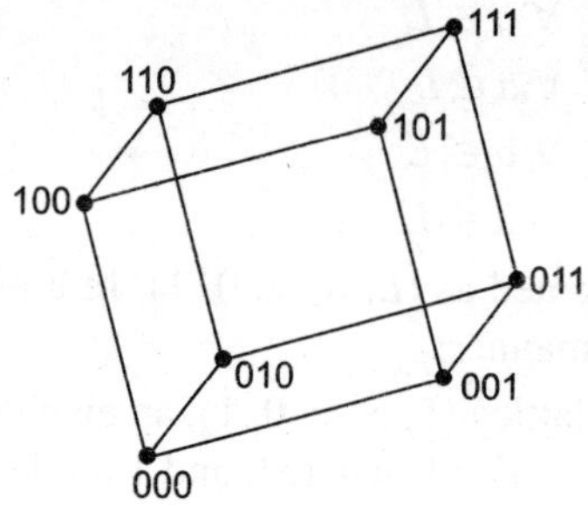

Fig. 3.18

Let us denote $(a_1, a_2, a_3) \in L^3$ by $a_1\ a_2\ a_3$. The bounds of the lattice are 000 and 111. This is a complemented lattice in which every element has a unique complement. The complement of 101 is 010 and the complement of an element of L^3 can be obtained by interchanging 1 by 0 and 0 by 1.

Note: In general, the lattice $(L^n, \leq_n)$ is a complemented lattice where the complement of an element of L^n is obtained by interchanging 1 by 0 and 0 by 1.

Example 39: *Consider the lattice $(L, \leq)$ where $L = \{1, 2, 3, 4, 6, 12\}$ and $a \leq b$ iff $a \mid b$? Find the least and greatest element of $(L, \leq)$. Is $(L, \leq)$ a complemented lattice.*

Solution: The Hasse diagram of the lattice $(L, \leq)$ is shown below (Fig. 3.19):

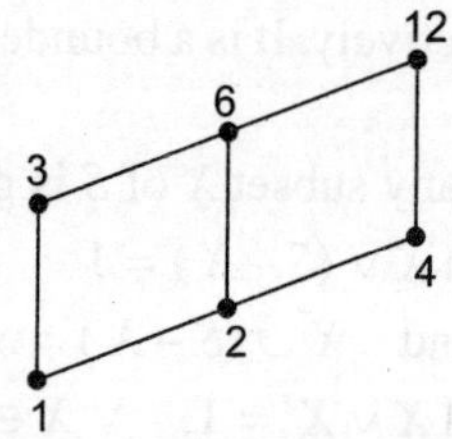

Fig. 3.19

From the Hasse diagram it is evident that 1 is the least element and 12 is the greatest element of $(L, \leq)$. Complements of 1, 2, 3, 4, 6, 12 are respectively 12, 6, 4, 3, 2, 1. Hence $(L, \leq)$ is a complemented lattice.

Note: Here L is the set of all positive divisors of 12.

Definition 23: A lattice $(L, \wedge, \vee)$ is said to be a **distributive lattice** if for any $a, b, c \in L$ the following distributive laws hold:

$$a \wedge (b \vee c) = (a \wedge b) \vee (a \wedge c) \text{ and } a \vee (b \wedge c) = (a \vee b) \wedge (a \vee c) \quad ...(1)$$

If these distributive laws are not satisfied in L, then $(L, \wedge, \vee)$ is said to be non-distributive.

Notes: (*i*) By the principle of duality, which interchanges $\wedge$ and $\vee$, it is evident that both the equalities given in (1) are equivalent to one another. Thus to check (or examine) whether a lattice is distributive or not, it is sufficient to verify one of the two equalities given in (1).

(*ii*) A complemented, distributive lattice is also called a **Boolean Algebra.**

Example 40: Let S be any set and $P(S)$ be the power set of S, then the poset $(P(S), \subseteq)$ is a lattice in which the g.l.b. (meet) and the l.u.b. (join) of $A, B \in P(S)$ are $A \cap B$ and $A \cup B$ respectively. It is a distributive lattice since

$$A \cap (B \cup C) = (A \cap B) \cup (A \cap C) \text{ and } A \cup (B \cap C) = (A \cup B) \cap (A \cup C),\ \forall\ A, B, C \in P(S).$$

Example 41: *Show that the lattices whose Hasse diagrams are given in Figs. 3.20 (a), (b) are not distributive.*

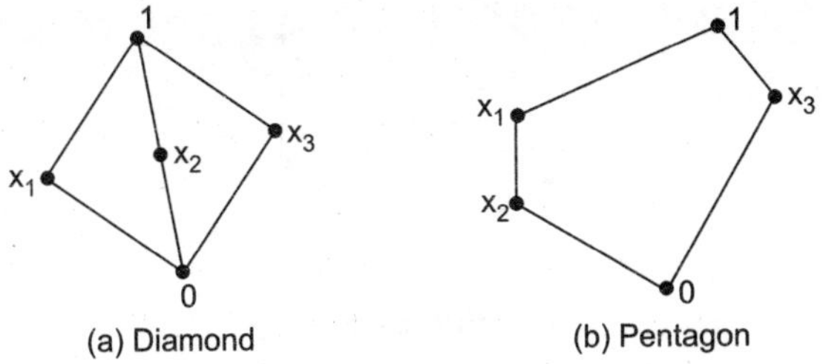

Fig. 3.20

Solution: (*i*) Consider the lattice whose Hasse diagram is shown in Fig. 3.20 (*a*). Observe that

$$x_1 \wedge (x_2 \vee x_3) = x_1 \wedge 1 = x_1 \text{ but } (x_1 \wedge x_2) \vee (x_1 \wedge x_3) = 0 \vee 0 = 0.$$

$$\therefore \quad x_1 \wedge (x_2 \vee x_3) \neq (x_1 \wedge x_2) \vee (x_1 \wedge x_3)$$

and hence the lattice shown in Fig. 3.20(*a*) is not distributive.

(*ii*) Consider the lattice whose Hasse diagram is shown in Fig. 3.20 (*b*). Observe that

$$x_1 \wedge (x_2 \vee x_3) = x_1 \wedge 1 = x_1 \text{ but } (x_1 \wedge x_2) \vee (x_1 \wedge x_3) = x_2 \vee 0 = x_2.$$

$$\therefore \quad x_1 \wedge (x_2 \vee x_3) \neq (x_1 \wedge x_2) \vee (x_1 \wedge x_3)$$

and so the lattice shown in Fig. 3.20 (*b*) is not distributive.

Notes: (*i*) The distributive equalities may be satisfied for a particular combination of elements (of a lattice) in a particular order but this does not guarantee that the lattice is distributive. For example in Fig. 3.20 (*b*):

$$x_3 \wedge (x_1 \vee x_2) = x_3 \wedge x_1 = 0 \text{ and } (x_3 \wedge x_1) \vee (x_3 \wedge x_2) = 0 \vee 0 = 0.$$

$$\therefore \quad x_3 \wedge (x_1 \vee x_2) = (x_3 \wedge x_1) \vee (x_3 \wedge x_2).$$

(*ii*) If any lattice contains at least one of these five elements lattices [shown in Figs. 3.20 (*a*), (*b*)] as a sublattices, then it cannot be distributive. The lattices shown in Figs. 3.20 (*a*), (*b*) are the smallest non-distributive lattices. The next theorem, which is stated without proof, states this result.

Theorem 13: A lattice $(L, \wedge, \vee)$ is distributive if and only if it does not contain the five element pentagonal or the diamond lattice [shown in Figs. 3.20 (*a*), (*b*)] as one of its sublattices.

Example 42: *Show that every chain is a distributive lattice.*

Solution: Let $(L, \leq)$ be a chain and let $a, b, c \in L$. Consider the following cases:

(*i*) $a \leq b$ or $a \leq c$ and (*ii*) $a \geq b$ or $a \geq c$

Now we shall show that distributive law is satisfied by $a, b, c \in L$.

For case (*i*), we have

$$a \wedge (b \vee c) = a \quad \text{and} \quad (a \wedge b) \vee (a \wedge c) = a$$

$$\therefore \quad a \wedge (b \vee c) = (a \wedge b) \vee (a \wedge c).$$

For case (*ii*), we have

$$a \wedge (b \vee c) = b \vee c \text{ and } (a \wedge b) \vee (a \wedge c) = b \vee c$$

$$\therefore \quad a \wedge (b \vee c) = (a \wedge b) \vee (a \wedge c).$$

Therefore, a chain is a distributive lattice.

Example 43: *Consider the lattice L whose Hasse diagram is given in Fig. 3.21.*

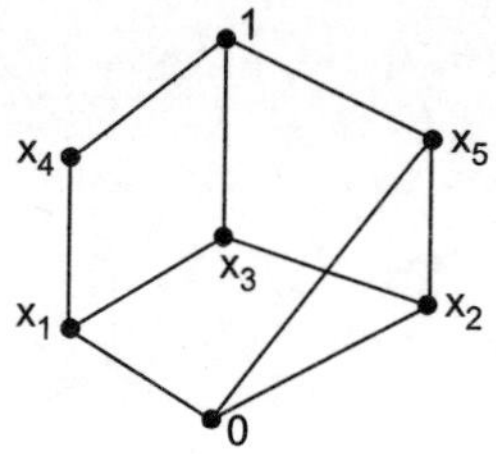

Fig. 3.21

(*i*) *Which of the followings are sublattices of L?*

$L_1 = \{0, x_1, x_2, 1\}$, $L_2 = \{0, x_1, x_5, 1\}$, $L_3 = \{x_1, x_3, x_4, 1\}$

(*ii*) *Find complements, if exist, for elements x_1, x_2 and x_3.*

(*iii*) *Is L distributive?*

(*iv*) *Is L a complemented lattice?*

Solution: (*i*) A non-empty subset L' is a sublattice of L if $a, b \in L' \Rightarrow a \wedge b, a \vee b \in L'$.

L_1 is not a sublattice since $x_1 \vee x_2 = x_3 \notin L_1$.

L_2 is a sublattice since $x_1 \wedge x_5 = 0 \in L_2$ and $x_1 \vee x_5 = 1 \in L_2$.

L_3 is a sublattice since $x_1 \wedge x_3 = x_1$, $x_1 \vee x_3 = x_3$, $x_1 \wedge x_4 = x_1$, $x_1 \vee x_4 = x_4$, $x_3 \wedge x_4 = x_1 \in L_3$ and $x_3 \vee x_4 = 1 \in L_3$.

(*ii*) We have $x_1 \wedge x_5 = 0$, $x_2 \vee x_5 = 1$, $x_2 \wedge x_4 = 0$, $x_2 \vee x_4 = 1$

So, the complements of x_1 and x_2 are x_5 and x_4 respectively. But x_3 has no complement.

(*iii*) L is not distributive since $K = \{0, x_1, x_4, x_5, 1\}$ is a sublattice which forms pentagon.

(*iv*) L is not complemented since x_3 has no complement.

Example 44: *Show that the poset given in the following Hasse diagram is a lattice. Is it distributive and complemented? Justify your answer.* [*WBUT 2012*]

Solution: The poset represented by the given Hasse diagram is a lattice since every pair of elements of it has a least upper bound as well as a greatest lower bound.

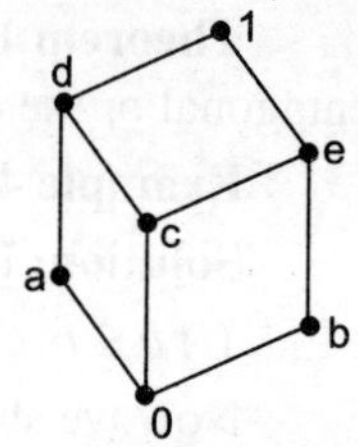

Fig. 3.22

The lattice represented by the given Hassse diagram is distributive because it does not contain the five element pentagonal or the diamond lattice as one of its sublattices (using Theorem 13).

The lattice represented by the given Hasse diagram is not complemented because the element c has no complement.

Theorem 14: *In a distributive lattice if an element has a complement then this complement is unique.*

Proof: Suppose that an element x has two complements y and z. Then

$$x \wedge y = 0, x \vee y = 1 \quad ...(1)$$

$$x \wedge z = 0, x \vee z = 1 \quad ...(2)$$

We have
$$y = y \wedge 1$$
$$= y \wedge (x \vee z) \quad \text{[By (2)]}$$
$$= (y \wedge x) \vee (y \wedge z) \quad \text{[By distributivity]}$$

$= 0 \vee (y \wedge z)$ [By (1) : $x \wedge y = 0$]

$= (x \wedge z) \vee (y \wedge z)$ [By (2)]

$= (x \vee y) \wedge z$ [By distributivity]

$= 1 \wedge z$ [By (1)]

$= z.$

Theorem 15: *If $(L, \wedge, \vee)$ be a complemented distributive lattice, then De Morgan's laws:*

$$(a \vee b)' = a' \wedge b' \text{ and } (a \wedge b)' = a' \vee b'$$ *[WBUT 2012]*

hold for all $a, b \in L$.

Proof: Given $(L, \wedge, \vee)$ be a complemented lattice and so it is bounded and every element in L has a complement. Since $(L, \wedge, \vee)$ is distributive, so each element in L has unique complement.

To prove $(a \vee b)' = a' \wedge b'$, we have to prove that $(a' \wedge b')$ is the complement of $(a \vee b)$, *i.e.*,

$$(a \vee b) \vee (a' \wedge b') = 1 \text{ and } (a \vee b) \wedge (a' \wedge b') = 0.$$

We have $(a \vee b) \vee (a' \wedge b')$

$= [(a \vee b) \vee a'] \wedge [(a \vee b) \vee b']$ (By distributivity)

$= [(a \vee a') \vee b] \wedge [a \vee (b \vee b')]$ (By commutativity and associativity)

$= (1 \vee b) \wedge (a \vee 1) = 1 \wedge 1 = 1$

and $(a \vee b) \wedge (a' \wedge b')$

$= [a \wedge (a' \wedge b')] \vee [b \wedge (a' \wedge b')]$ (By distributivity)

$= [(a \wedge a') \wedge b'] \vee [a' \wedge (b \wedge b')]$ (By associativity and commutativity)

$= (0 \wedge b') \vee (a' \wedge 0) = 0 \vee 0 = 0.$

Thus $(a' \wedge b')$ behaves as a complement of $(a \vee b)$ and by uniqueness of complements, it is the only complement of $a \vee b$.

$\therefore \quad (a \vee b)' = a' \wedge b'.$

The other result follows from the principle of duality.

MULTIPLE CHOICE QUESTIONS

1. The usual 'less than (<)' relation on the set of all real numbers is
   (*a*) a partial ordering relation because it is asymmetric and reflexive
   (*b*) a partial ordering relation because it is antisymmetric and reflexive
   (*c*) not a partial ordering relation because it is not antisymmetric and not reflexive.
   (*d*) not a partial ordering relation because it is not asymmetric and not reflexive.

   Note: A relation R on a non-empty set A is said to be **asymmetric** if $(a, b) \in R \Rightarrow (b, a) \notin R$. It means that R is not asymmetric if for some $a, b \in R$, both $(a, b) \in R$ and $(b, a) \in R$.
2. Among the relations usual 'less than (<)' and usual 'not equal to (≠)' defined on the set Z of all integers
   (*a*) both are partial ordering relations.
   (*b*) the first is a partial ordering relation but not the second.
   (*c*) the second is a partial ordering relation but not the first.
   (*d*) none of them is a partial ordering relation.
3. The relation R on Z defined by $a\,R\,b$ iff $a + b$ is even is not a partial ordering relation on Z since R is
   (*a*) not reflexive (*b*) not antisymmetric
   (*c*) not transitive (*d*) none of these

4. The relation $\leq$ on the set Z^+ of all positive integers defined by $m \leq n$ iff $m + n$ is odd is not a partial ordering relation because R is
   (*a*) not reflexive (*b*) not antisymmetric
   (*c*) not transitive (*d*) not reflexive, not antisymmetric, not transitive
5. The relation $>$ defined on the set R of all real numbers is
   (*a*) reflexive (*b*) antisymmetric
   (*c*) not a partial ordering (*d*) none of these
6. If the relations R and S are defined on the set N of natural numbers (*i.e.*, positive integers) by mRn if $m = 2n$ and mSn if $m = n^k$ for some $k \in N$, then
   (*a*) only R is a partial ordering (*b*) only S is a partial ordering
   (*c*) R and S both are partial orderings (*d*) neither R nor S is a partial ordering
7. The relation R on the set Z of all integers defined by $a\,R\,b$ iff $a + b$ is even is reflexive.
   (*a*) True (*b*) False
8. Two elements a, b in a poset $(S, \leq)$ are comparable if $a \leq b$ or $b \leq a$ and incomparable if they are not comparable. The relation 'incomparability' is
   (*a*) reflexive (*b*) antisymmetric (*c*) transitive (*d*) none of these
9. In $(P, \subseteq)$, where P is the power set of $\{a, b, c\}$, the following subsets are incomparable:
   (*a*) $\{a\}$ and $\{a, b, c\}$ (*b*) $\{b, c\}$ and $\{a, b, c\}$ (*c*) $\{b\}$ and $\{a, c\}$ (*d*) $\{a, b, c\}$ and $\{c\}$
10. In $(P, \subseteq)$, where P is the power set of $\{a, b, c\}$, the number of subsets incomparable to $\{a\}$ is
   (*a*) 1 (*b*) 2 (*c*) 3 (*d*) 4
11. The number of possible partial ordering on $(\{a, b, c\}, \leq)$ is
   (*a*) 3 (*b*) 4 (*c*) 5 (*d*) 6
12. If $(S, \leq)$ be a lattice and $a \leq b \leq c$ for $a, b, c \in S$, then
   (*a*) $a \vee b = b \wedge c$ (*b*) $a \vee b = b \vee c$ (*c*) $a \wedge b = b \vee c$ (*d*) $a \wedge b = b \wedge c$
13. In a lattice $(S, \leq)$ if $a \vee b = b$, then
   (*a*) $b \leq a$ (*b*) $a \wedge b = a$ (*c*) $a \geq b$ (*d*) none of these
14. In a lattice $(S, \leq)$, $((a \vee b) \wedge a) \vee b =$
   (*a*) $a \wedge b$ (*b*) $((a \wedge b) \vee a) \wedge b$ (*c*) $a \vee b$ (*d*) $(a \vee b) \wedge a$
15. In a lattice $(S, \leq)$, if $a \leq b$ and $c \leq d$, then
   (*a*) $a \leq d$ (*b*) $b \leq c$ (*c*) $a \wedge c \leq b \wedge d$ (*d*) none of these
16. If $(\{1, 2, 3, 10, 15, x\}, \leq)$ be a lattice where $a \leq b$ iff $a \mid b$, then the smallest value for x is
   (*a*) 150 (*b*) 90 (*c*) 60 (*d*) 30
17. In the lattice defined by the adjacent Hasse diagram, how many complements does the element 'b' have?
   (*a*) 3 (*b*) 2 (*c*) 1 (*d*) 0

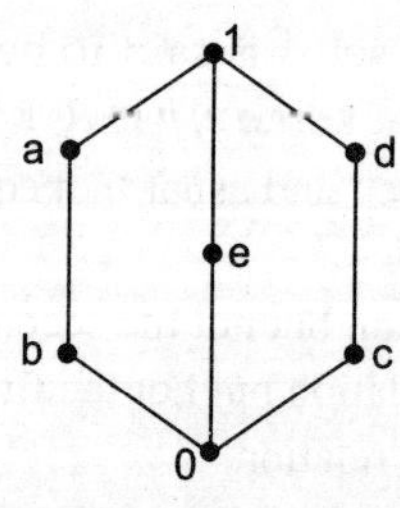

18. Let $S = \{2, 3, 6, 12, 24\}$ and let $\leq$ be a partial ordering relation on S defined by $a \leq b$ iff a divides b, then the number of edges in the Hasse diagram of $(S, \leq)$ is
   (*a*) 3 (*b*) 4 (*c*) 5 (*d*) 6 (*GATE 1996*)

19. Every finite subset of a lattice has a least upper bound and a greatest lower bound.

(*a*) True (*b*) False

20. A complemented, distributive lattice is called a Boolean Algebra

(*a*) True (*b*) False

21. Consider the set *A* of all integers greater than 1. Let *D* be a relation defined on *A* by $(x, y) \in D$ iff *x* divides *y*. Then which of the following is true?

(*a*) *D* is both a lattice and a partial ordering
(*b*) *D* is a lattice but not a partial ordering
(*c*) *D* is neither a lattice nor a partial ordering
(*d*) *D* is a partial ordering but not a lattice. [*WBUT 2012*]

22. The set of natural numbers *N* with the relationship '|' (divides) is a poset. How many minimal and maximal elements does it have?

(*a*) 1 minimal and 1 maximal
(*b*) 1 minimal and 0 maximal
(*c*) 1 minimal and more than one maximal
(*d*) 0 minimal and 0 maximal. [*WBUT 2012*]

23. A complemented, distributive lattice is also called a Boolean Algebra. Consider a set $S = \{a, b, c\}$ and let $M = P(S)$ be the power set of *S*. Consider the inclusion (subset) relation '$\subseteq$'. The $(M, \subseteq)$ is

(*a*) not a partial ordering
(*b*) a partial ordering but not a lattice
(*c*) a lattice but not a boolean algebra
(*d*) a boolean algebra. [*WBUT 2012*]

24. The maximal and minimal elements of the poset described by the following Hasse diagram are

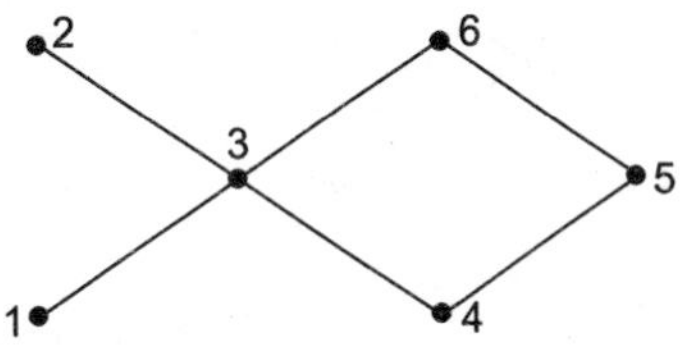

(*a*) maximal 5; minimal 1, 2
(*b*) maximal 5; minimal 1
(*c*) maximal 2, 6; minimal 1, 4
(*d*) none of these

25. If the posets P_1 and P_2 are described by the following Hasse diagrams:

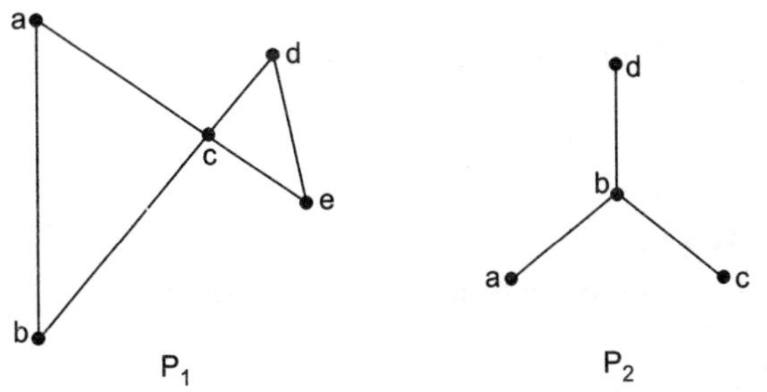

P_1 P_2

Then

(*a*) P_1 is a lattice
(*b*) P_2 is a lattice
(*c*) P_1 and P_2 both are lattices
(*d*) None of them is a lattice

26. In the lattice $(R, \geq)$, where *R* is the set of all real numbers and $\geq$ is the usual relation 'greater than or equal to', lub $\{a, b\}$ =

(*a*) min $\{a, b\}$
(*b*) max $\{a, b\}$
(*c*) greatest common divisor of *a* and *b*
(*d*) least common multiple of *a* and *b*

27. Among $L_1 = (\{7, 13, 91\}, 1)$, $L_2 = (\{1, 7, 13, 91\}, |)$ and $L_3 = (\{1, 7, 13, 49, 91\}, |)$

(*a*) L_1 alone is a lattice
(*b*) L_2 alone is a lattice
(*c*) L_1 and L_2 both are lattices
(*d*) L_2 and L_3 both are lattices

28. Consider the set of integers {1, 2, 3, 4, 6, 8, 12, 24} together with the two binary operations LCM (least common multiple) and GCD (greatest common divisor). Which of the following algebraic structures does this represent?
(*a*) Group (*b*) Ring
(*c*) Field (*d*) Lattice [*GATE 1992*]

29. The less than relation, <, on reals is
(*a*) a partial ordering since it is asymmetric and reflexive
(*b*) a partial ordering since it is not asymmetric and reflexive
(*c*) not a partial ordering because it is not antisymmetric and not reflexive
(*d*) not a partial ordering because it is not symmetric and reflexive
(*e*) none of the above. [*GATE 1992*]

30. The inclusion of which of the following sets into $S = \{\{1, 2\}, \{1, 2, 3\}, \{1, 3, 5\}, \{1, 2, 4\}, \{1, 2, 3, 4, 5\}\}$ is necessary and sufficient to make S a complete lattice and the partial ordering defined by set containment?
(*a*) {1} (*b*) {1}, {2, 3}
(*c*) {1}, {1, 3} (*d*) {1}, {1, 3}, {1, 2, 3, 4}, {1, 2, 3, 5}
[*GATE 2004*]

31. The following is the Hasse diagram of the poset $(\{a, b, c, d, e\}, \leq)$:

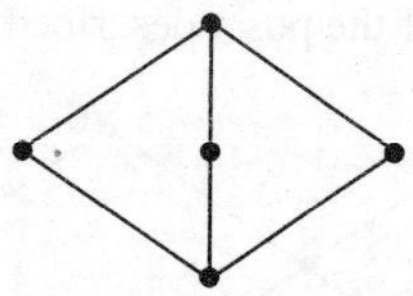

The poset is
(*a*) not a lattice (*b*) a lattice but not a distributive lattice
(*c*) a distributive lattice but not a Boolean algebra (*d*) a Boolean algebra [*GATE 2005*]

32. A relation R is defined on ordered pair of integers as follows:
$(x, y)\ R\ (u, v)$ if $x < u$ and $y > v$.
Then R is
(*a*) neither a partial order nor an equivalence relation
(*b*) a partial order but not a total order
(*c*) a total order
(*d*) an equivalence relation [*GATE 2006*]

33. The set of natural numbers has
(*a*) maximal element (*b*) upper bound (*c*) lower bound (*d*) none of these
[*SET 2010*]

ANSWERS

1. (*c*) **2.** (*d*) **3.** (*b*) **4.** (*d*)
5. (*c*) **6.** (*b*) **7.** (*a*) **8.** (*d*)
9. (*c*) **10.** (*c*) **11.** (*c*) [Hint: $a \leq a, a \leq b, b \leq c, b \leq b, c \leq c$]
12. (*a*) **13.** (*b*) **14.** (*c*) (by absorption law)
15. (*c*) **16.** (*d*) **17.** (*a*) **18.** (*b*)
19. (*a*) **20.** (*a*) **21.** (*d*) **22.** (*b*)
23. (*d*) **24.** (*c*) **25.** (*d*) **26.** (*a*)
27. (*b*) **28.** (*d*) **29.** (*c*) **30.** (*d*)
31. (*b*) **32.** (*a*) **33.** (*c*)

PROBLEMS

1. Define partial ordering relation on a non-empty set. Show that the divisibility relation on the set of all positive integers is a partial ordering relation.
2. Define a relation R on the set Z of all integers as follows:
$$mRn \Leftrightarrow m + n \text{ is odd}, \forall\, m, n \in Z.$$
Is R a partial ordering relation? Prove or give a counter example.
3. Show that the relation S on the set R of all real numbers defined by $S = \{(a, b) \in R \times R : a - b \le 0\}$ is a partial ordering relation.
4. What is a poset? Explain the rules to draw the Hasse diagrams. Draw the Hasse diagram for poset $(P(A), \subseteq)$, where $A = \{a, b, c, d\}$, $P(A)$ is the power set and '$\subseteq$' stand for 'subset of '.
5. Define the following terms with examples:
   (*i*) comparable (*ii*) incomparable (*iii*) chain (*iv*) anti-chain
   (*v*) totally ordered set (*vi*) least element and greatest element (*vii*) bounded poset
   (*viii*) minimal and maximal elements (*ix*) greatest lower bound and least upper bound
   (*x*) Lattice (*xi*) sublattice (*xii*) complete lattice (*xiii*) bounded lattice
   (*xiv*) complemented lattice (*xv*) distributive lattice (*xvi*) Boolean algebra
6. Find the maximal and minimal elements of the poset $(S, \le)$, where $S = \{2, 4, 5, 10, 12, 20, 25\}$ and $a \le b$ iff a divides b.
7. Find the least upper bound and the greatest lower bound of the set $\{1, 2, 4, 5, 15\}$ in the poset $(Z^+, \le)$ where Z^+ denotes the set of all positive integers and $a \le b$ iff a divides b.
8. Show that $(S, \le)$ is a lattice where $S = \{1, 2, 4, 8, 16, 32\}$ and $a \le b$ iff a divides b.
9. Draw the Hasse diagram of the poset $(S, \le)$ where $S = \{1, 2, 3, 5, 7, 11, 13, 17\}$ and $a \le b$ iff a divides b.
10. Which of the following posets whose Hasse diagrams are given below are lattices?

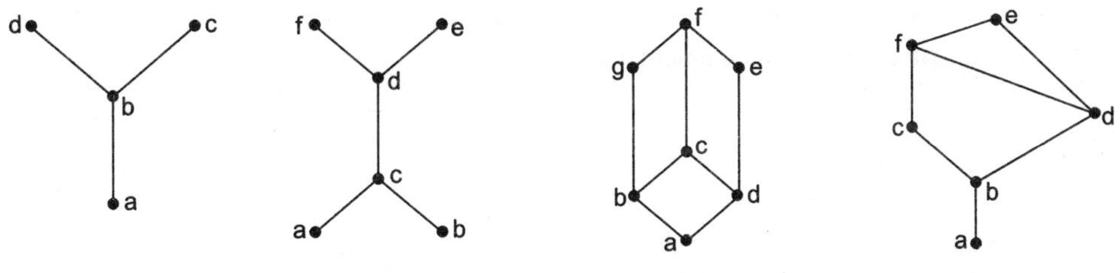

(*i*) (*ii*) (*iii*) (*iv*)

11. In any bounded distributive lattice, show that the elements having complements form a sublattice.
12. Consider the lattice $(L, \le)$ whose Hasse diagram is shown below:

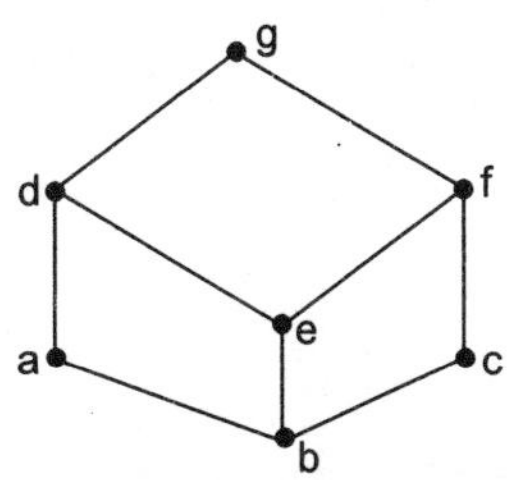

(*i*) Find all sublattices having five elements (*ii*) Find complements of a and e if exist
(*iii*) Is $(L, \le)$ complemented? Distributive?

13. Show that the lattice represented by the following Hasse diagram is complemented but not distributive.

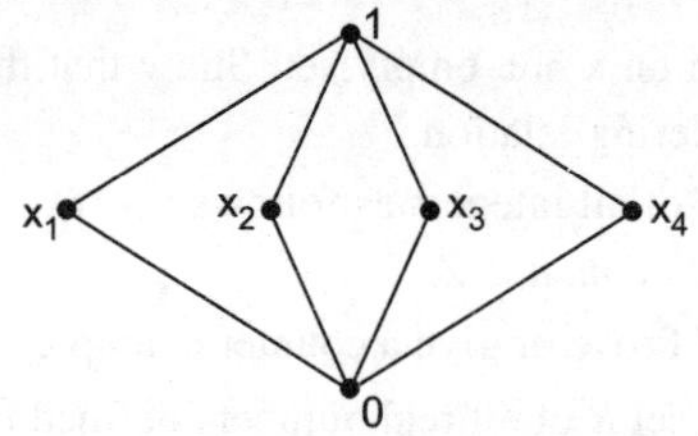

14. Show that the lattice represented by the following Hasse diagram is distributive but not complemented.

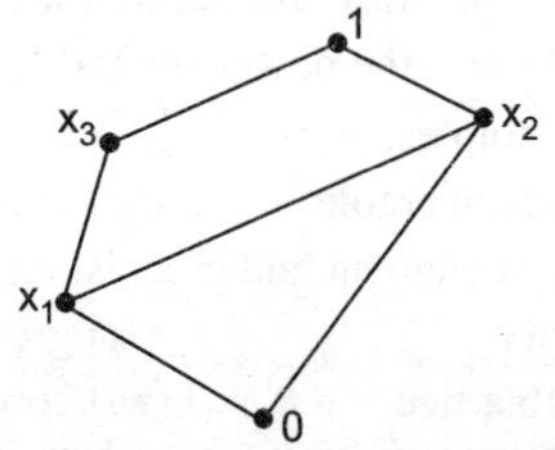

ANSWERS

2. No, R is not reflexive. Counter example: $1 \not R 1$, because 1 + 1 is even.

6. Minimal elements are 2, 5. Maximal elements are 12, 20, 25.

7. lub = 60, glb = 1.

10. (*i*) No (*ii*) No (*iii*) Yes (*iv*) Yes

12. (*i*) Six: *abdeg* ; *abcdg, badfg, becfg, bacfg, bcdfg*

(*ii*) *c* and *f* are complements of *a*. *e* has no complement.

(*iii*) $(L, \leq)$ is distributive but not complemented.

Module III

CHAPTER 4

Combinatories

4.1 INTRODUCTION

Combinatories is the study of counting and enumeration of specified objects, patterns or designs. Techniques of counting are important in discrete mathematics and computer science especially in probability theory and to determine the complexity of algorithms. Counting is also needed to find whether there are enough telephone numbers or internet protocol addresses to meet the demand.

In this chapter, we shall study the basic rules of counting, the permutation, the combination, binomial coefficients, the pigeon-hole principle and the principles of inclusion and exclusion.

4.2 TWO BASIC PRINCIPLES OF COUNTING

In this section, we will present two basic counting principles which will show how they can be used to solve many different counting problems.

4.2.1 Sum Rule (Principle of Disjunctive Counting)

First form: Let S be a non-empty set and let $|S|$ denotes the number of elements in S. If S is a union of disjoint non-empty subsets $X_1, X_2, ..., X_n$, then $|S| = |X_1| + |X_2| + ... + |X_n|$. Here the subsets X_i (i = 1, 2, ..., n) of S are all disjoint, *i.e.*, they have no element in common. If X_i and X_j are two distinct subsets of S, then $X_i \cap X_j = \phi$ (for $i \neq j$). Therefore, each element of

$$S = X_1 \cup X_2 \cup X_3 \cup ... \cup X_n$$

is exactly in one of the subsets X_i (i = 1, 2, ..., n).

In other words, the subsets $X_1, X_2, ..., X_n$ is a partition of S.

For this case, **the sum rule states that** if X_i has r_i (i = 1, 2, ..., n) elements, the number of possible elements that can be selected from X_1 or X_2 or X_3 or ... or X_n is $r_1 + r_2 + ... + r_n$.

Second form: Let S be a sample space (or state space). Two events $E_1 \subseteq S$ and $E_2 \subseteq S$ are said to be mutually exclusive if these two events have no elements in common. If $E_1, E_2, E_3, ..., E_n$ are mutually exclusive events of S, then the sum rule for counting events can be stated as follows:

If $E_1, E_2, E_3, ..., E_n$ are mutually exclusive events of a sample space S and event E_i can happen in m_i ways for $i = 1, 2, ..., n$, then E_1 or E_2 or ... or E_n can happen in $m_1 + m_2 + ... + m_n$ ways.

Example 1: *Suppose there are 38 boys and 13 girls in a class, find the number of ways of selecting one student as class representative.*

Solution: Using sum rule, there are 38 + 13 = 51 ways of selecting one student (either a boy or a girl) as class representative.

Example 2: *If a student is getting admission in 5 different Engineering Institutes and 2 Medical Colleges, find the number of ways of choosing one of these Institutes/Colleges.*

Solution: Using sum rule, there are 5 + 2 = 7 ways of choosing one of these Institutes/Colleges.

Example 3: *In how many ways one can get a sum of 8 or 11 when two distinguishable dice are rolled?*

Solution: Given two dice are distinguishable and so the ordered pairs (a, b) and (b, a) are distinct when $a \neq b$, *i.e.*, $(a, b) \neq (b, a)$ for $a \neq b$.

The distinct ordered pairs in which the sum is 8 are (2, 6), (6, 2), (3, 5), (5, 3), (4, 4). Thus there are 5 ways to obtain the sum 8. Similarly, the distinct ordered pairs in which the sum is 11 are (5, 6), (6, 5). So, the number of ways in which one get a sum 11 is 2.

Therefore one get a sum 8 or 11 with two distinguishable dice is 5 + 2 = 7 ways.

Example 4: *How many 10-bit strings begin with 01010 or 1001?*

Solution: The number of 10-bit strings that begin with 01010 is equal to the number of ways bit six to bit ten can be chosen. This number is $2^5 = 32$. Similarly, the number of 10-bit strings that begin with 1001 is $2^6 = 64$. Since the set of strings begin with 01010 is disjoint from the set of strings begin with 1001, so using sum rule the number of strings begin with 01010 or 1001 is 32 + 64 = 96.

Third Form:

The sum rule: If a first task can be done in n_1 ways and a second task in n_2 ways, and if these tasks cannot be done simultaneously, then there are $n_1 + n_2$ ways to perform either task.

It can be extended to more than two tasks. Suppose that the tasks $T_1, T_2, ..., T_m$ can be done in n_1, $n_2, ..., n_m$ ways respectively, and no two of these tasks can be done simultaneously. Then the number of ways to perform one of these tasks is $n_1 + n_2 + ... + n_m$.

Example 5: *A student can choose a computer project from one of the five given lists. The five lists contain respectively 5, 7, 9, 12, 13 possible projects. Find the number of possible projects to choose by a student.*

Solution: A student can choose a project from the first list in 5 ways, from the second list in 7 ways, from the third list in 9 ways, from the fourth list in 12 ways and from the fifth list in 13 ways. Hence a student can choose a project in 5 + 7 + 9 + 12 + 13 = 46 ways.

4.2.2 Product Rule (Principle of Sequential Counting)

First form: Suppose a procedure can be split up into two tasks. If there are n_1 ways to perform the first task and n_2 ways to perform the second task after the first task has been done, then there are $n_1 n_2$ ways to perform the procedure.

It can be extended to more than two tasks as follows:

Suppose a procedure can be performed by doing the tasks $T_1, T_2, ..., T_m$. If task T_i can be performed in n_i ways after tasks $T_1, T_2, ...,$ and T_{i-1} have been done, then there are $n_1\, n_2 ... n_m$ ways to perform the procedure.

Second form: If $A_1, A_2, ..., A_m$ are finite non-empty sets, then the number of elements in the Cartesian product $A_1 \times A_2 \times ... \times A_m$ is the product of the number of elements in each set.

To relate this to the first form, observe that the task of choosing an element in the Cartesian product $A_1 \times A_2 \times ... \times A_m$ is performed by choosing an element in A_1, an element in A_2, ..., and an element in A_m. Using the product rule, we get

$$|A_1 \times A_2 \times ... \times A_m| = |A_1||A_2| ... |A_m|.$$

Third form: If $E_1, E_2, ..., E_i$ are events and E_1 can happen in m_1 ways, E_2 can happen in m_2 ways, ..., E_i can happen in m_i ways, then the sequence of events E_1 first, followed by E_2, followed by E_3, .., followed by E_i can happen in $m_1 m_2 m_3 ... m_i$ ways.

Example 6: *In how many different ways a candidate can answer all the questions of a true-false test consisting of 8 questions?*

Solution: There are two ways of answering each of the 8 questions and so by product rule the number of ways in which all the 8 questions can be answered = $2^8 = 256$.

Example 7: *How many different bit strings are there of length ten?*

Solution: Each of the ten bits can be chosen in two ways since each bit is either 0 or 1. Using product rule, there are a total of $2^{10} = 1024$ different bit strings of length ten.

Example 8: *How many possible outcomes are there when a pair of dice, one white and one yellow are rolled?*

Solution: The white die can land in any one of six ways and for each of these six ways, the yellow die can also land in six ways.

∴ Number of possible outcomes when two dice are rolled = $6 \times 6 = 36$.

Example 9: *A new flag is to be designed with 6 vertical strips using 4 colours. In how many ways can this be done so that no 2 adjacent strips have the same colour?* *[WBUT 2012]*

Solution: To colour first strip, any one of the 4 colours can be used whereas the second strip can be coloured using 3 colours excluding first one. The third strip can also be coloured using 3 colours excluding the colour of the second strip and so on.

∴ Required number of ways = $4 \times 3^5 = 972$.

Example 10: *The chairs of an Institute auditorium are to be labelled with a letter followed by a positive integer not exceeding 100. Find the largest number of chairs that can be labelled differently.*

Solution: The process of labelling a chair consists of two tasks:

(*i*) Assigning one of the 26 letters and then

(*ii*) Assigning one of the 100 integers.

Using product rule, there are $26 \times 100 = 2600$ different ways that a chair can be labeled. Thus the largest number of chairs that can be labelled differently is 2600.

Example 11: *In how many ways can a student select two books from different subjects among eight distinct computer science books, five distinct mathematics books and three distinct physics books?*

Solution: Using product rule, a student can select two books from different subjects as follows:

(*i*) one from computer science and one from mathematics in $8 \times 5 = 40$ ways,

(*ii*) one from computer science and one from physics in $8 \times 3 = 24$ ways,

(*iii*) one from mathematics and one from physics in $5 \times 3 = 15$ ways.

Since these sets of selections are pairwise disjoint, by sum rule there are $40 + 24 + 15 = 79$ ways of selecting two books from different subjects among computer science, mathematics and physics.

Example 12: *In a certain state automobile 'license plate numbers' are made such that each plate contains three letters followed by three digits.*

(*i*) *How many different 'license plate numbers' can be made if different letters and different digits are used?*

(*ii*) *How many different 'license plate numbers' can be made if letters and digits are allowed to repeat?*

(*iii*) *How many different 'license plate numbers' can be made if only letters are allowed to repeat?*

(*iv*) *How many different 'license plate numbers' can be made if only digits are allowed to repeat?*

Solution: (*i*) First letter can be printed in 26 different ways, the second letter must be chosen from any one of the 25 letters not previously used and the third letter must be chosen from 24 letters not previously printed. Similarly, the first digit can be printed in 10 ways, the second digit in 9 ways and the third in 8 ways.

Using product rule, the number of license plates that can be manufactured so that each plate contains three distinct letters followed by three different digits is $26 \times 25 \times 24 \times 10 \times 9 \times 8 = 11232000$.

(*ii*) Using product rule, there are a total of $26 \times 26 \times 26 \times 10 \times 10 \times 10 = 17576000$ possible license plates can be made.

Similarly,

(*iii*) the number of different license plates are $26^3 \times 10 \times 9 \times 8$,

(*iv*) the number of different license plates are $26 \times 25 \times 24 \times 10^3$.

4.3 PERMUTATIONS AND COMBINATIONS

4.3.1 Fundamental Principle of Counting

If an event can occur in n_1 ways and thereafter for each of these events a second event can occur in n_2 ways and thereafter for each of these events a third event can occur in n_3 ways, then the number of ways these three events can happen is given by $n_1 n_2 n_3$. This result can be extended for any number of events.

Illustration: If a coin is tossed, then we know that there are two outcomes, namely 'head' and 'tail'. Suppose that three coins are tossed at random. By Principle of Counting, the total number of outcomes is $n = 2 \times 2 \times 2 = 8$.

4.3.2 Factorial *n*

If n be a positive integer, then $n\,!$ or $|\underline{n}$ (read as factorial n) stands for the product of first n natural numbers (*i.e.*, positive integers).

$$\therefore \quad n\,! = |\underline{n} = 1 \times 2 \times 3 \times \ldots \times (n-2) \times (n-1) \times n$$

$$= n \times \{(n-1) \times (n-2) \times \ldots \times 3 \times 2 \times 1\}$$

$$= n \times (n-1) \times \{(n-2) \times (n-3) \times \ldots \times 3 \times 2 \times 1\}$$

$$\therefore \quad |\underline{n} = n\,|\underline{n-1} = n\,(n-1)\,|\underline{n-2} = n\,(n-1)(n-2)\,|\underline{n-3}\,.$$

Notes: (*i*) $|\underline{0} = 0\,! = 1$ (by definition)

(*ii*) $|\underline{1} = 1\,! = 1;\ |\underline{2} = 2\,! = 2 \times 1 = 2;\ |\underline{3} = 3\,! = 3 \times 2 \times 1 = 6.$

4.3.3 Permutations

Definition. Each of the different **arrangements** that can be made out of a given set of objects by taking some or all of them at a time is called a **permutation.**

Illustration (*i*) The different permutations of two letters a, b are ab, ba.

(*ii*) The different permutations of three letters a, b, c **taken two at a time** are ab, ba, ac, ca, bc, cb.

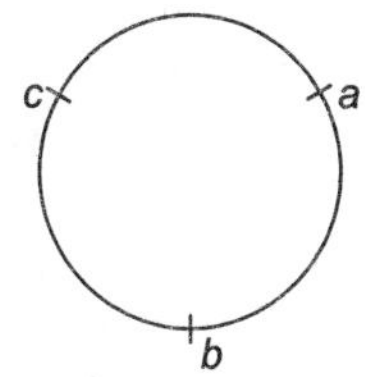

(*iii*) The different permutations of three letters a, b, c **taken all at a time** are abc, acb, bca, bac, cab, cba.

The number of permutations of n different objects taking r at a time is usually denoted by the symbol ${}^nP_r$ or $P(n, r)$.

Note: Here the word **object** is being used in a very general sense to include individuals as well.

4.3.4 Permutations of Objects all Different

To find the number of permutations of n different objects taken r (≤ n) at a time.

The number of permutations of n different objects taken r ($\le n$) at a time is the same as the number of different ways in which r places can be occupied by the n different objects.

The first place can be occupied in n ways, because any one of the n different objects can be placed there. When the first place has been occupied in **any one** of these n ways, the second place can be occupied in $(n-1)$ ways, because any one of the remaining $(n-1)$ objects can be placed there. Since each way of filling up the first place can be associated with each way of filling up the second place, so the first-two places can be occupied in $n\ (n-1)$ ways by using principle of counting.

When the first two places have been occupied in **any one** of these $n(n-1)$ ways, the third place can be occupied in $(n-2)$ ways, because there are now $(n-2)$ objects left to fill up the third place. Each of these $(n-2)$ ways can be associated with each of the $n(n-1)$ arrangements of the first-two places. Using principle of counting, the first three places can be occupied in $n(n-1)(n-2)$ ways.

Proceeding in this way and observing that at any stage the number of factors is the same as the number of places occupied, so, the total number of ways in which r ($\le n$) places can be filled up

$$= n(n-1)(n-2)(n-3) \ldots \text{ to } r \text{ factors}$$
$$= n(n-1)(n-2) \ldots \{n-(r-1)\}$$
$$= n(n-1)(n-2) \ldots (n-r+1)$$

$$\therefore \quad \mathbf{{}^nP_r = n(n-1)(n-2) \ldots (n-r+1)} \qquad \ldots(1)$$

Notes: (*i*) The number of permutations of n different objects taken all at a time = Number of permutations of n different objects among themselves

$$= {}^nP_n = n(n-1)(n-2) \ldots 3.2.1. \qquad \text{[putting } r = n \text{ in (1)]}$$
$$= n!$$

(*ii*) ${}^nP_r = n(n-1)(n-2) \ldots (n-r+1)$

$$= \frac{n(n-1)(n-2)\ldots(n-r+1)(n-r)(n-r-1)\ldots 3.2.1}{(n-r)(n-r-1)\ldots 3.2.1} = \frac{n!}{(n-r)!}$$

$$\therefore \quad {}^nP_0 = 1,\ {}^nP_1 = \frac{(n)!}{(n-1)!} = \frac{n(n-1)!}{(n-1)!} = n,$$

$${}^nP_2 = \frac{(n)!}{(n-2)!} = \frac{n(n-1)\,(n-2)!}{(n-2)!} = n(n-1) \text{ etc.}$$

(*iii*) Meaning of $\lfloor 0 = 0!$:

Putting $r = n$ in (*ii*), we get $^nP_n = \dfrac{n!}{0!}$. Also, from (*i*), $^nP_n = n!$

$$\therefore \quad n! = \frac{n!}{0!} \Rightarrow 0! = 1.$$

So, we define $0! = \lfloor 0 = 1$.

(*iv*) It is evident that $^nP_r = \dfrac{n!}{(n-r)!}$ is greatest when $r = n$ or $n - 1$.

Obviously, $^nP_n = {^nP_{n-1}}$.

Example 13: *A salesman has to visit seven different cities. He must start his trip in a specified city, but he can visit the other six cities in any order he wishes. Determine the number of possible ways the salesman can use when visiting these cities.*

Solution: Clearly the number of possible paths between the cities is the number of permutations of six elements, since the first city is specified, but the remaining six can be ordered arbitrarily. Consequently, there are $6! = 6 \times 5 \times 4 \times 3 \times 2 \times 1 = 720$ possible ways for the salesman to choose his trip.

4.3.5 Permutations of Objects not all Different

To find the number of permutations of n objects taken all together when p of the objects are alike of one kind, q of the objects are alike of another kind, r of the objects are alike of a third kind and the rest all different.

Let the n objects be represented by n letters and suppose p of them are a's, q of them are b's, r of them are c's and the rest are all different.

Let the required number of permutations be x.

Step 1. Let p number of a's are changed into p new letters $a_1, a_2, ..., a_p$ different from each other and from the rest. Then by changing the arrangement of the p new letters only among themselves without changing the position of any other letter, each permutation would give rise to $p!$ different permutations. Hence, if this change were made in each of the x permutations, then the number of permutations would become $x \times p!$ (by principle of counting).

Step 2. Similarly, if in any one of these new permutations q number of b's are changed into q new letters $b_1, b_2, ..., b_q$ different from each other and from the rest, then we should get from each permutation $q!$ permutations by changing the order of $b_1, b_2, ..., b_q$ only. Therefore, the total number of permutations would now become $x \times p! \times q!$.

Step 3. Again, if r number of c's are changed into r new letters $c_1, c_2, ..., c_r$ different from each other and from the rest, then the total number of permutations (arrangements) would similarly become $x \times p! \times q! \times r!$.

Step 4. Now there are n different objects and the number of permutations of n different objects taken all together is $n!$

$$\therefore \quad x \times p! \times q! \times r! = n!$$

$$\Rightarrow x = \text{Required number of permutations} = \frac{n!}{p!q!r!}$$

Note: This result can be extended when more than three kinds of objects are repeated.

4.3.6 Permutations of Objects when Repetition are Allowed

To find the number of permutations of n different objects taken r at a time when each object may be repeated up to r times in any arrangement.

Let the n different objects be represented by n distinct letters.

Obviously, the number of permutations in this case is the same as the number of ways in which r places can be occupied by these n distinct letters, each letter being repeated once, twice, thrice, ... up to r times.

The first place can be occupied in n ways, since any one of the n letter can be placed there. When the first place has been occupied in any one way, the second place can also be occupied in n ways, because the same letter can be repeated again. Therefore, by principle of counting, the first-two places can be filled up in $n \times n = n^2$ ways. Similarly, the first-three places can be occupied in n^3 ways.

Proceeding in this way and observing that at any stage the index (power) of n is the same as the number of places filled up, the number of ways in which r places can be occupied is evidently equal to n^r.

$\therefore$ Required number of permutations $= n^r$.

4.3.7 Restricted Permutations

1. The number of permutations (or arrangements) of n distinct objects taken r at a time in which m particular objects never occur is ${}^{n-m}P_r$, provided $n - m \geq r$.

2. The number of permutations of n distinct objects taken r at a time in which m particular objects always occur is ${}^{r}P_m \times {}^{n-m}P_{r-m}$ (by using principle of counting).

3. The number of permutations of n distinct objects taken r at a time in which m particular objects are placed in m given positions (*i*) in a given order is ${}^{n-m}P_{r-m}$ (*ii*) in any order is $m\,!\,{}^{n-m}P_{r-m}$.

4. The number of permutations of n distinct objects taken r at a time in which m particular objects occur together (with fixed order) is $(r - m + 1)\,{}^{n-m}P_{r-m}$.

4.3.8 Summary

1. $n\,! = \underline{|n} = n\,(n-1)(n-2)\,...\,3.2.1$
2. $0! = \underline{|0} = 1$ (by definition)
3. $n\,! = \underline{|n} = n\,(n-1)\,! = n(n-1)\{(n-2)\,!\} = n(n-1)(n-2)\{(n-3)\,!\}$
4. The number of permutations of n different objects taken r at a time $= {}^{n}P_r = \dfrac{n\,!}{(n-r)\,!}$

$$= n(n-1)(n-2)\,...\,(n-r+1).$$

5. ${}^{n}P_1 = n,\ {}^{n}P_2 = n(n-1),\ {}^{n}P_3 = n(n-1)(n-2),\ {}^{n}P_r = n(n-1)(n-2)\,...$

$$(n-r+1),\ {}^{n}P_{n-1} = {}^{n}P_n = n\,!$$

6. The number of permutations of n different objects taken all at a time $= {}^{n}P_n = n\,!$
7. The number of permutations of n objects taken all at a time when p of them are alike of one kind, q of them are alike of another kind, r of them are alike of a third kind and the rest all different $= \dfrac{n\,!}{p\,!\,q\,!\,r\,!}$.
8. The number of permutations of n different objects taken r at a time when each object may be repeated up to r times is n^r.

4.3.9 Circular Permutations

Definition: A **circular permutation** of n objects is an arrangement of the objects in a circle and so relative positions of different objects are to be considered.

Number of circular permutations of n different objects taken all at a time

Here having fixed one object, the remaining $(n - 1)$ objects can be arranged round the circle in $(n - 1)$! ways.

Notes: *(i)* If n persons are to be arranged in a row, then the number of arrangements (permutations) is $n!$, whereas for a round (circular) table the number of arrangements is $(n - 1)!$.

(ii) **Necklace:** Number of arrangements (permutations) of n different beads to form a necklace or placed on a circular wire will be $\frac{1}{2}(n - 1)!$ as explained below:

R = Rakhi, G = Gopal, B = Bijoy
R = Red, G = Green, B = Black

The two seating arrangements (see Fig. 4.1) of three persons in a round table are different and that is why we take $(n - 1)$!. Here clockwise and anticlockwise make different arrangements.

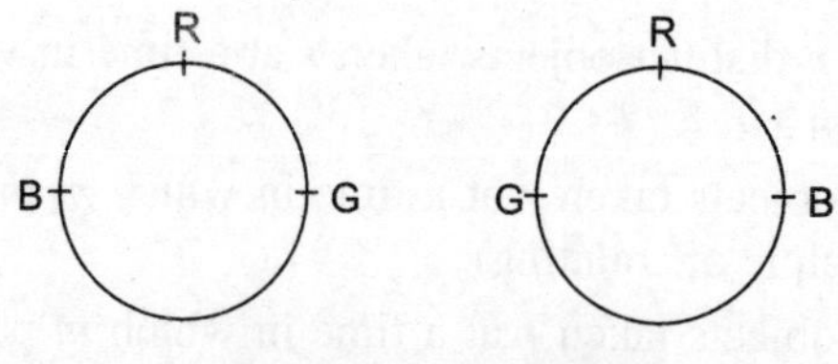

Fig. 4.1. Round table

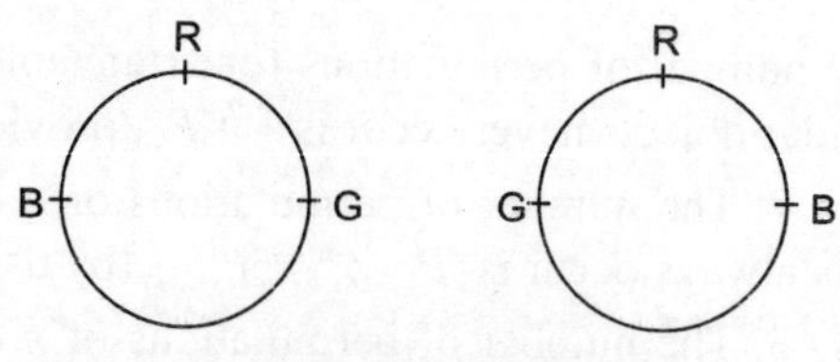

Fig. 4.2. Necklace

The same two arrangements of three colour beads to form a necklace is identical (see Fig. 4.2) because clockwise or anticlockwise does not change the character of the necklace. This is why we say the total number of arrangements of n beads to form a necklace is $(n - 1)!/2$.

ILLUSTRATIVE EXAMPLES—I

Example 1: *If* ${}^{10}P_6 + 6\,{}^{10}P_5 = {}^{11}P_r$, *find r.*

Solution: Given : ${}^{10}P_6 + 6\,{}^{10}P_5 = {}^{11}P_r$ ...(1)

Now, $${}^{10}P_6 + 6\,{}^{10}P_5 = \frac{10!}{(10-6)!} + 6.\ \frac{10!}{(10-5)!}$$

$$= \frac{10!}{4!} + \frac{6.10!}{5!} = \frac{5.10!}{5!} + \frac{6.10!}{5!} = \frac{11.10!}{5!} = \frac{11!}{5!}.$$

So, from (1), we get

$$\frac{11!}{5!} = \frac{11!}{(11-r)!}. \quad \therefore \quad 11 - r = 5, \quad \text{or,} \quad r = 11 - 5, \quad \textit{i.e.,} \quad r = 6.$$

Example 2: *Prove that (i)* $(2n)! = n!\,2^n\{1.\,3.\,5 \ldots (2n - 1)\}$

(ii) ${}^nP_r = n\,{}^{n-1}P_{r-1} = (n - r + 1)\,{}^nP_{r-1}$

(iii) ${}^nP_r = {}^{n-1}P_r + r\,{}^{n-1}P_{r-1}$.

Solution: (*i*) L.H.S. = $(2n)! = 1.2.3.4.5.6. \ldots (2n-1).2n$

$$= \{1.3.5. \ldots (2n-1)\}\{2.4.6. \ldots 2n\}$$
$$= \{1.3.5. \ldots (2n-1)\}\{(2.1).(2.2).(2.3). \ldots (2n)\}$$
$$= \{1.3.5. \ldots (2n-1)\}2^n\{1.2.3. \ldots n\}$$
$$= n!\, 2^n\{1.3.5. \ldots (2n-1)\} = \textbf{R.H.S.} \quad \textbf{(Proved)}$$

(*ii*) Now, $${}^nP_r = \frac{n!}{(n-r)!} = \frac{n.(n-1)!}{\{(n-1)-(r-1)\}!} = n\,{}^{n-1}P_{r-1}$$

Also, $${}^nP_r = \frac{n!}{(n-r)!} = (n-r+1).\frac{n!}{(n-r+1).(n-r)!}$$

$$= (n-r+1).\frac{n!}{(n-r+1)!} = (n-r+1)^n P_{r-1}$$

$\therefore$ $${}^nP_r = n^{n-1}P_{r-1} = (n-r+1)\,{}^nP_{r-1} \quad \textbf{(Proved)}$$

(*iii*) $$\text{R.H.S.} = {}^{n-1}P_r + r\,{}^{n-1}P_{r-1} = \frac{(n-1)!}{(n-1-r)!} + r.\frac{(n-1)!}{(n-1-r+1)!}$$

$$= \frac{(n-1)!}{(n-r-1)!} + r.\frac{(n-1)!}{(n-r)!} = \frac{(n-1)!}{(n-r-1)!} + r.\frac{(n-1)!}{(n-r).(n-r-1)!}$$

$$= \frac{(n-1)!}{(n-r-1)!}\left\{1+\frac{r}{n-r}\right\} = \frac{(n-1)!}{(n-r-1)!}\cdot\frac{n}{n-r} = \frac{n!}{(n-r)!} = {}^nP_r = \textbf{L.H.S.} \quad \textbf{(Proved)}$$

Example 3: (*i*) *If* ${}^{4-x}P_2 = 6$, *find the value of x.*
(*ii*) *If* ${}^nP_5 : {}^nP_3 = 2:1$, *then find the value of n.*
(*iii*) *If* ${}^{2n+1}P_{n-1} : {}^{2n-1}P_n = 3:5$, *then find n.*
(*iv*) *If* ${}^{n+r+1}P_2 = 56$ *and* ${}^{n-r}P_2 = 20$, *find n and r.*

Solution:

(*i*) Given ${}^{4-x}P_2 = 6$ or, ${}^nP_2 = 6$ (where $n = 4-x$)

$\therefore$ $n(n-1) = 6$, or, $n^2 - n - 6 = 0$, or, $n^2 - 3n + 2n - 6 = 0$

$\therefore$ $(n-3)(n+2) = 0$. So, either $n - 3 = 0$, *i.e.*, $n = 3$

or $n + 2 = 0$, *i.e.*, $n = -2$.

Since $n = 4 - x$ is positive, rejecting the inadmissible value ($n = -2$), we get $n = 3$, or, $4 - x = 3$, or, $x = 1$.

(*ii*) Given, ${}^nP_5 : {}^nP_3 = 2:1$.

$\therefore$ ${}^nP_5 = 2\,{}^nP_3$ or, $n(n-1)(n-2)(n-3)(n-4) = 2n(n-1)(n-2)$

Obviously, $n = 0, 1, 2$ cannot be the solutions, so cancelling the common factor $n(n-1)(n-2)$ from both sides, we get

$$(n-3)(n-4) = 2, \text{ or } n^2 - 7n + 10 = 0, \text{ or } n^2 - 5n - 2n + 10 = 0,$$

or $(n-5)(n-2) = 0$. $\therefore$ $n = 5$, or 2.

Rejecting the inadmissible value $n = 2$, we get $n = 5$.

Note: ${}^nP_r$ is defined only when n, r are positive integers such that $r \le n$.

(*iii*) Given, ${}^{2n+1}P_{n-1} : {}^{2n-1}P_n = 3 : 5$, or $5\{{}^{2n+1}P_{n-1}\} = 3\{{}^{2n-1}P_n\}$

$$\therefore \quad 5 \cdot \frac{(2n+1)!}{(2n+1-n+1)!} = 3\frac{(2n-1)!}{(2n-1-n)!}, \quad \text{or} \quad 5\,\frac{(2n+1)!}{(n+2)!} = 3\,\frac{(2n-1)!}{(n-1)!}$$

$$\text{or} \quad \frac{5(2n+1)\,2n}{(n+2)\,(n+1)\,n}\cdot\frac{(2n-1)!}{(n-1)!} = 3\frac{(2n-1)!}{(n-1)!}, \text{ or } 10(2n+1) = 3\,(n+2)(n+1)$$

$$\text{or} \quad 20n + 10 = 3n^2 + 9n + 6, \text{ or } 3n^2 - 11n - 4 = 0$$

$$\text{or} \quad 3n^2 - 12n + n - 4 = 0, \text{ or } 3n(n-4) + (n-4) = 0, \text{ or } (n-4)(3n+1) = 0$$

So, either $n - 4 = 0$, *i.e.*, $n = 4$ or $3n + 1 = 0$, *i.e.*, $n = -1/3$.

Rejecting the inadmissible value ($n = -1/3$), we get $n = 4$.

(*iv*) Given, ${}^{n+r+1}P_2 = 56$, or $(n+r+1)(n+r) = 56$ $\quad [\because \; {}^nP_2 = n(n-1)]$

or $(x+1)x = 56$ (where $x = n + r$)

or $x^2 + x - 56 = 0$, or $x^2 + 8x - 7x - 56 = 0$, or $x\,(x+8) - 7\,(x+8) = 0$

or $(x+8)\,(x-7) = 0$.

$\therefore$ $x - 7 = 0$ $\quad [\because \; n, r$ are both positive integers, so $x + 8 = n + r + 8 \neq 0.]$

or $x = 7$, *i.e.*, $n + r = 7$ ...(1)

Also, it is given

$${}^{n-r}P_2 = 20, \text{ or } (n-r)\,(n-r-1) = 20,$$

or $y\,(y-1) = 20$ (where $y = n - r$)

or $y^2 - y - 20 = 0$, or $y^2 - 5y + 4y - 20 = 0$

or $y(y-5) + 4(y-5) = 0$, or $(y-5)\,(y+4) = 0$

Since, n, r are both positive integers with $r \leq n$, so

$y + 4 = n - r + 4 \neq 0$. Hence $y - 5 = 0$, or $n - r - 5 = 0$.

$\therefore$ $n - r = 5$ ...(2)

Solving (1) and (2), we get $n = 6$, $r = 1$.

Example 4: *How many numbers lying between 100 and 1000 can be formed with the digits 0, 3, 4, 6, 8, 9?*

Solution: Given digits are 0, 3, 4, 6, 8, 9, *i.e.*, six in all. We have to form numbers between 100 and 1000. Clearly they will be of three digits and their number will be

$${}^6P_3 = \frac{6!}{3!} = 6 \times 5 \times 4 = 120.$$

Out of these we have to exclude those numbers of 3 digits which have zero in the first place.

Their number is $\quad {}^5P_2 = \dfrac{5!}{(5-2)!} = 5 \times 4 = 20.$

$\therefore$ Required number of numbers = 120 – 20 = 100.

Example 5: *Without taking any digit more than once, how many odd numbers of 6 digits can be formed with the digits 0, 1, 4, 5, 6, 7?*

Solution: In the problem, there are two restrictions:

(*i*) the numbers to be formed are of 6 digits,

(*ii*) the numbers to be formed with the given digits must be odd.

To satisfy the restriction (*ii*), we must put an odd digit in the unit's place (*i.e.*, in the last place) and this can be made in 3 ways (viz. 1, 5, 7). If the unit's place is filled up by 1, then the remaining

5 places must be filled up by remaining 5 digits as per restriction (*ii*). Obviously, this can be done in 5! ways out of which 4! cases have zero in the first place and so we have to exclude these cases.

∴ Total number of numbers of 6 digits with 1 in the last place = 5 ! – 4 ! = 120 – 24 = 96.

Since the last place can be filled up in 3 ways and for each case the remaining 5 places can be filled up in 96 ways, so the required number of 6 digit odd numbers = 3 × 96 = 288.

Example 6: *Without taking any digit more than once, how many numbers divisible by 5 and lying between 3000 and 4000 can be formed from the digits 3, 4, 5, 6, 7, 8?*

Solution: Given: 3, 4, 5, 6, 7, 8.

Arrangements: 3 | any 2 out of remaining 4 | 5.

Interpretation: Since the numbers are to be divisible by 5, we must put 5 in the last place. Also, as they lie between 3000 and 4000, we must put 3 in the first place. Clearly they will be of 4 digits and the first and last digits have been fixed by 3 and 5 respectively. So, we have to select 2 digits out of the remaining 4.

$$\therefore \quad \text{Required number of numbers} = {}^4P_2 = \frac{4!}{(4-2)!} = 4 \times 3 = 12.$$

Example 7: *How many different numbers of six digits each (without repetition of digits) can be formed from the digits 2, 4, 5, 6, 7, 9? How many of these are not divisible by 5?*

Solution: Here we have to take all six digits 2, 4, 5, 6, 7, 9. So, number of 6 digit numbers = 6 ! = 720. Now 5! numbers are divisible by 5 as 5 is fixed in the last place and the remaining 5 places can be filled up by remaining 5 digits.

∴ Total number of numbers which are not divisible by 5 is = 6! – 5! = 720 – 120 = 600.

Example 8: *Find the total number of 9 digit numbers which have all different digits.*

Solution: Now, the number of 9 digit numbers which have all different digits taken from 0, 1, 2, ..., 9 is ${}^{10}P_9$. Out of which we have to exclude those numbers of 9 digits which have zero in the first place. Their number is ${}^9P_8$.

∴ Required number of numbers

$$= {}^{10}P_9 - {}^9P_8 = \frac{10!}{(10-9)!} - \frac{9!}{(9-8)!} = 10\,! - 9\,! = 9!\,(10-1)$$

$$= 9 \times (9 \times 8 \times 7 \times 6\,!) = 81 \times 56 \times 720 = 3265920.$$

Example 9: *How many numbers greater than 24000 can be formed from the digits 1, 2, 3, 4, 5?*

Solution: Given digits are 1, 2, 3, 4, 5 and we have to form numbers greater than 24000.

Required number of numbers

= Total – (those beginning with 1) – (those beginning with 21) – (those beginning with 23)

= 5 ! – 4 ! – 3 ! – 3 ! = 120 – 24 – 6 – 6 = 84.

Example 10: *How many numbers can be formed by using all the digits 1, 2, 3, 4, 3, 2, 1 so that the odd digits always occupy the odd places.*

Solution: There are 4 odd places and there are 4 odd integers. Observe that

1, 1 → alike, 2, 2 → alike, 3, 3 → alike.

$$\text{So, odd integers can be arranged in odd places in } \frac{4!}{2!2!} = 6 \text{ ways.}$$

There are 3 even places namely 2nd, 4th and 6th in which 3 even numbers (of which 2, 2 are alike) can be arranged in

$$\frac{3!}{2!} = 3 \text{ ways.}$$

Hence, by principle of counting, the required number of numbers is $6 \times 3 = 18$.

Example 11: *A tea party is arranged for 16 people along two sides of a long table with 8 chairs on each side. Four men wish to sit on one particular side and two on the other side. In how many ways can they be seated?*

Solution: The 4 persons who wish to occupy side 1 can be seated on 8 chairs in

$$^8P_4 = \frac{8!}{(8-4)!} = 8 \times 7 \times 6 \times 5 = 1680 \text{ ways.}$$

The 2 persons who wish to occupy side 2 can be seated on 8 chairs in

$$^8P_2 = \frac{8!}{(8-2)!} = 8 \times 7 = 56 \text{ ways.}$$

Having seated 6 persons on 6 chairs, remaining 10 persons can be seated on remaining 10 chairs in 10 ! ways.

Using principle of counting, the required number of arrangements is $= 10! \times 56 \times 1680$.

Example 12: (*i*) *In a 14 storey building 3 persons enter a lift cabin. It is known that they will leave the lift at different storeys. In how many ways can they do so if the lift does not stop at the second storey?*

(*ii*) *Find the number of positive integers which can be formed by any number of digits from 0, 1, 2, 3, 4, 5, by using each digit not more than once in each number. How many of these integers are greater than 3000?*

Solution: (*i*) Each of the 3 persons can leave the lift at one of the twelve floors (excluding second storey and one at which they enter the lift). Since they leave the lift at different storeys, the required number of ways is

$$^{12}P_3 = 12\,(12-1)\,(12-2) = 12 \times 11 \times 10 = 1320.$$

(*ii*) Now, $^6P_6$ numbers of positive integers can be formed by using given six digits taking all at a time such that each digit does not occur more than once in each number. Rejecting those which have 0 in the left most place (*i.e.*, $^5P_5$ numbers), we get

number of 6 digit positive integers $= {^6P_6} - {^5P_5} = 6! - 5!$

$$= (6-1)\,5! = 5 \times 5 \times 4 \times 3 \times 2 = 600.$$

Similarly, no. of 5 digit positive integers

$$= {^6P_5} - {^5P_4} = \frac{6!}{(6-5)!} - \frac{5!}{(5-4)!} = 6! - 5! = 600,$$

number of 4 digit positive integers $= {^6P_4} - {^5P_3}$

$$= \frac{6!}{(6-4)!} - \frac{5!}{(5-3)!} = \frac{1}{2}(6! - 5!) = 300,$$

number of 3 digit positive integers $= {^6P_3} - {^5P_2}$

$$= \frac{6!}{(6-3)!} - \frac{5!}{(5-2)!} = \frac{1}{3!}(6! - 5!) = \frac{1}{6} \times 600 = 100,$$

number of 2 digit positive integers = ${}^6P_2 - {}^5P_1$

$$= \frac{6!}{(6-2)!} - 5 = \frac{6!}{4!} - 5 = 6 \times 5 - 5 = 25,$$

number of 1 digit positive integers = ${}^6P_1 - 1 = 6 - 1 = 5.$

$\therefore$ Required number of positive integers formed

$$= 600 + 600 + 300 + 100 + 25 + 5 = 1630.$$

For positive integers greater than 3000, we consider all 6, 5 digits positive integers and all those integers of 4 digit positive integers which have 5, 4 or 3 in the first place.

No. of such 4 digit positive integers = ${}^5P_3 + {}^5P_3 + {}^5P_3$

$$= 3 \times \frac{5!}{(5-3)!} = 3 \times 5 \times 4 \times 3 = 180.$$

$\therefore$ Required number of positive integers greater than 3000

$$= 600 + 600 + 180 = 1380.$$

Note: Where to multiply and where to add

When the occurrence of event B depends on the occurrence of event A then we multiply the corresponding number of ways by principle of counting to get the total number of ways. But when the events are independent we just add the corresponding number of ways.

Example 13: *How many words of 6 letters can be formed by using all the letters of the word MONDAY which begins with M but does not end with Y?*

Solution: Having fixed up M at the first place we are left with the permutations of remaining 5 distinct letters which can be arranged in ${}^5P_5 = 5\,!$ ways.

Having fixed up M at the first place and Y at the end we are left with the permutations of remaining 4 distinct letters which can be arranged in ${}^4P_4 = 4\,!$ ways.

$\therefore$ Required number of words is $5\,! - 4\,!$

$$= (5-1)\,4\,! = 4 \times 4 \times 3 \times 2 = 96.$$

Example 14: *Ten different letters of an alphabet are given. Words with five letters are formed from these given letters. Then find the number of words which have at least one letter repeated.*

Solution: Number of words in which all the 5 letters are repeated = 10^5 (by Art. 4.3.6).

Again, the number of words in which no letter is repeated = ${}^{10}P_5$.

Therefore, the number of words which have at least one letter repeated = $10^5 - {}^{10}P_5$

$$= 100000 - 10 \times 9 \times 8 \times 7 \times 6 = 100000 - 30240 = 69760.$$

Example 15: *How many different words can be formed from the letters of the word BILASHPUR when*

(i) all the letters are taken

(ii) the letter B always occupies the first place

(iii) the letters A and R occupy the first and last places respectively

(iv) the vowels are always together

(v) the letters L, H, P are never together

(vi) the vowels always occupy even places (i.e., 2nd, 4th etc.)

(vii) how many words of 5 letters each can be formed each containing 2 vowels and 3 consonants?

Solution: BILASHPUR : No. of letters = 9 ; vowels are AIU, *i.e.*, 3 and consonants 6.

(*i*) When all the letters are taken at a time, then the number of different words formed is $^9P_9 = 9\,!$

(*ii*) Having fixed up *B* at the first place we are left with the permutations of remaining 8 distinct letters which can be arranged in 8 ! ways.

∴ Required number of different words = 8 !

(*iii*) *A* occupies the first place and *R* occupies the last place and so we have to arrange the remaining 7 letters which can be done in 7 ! ways.

∴ Required number of different words = 7 !

(*iv*) The three vowels *AIU* are to be together. Take these three letters as one letter and so we have in all 9 – 3 + 1 = 7 letters which can be arranged in 7 ! ways. In each of these 7 ! arrangements the three vowels are together and these 3 vowels can be arranged amongst themselves in 3 ! ways. Hence by using principle of counting, the number of different words will be 7 ! 3 ! in which the vowels will be together.

(*v*) Take *L, H, P* as one letter and so the number of letters will be 9 – 3 + 1 = 7. As in part (*iv*), the number of different words in which *L, H, P* are together will be 7 ! 3 !. The total number of different words, subject to no restriction, is 9 !, by (*i*). Hence the number of different words when *L, H, P* are never together is 9 ! – 7 ! 3 ! = 7 ! (9 × 8 – 3 × 2) = 7 ! × 66.

Note: Some students may try to solve this question using gap method. Arrange the six letters (excluding *L, H, P*) in 6 ! ways. There are 7 gaps in which these three can be put in

$$^7P_3 = \frac{7\,!}{(7-3)\,!} = 7 \times 6 \times 5 = 210 \text{ ways.}$$

Hence the required number of words is 6 ! × 210 = 7 ! × 30. This is less than the number of ways calculated in (*v*).

Reason: The condition of question (*v*) is that the three letters *L, H, P* all are never together. But any two of them could be together, or all of them could be separated.

(*vi*) We have nine places out of which 5 places are odd, *i.e.*, 1st, 3rd, 5th, 7th, 9th and four are even, *i.e.*, 2nd, 4th, 6th, 8th. In the four even places we have to put up 3 vowels which can be done in $^4P_3$ ways. We will be left with 6 places (5 odd and 1 even) after fixing the three vowels. In these 6 places we have to put 6 consonants which can be done in $^6P_6$ ways. By using principle of counting, the required number of different words is $^4P_3 \times {}^6P_6 = 4\,!\,6\,! = 24 \times (6 \times 5 \times 24) = 24 \times 720$.

(*vii*) Here each word is to contain 3 consonants out of 6 and 2 vowels out of 3. We can **select** them in $^6C_3$ and $^3C_2$ ways respectively (see Art. 4.3.10). Thus the total number of combinations (groups) of 5 letters having 2 vowels and 3 consonants is $^6C_3 \times {}^3C_2$, by the principle of counting. Now the 5 letters in each group can be arranged amongst themselves in $^5P_5 = 5\,! = 120$ ways. So, using the principle of counting, the required number of different words is

$$= {}^6C_3 \times {}^3C_2 \times 120 = \frac{6\,!}{3\,!(6-3)\,!} \times \frac{3\,!}{2\,!(3-2)\,!} \times 120$$

$$= \frac{1}{6} \times 6 \times 5 \times 4 \times 3 \times 120 = 7200$$

Note: From (*i*), (*iv*) and (*v*), we conclude that the number of permutations of *n* different objects taken all at a time in which *p* particular objects are never together is $n\,! - (n - p + 1)\,!\,p\,!$.

Example 16: *Find how many different words can be formed from the letters of the word PEOPLE taken all together in which two P's would not remain side by side.*

Solution: Since there are two *P*'s, two *E*'s, the number of words that can be formed from the letters of the given word taken all together, subject to no restriction, is (by Art. 4.3.5).

$$= \frac{6!}{2!2!} = \frac{1}{2} \times 6 \times 5 \times 4 \times 3 = 180.$$

Now, the number of words in which two *P*'s remain side by side is

$$= \frac{5!}{2!} = 5 \times 4 \times 3 = 60$$

$\therefore$ Required number of words in which two *P*'s would not remain side by side is

$$= 180 - 60 = 120.$$

Example 17: *How many different words can be formed from the letters of the word INTERMEDIATE? In how many of them, two vowels never come together?*

Solution: The different letters and their numbers in the word INTERMEDIATE are as follows:

Letter :	*E*	*I*	*T*	*A*	*D*	*M*	*N*	*R*	*Total*
Number :	3	2	2	1	1	1	1	1	12

No. of letters = 12

No. of vowels = 6 (3 *E*, 2 *I*, 1 *A*)

No. of consonants = 6 (2 *T*, 1 *D*, 1 *M*, 1 *N*, 1 *R*)

The numbers of words that can be formed from the letters of the given word taken all together, subject to no restriction, is (by Art. 4.3.5)

$$= \frac{12!}{3!2!2!}$$

Gap Method

No. of ways of arranging 6 consonants (2 alike) is

$$\frac{6!}{2!} = \frac{1}{2} \times 6 \times 5 \times 4 \times 3 \times 2 = 360 \quad ...(1)$$

There will be 7 gaps in which 6 vowels (if all different) can be arranged in ${}^7P_6 = 7!$ ways. But 3 are alike of one kind and 2 of other kind.

$\therefore$ Number of arranging the vowels is

$$= \frac{7!}{3!2!} = \frac{1}{2} \times 7 \times 6 \times 5 \times 4 = 420 \quad ...(2)$$

By the principle of counting, the total number of words when the two vowels never come together is $360 \times 420 = 151200$.

Example 18: (*i*) *How many different words can be formed by taking all the letters in the word MATHEMATICS? How many of them begin with S? How many of them begin with M?*

(*ii*) *Find the number of different words each containing five letters such that the letters appearing in the odd positions are taken from the unrepeated letters of the word MATHEMATICS whereas the letters which occupy even places are taken from the repeated letters.*

Solution: (*i*) The different letters and their numbers in the word MATHEMATICS are as follows:

Letter :	*M*	*A*	*T*	*H*	*E*	*I*	*C*	*S*	Total
Number :	2	2	2	1	1	1	1	1	11

The number of words that can be formed from the letters of the given word taken all together, subject to no restriction, is (by Art. 4.3.5)

$$= \frac{11!}{2!2!2!} = \frac{11 \times 10 \times 9 \times 8}{2 \times 2 \times 2}(7!) = 990\ (7!)$$

To begin with S

Having fixed *S* at the first place, we have 10 letters in which 2 are *M*, 2 are *A*, 2 are *T* and rest 4 different. So, the number of different words starting with *S* is

$$= \frac{10!}{2!2!2!}$$

To begin with M

Having placed *M* at the first place, we have 10 letters out of which 2 are *A*, 2 are *T* and rest six are *M, H, E, I, C* and *S*. Hence the number of different words starting with *M* is

$$= \frac{10!}{2!2!}$$

(*ii*) There are 3 odd places namely 1st, 3rd and 5th which are to be fixed by five unrepeated letters *H, E, I, C, S*. This can be made in ${}^5P_3 = 5 \times 4 \times 3 = 60$ ways.

There are 2 even places namely 2nd and 4th which are to be fixed by three repeated letters (2*M*, 2*A*, 2*T*), *i.e.*, 6 letters. These two even places can be fixed by 3 different types of letters as follows:

(*a*) **All different**

No. of ways is ${}^3P_2 = 3 \times 2 = 6$.

(*b*) **Both alike**

No. of ways is 3.

Thus even places can be fixed in 6 + 3 = 9 ways.

Hence by principle of counting, the required number of words is 60 × 9 = 540.

Example 19: *In how many paths can one move from (0, 0) to (a, b) by taking unit steps in positive X, Y directions only?*

Solution: To move from (0, 0) to (*a*, *b*) one has to move *a* steps of 1 unit each along *X*-axis and *b* steps of 1 unit each along *Y*-axis. Each such movement can be represented as a permutation *xyyxxyxy* ... *xxy* where there are *a* number of '*x*'s and *b* number of '*y*'s. Each *x* denotes a 1-unit movement along the *X*-axis and each *y* denotes a 1-unit movement along the *Y*-axis. Therefore, the number of distinct paths from (0, 0) to (*a*, *b*) is equal to the number of permutations of (*a* + *b*) things, '*a*' of which are alike of one kind (*viz*., they are *x*) and '*b*' of which are alike of another kind (viz., they are *y*).

$$\therefore \quad \text{Required number of paths} = \frac{(a+b)!}{a!b!}.$$

Notes: (*i*) The number of distinct paths from (0, 0, 0) to (*a*, *b*, *c*) is $\frac{(a+b+c)!}{a!b!c!}$, by taking unit steps in positive *x*, *y*, *z* directions only.

(*ii*) These type of walks is called **Lattice-Walking** where one has to travel from A to B using the grid of streets and without 'doubling back' as shown in the following figure:

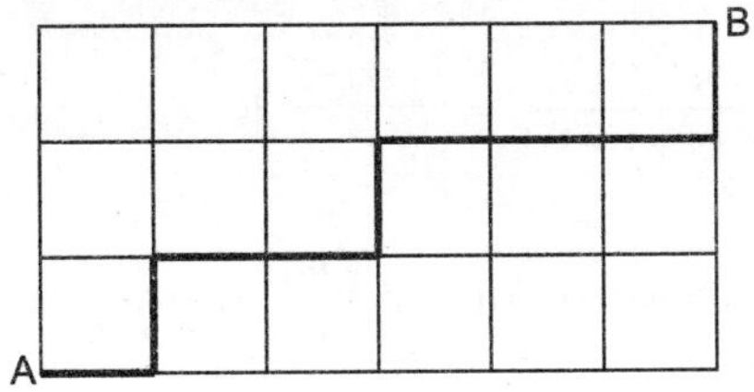

4.3.10 Combinations

Definition. The different **groups** or **collections** or **selections** that can be formed out of a given set of objects by taking some or all of them at a time (irrespective of order) are called their **combinations.**

For example, the combinations of three letters a, b, c taken two at a time are ab, bc, ca.

The number of combinations of n different objects taken r at a time is usually denoted by the symbol ${}^nC_r$ or $C(n, r)$.

4.3.11 Combinations of Objects all Different

To find the number of combinations of n different objects taken r (≤ n) at a time.

Let x denotes the number of combinations of n different objects taken r ($\leq n$) at a time. If the r different objects in each combination be arranged among themselves in all possible ways, then each combination will produce ${}^rP_r = r\,!$ permutations (arrangements). So, x combinations will produce $x \times r\,!$ permutations. When all the objects in each combination are arranged among themselves in all possible ways, all the possible permutations of n different objects taken r at a time are obtained.

$$\therefore \qquad x \times r\,! = {}^nP_r = \frac{n\,!}{(n-r)\,!}, \quad \text{or} \quad x = {}^nC_r = \frac{n\,!}{r\,!\,(n-r)\,!}$$

Notes: (*i*) ${}^nP_r = r\,! \times {}^nC_r$

(*ii*) ${}^nC_0 = {}^nC_n = 1$ (In reality, ${}^nC_0$ has no meaning)

$${}^nC_1 = {}^nC_{n-1} = n,\ {}^nC_2 = {}^nC_{n-2} = \frac{1}{2}\,n(n-1).$$

(*iii*) $${}^nC_{n-r} = \frac{n\,!}{(n-r)\,!\,\{n-(n-r)\}\,!} = \frac{n\,!}{r\,!\,(n-r)\,!} = {}^nC_r.$$

So, **the number of combinations of n different objects taken r at a time is equal to the number of combinations of n different objects taken (n – r) at a time.**

(*iv*) ${}^nC_r + {}^nC_{r-1} = {}^{n+1}C_r$.

$$\text{L.H.S.} = \frac{n\,!}{r\,!\,(n-r)\,!} + \frac{n\,!}{(r-1)\,!\,(n-r+1)\,!} = \frac{n\,!}{(r-1)\,!\,(n-r)\,!}\left\{\frac{1}{r} + \frac{1}{n-r+1}\right\}$$

$$= \frac{n\,!}{(r-1)\,!\,(n-r)\,!} \cdot \frac{n+1}{r(n-r+1)} = \frac{(n+1)\,!}{r\,!\,(n-r+1)\,!} = {}^{n+1}C_r = \text{R.H.S.}$$

This is known as **Pascal's identity.**

(*v*) $^nC_r = \dfrac{n!}{r!(n-r)!} = \dfrac{n-r+1}{r} \cdot \dfrac{n!}{(r-1)!(n-r)!(n-r+1)}$ $[\because \; r! = r.(r-1)!]$

$$= \frac{n-r+1}{r} \cdot \frac{n!}{(r-1)!(n-r+1)!} = \frac{n-r+1}{r} \, {}^nC_{r-1}$$

(*vi*) The number ^nC_r $\left[\text{also denoted as } C(n, r) \text{ or } \binom{n}{r}\right]$ is called a **binomial coefficient.** The name **binomial coefficient** is used since these numbers occur as coefficients in the expansion of powers of binomial expressions such as $(a + b)^n$.

(*vii*) **Pascal's triangle**

The binomial coefficients in the expansion of $(a + x)^n$ are $^nC_0, {}^nC_1, {}^nC_2, ..., {}^nC_r, ..., {}^nC_n$ which can be represented in the following pattern known as **Pascal's triangle:**

n = 0 1

n = 1 1 1

n = 2 1 2 1

n = 3 1 3 3 1

n = 4 1 4 6 4 1

n = 5 1 5 10 10 5 1

n = 6 1 6 15 20 15 6 1

n = 7 1 7 21 35 35 21 7 1

n = 8 1 8 28 56 70 56 28 8 1

1. Each row starts with 1 and ends in 1.
2. Each coefficient is given by sum of two coefficients, one just before it and other just after it, in the preceding row [using Pascal's identity given in (*iv*)].
3. Pascal's identity is the basis for geometric arrangement of binomial coefficients in the Pascal's triangle.

Illustration: The sum of the coefficients in the expansion of $(1 + x)^5$ is

(*a*) 16 (*b*) 32 (*c*) 64 (*d*) 128

Solution: Using Pascal's triangle for $n = 5$, the sum is $= 1 + 5 + 10 + 10 + 5 + 1 = 32$.

Ans. (*b*).

4.3.12 Combinations of Objects not all Different

To find the total number of combinations of (p + q + r + ...) objects of which p are alike of one kind, q are alike of second kind, r are alike of third kind and so on taken any number at a time.

The p like objects of the first kind can be selected in $(p + 1)$ ways, because we can take either 1, 2, 3, .., or p of them at a time or none at all. Similarly, q like objects of the second kind can be selected in $(q + 1)$ ways and so on. Now, for each way of selection of objects of the first kind, the second kind of objects can be selected in $(q + 1)$ ways and so, by principle of counting, the total number of combinations of the first and second kinds of objects is $(p + 1)(q + 1)$. Similarly, the total number of combinations of the first, second and third kinds of objects is $(p + 1)(q + 1)(r + 1)$ and so on. Therefore, total number of combinations is $\{(p + 1)(q + 1)(r + 1) \ldots\}$, which includes the case in which all the objects are left out. Rejecting this case, the total number of combinations $= \{(p + 1)(q + 1)(r + 1) \ldots\} - 1$.

Note: Let the total number of objects be $\{m + p + q + r + \ldots\}$ of which m objects are different from each other and from the rest and p are alike of one kind, q are alike of second kind, r are alike of third kind and so on. Now, each of m different objects can be selected in 2 ways, hence by the previous result and the principle of counting, the total number of combinations is

$$= \{2^m (p + 1)(q + 1)(r + 1) \ldots\} - 1.$$

4.3.13 Division into Groups

(*i*) To find the number of ways in which (m + n) different objects can be divided into two groups containing m and n objects respectively.

The number of ways in which m objects can be selected out of $(m + n)$ objects is ${}^{m+n}C_m$ and each time that m objects are taken, n objects are left out to form the other group. So, the required number of ways is

$$= {}^{m+n}C_m = \frac{(m+n)!}{m!(m+n-m)!} = \frac{(m+n)!}{m!n!} = {}^{m+n}C_n.$$

Note: If $m = n$, i.e., the two groups consist of equal number of objects, then no new subdivision is obtained by interchanging these two groups. So, in this case the number of different ways of subdivision is

$$= \frac{1}{2!}\,{}^{2m}C_m = \frac{(2m)!}{2!(m!)^2}$$

However, if $2m$ different objects are to be distributed equally among two persons, the number of ways will be $= {}^{2m}C_m = \dfrac{(2m)!}{(m!)^2}$.

(*ii*) To find the number of ways in which (m + n + p) different objects can be divided into three groups containing m, n and p objects respectively.

The m objects can be selected out of $(m + n + p)$ objects in ${}^{m+n+p}C_m$ ways; then n objects out of the remaining $(n + p)$ objects can be taken in ${}^{n+p}C_n$ ways and lastly the p objects are left out to form the third group. So, by principle of counting, the required number of ways is $= {}^{m+n+p}C_m \times {}^{n+p}C_n$

$$= \frac{(m+n+p)!}{m!(n+p)!}\cdot\frac{(n+p)!}{n!p!} = \frac{(m+n+p)!}{m!n!p!}.$$

Notes: (*i*) If $m = n = p$, *i.e.*, the three groups consist of equal number of objects, then no new subdivision is obtained by interchanging the three groups (which can be made in 3! ways). So, in this case the number of different ways of subdivision is

$$= \frac{1}{3!}\,{}^{3m}C_m \times {}^{2m}C_m = \frac{1}{3!}\,\frac{(3m)!}{m!(2m)!}\cdot\frac{(2m)!}{m!m!} = \frac{(3m)!}{3!(m!)^3}$$

However, if $3m$ different objects are to be distributed equally among three persons, the number of ways will be $= {}^{3m}C_m \times {}^{2m}C_m = \dfrac{(3m)!}{(m!)^3}$.

(*ii*) The previous results and proof may easily be generalized for any number of groups.

4.3.14 Restricted Combinations and Permutations

1. Number of combinations of n different objects taken r at a time when p particular objects always occur is $= {}^{n-p}C_{r-p}$

2. Number of combinations of n different objects taken r at a time when p particular objects never occur is $= {}^{n-p}C_r$.

3. Number of permutations of n different objects taken r at a time when p particular objects always occur is

$$= {}^{n-p}C_{r-p}\, r\,!.$$

Note: Compare this with 2 of Art. 4.3.7.

4. Number of permutations of n different objects taken r at a time when p particular objects never occur is

$$= {}^{n-p}C_r\, r\,!$$

Note: Compare this with 1 of Art. 4.3.7.

5. The total number of combinations of n different objects taken any number at a time is

$${}^nC_1 + {}^nC_2 + {}^nC_3 + \dots + {}^nC_n = 2^n - 1.$$

Proof. We know that

$$(1 + x)^n = {}^nC_0 + {}^nC_1 x + {}^nC_2 x^2 + {}^nC_3 x^3 + \dots + {}^nC_n x^n$$

This is an identity and it is true for all values of x.

Putting $x = 1$, we get

$$(1 + 1)^n = {}^nC_0 + {}^nC_1 + {}^nC_2 + {}^nC_3 + \dots + {}^nC_n$$

$$\therefore \quad {}^nC_1 + {}^nC_2 + {}^nC_3 + \dots + {}^nC_n = 2^n - 1 \ (\because \ {}^nC_0 = 1).$$

Notes: (*i*) **Where to use permutation and where to use combination**

The word arrangement is associated with permutation and selection (or collection or group) is associated with combination.

(*ii*) **Permutation via combination**

$${}^nP_r = {}^nC_r\, r\,!$$

(*iii*) Problems on letters of alphabet forming words will be of permutatión. Similarly, problems on numbers involving digits will be of permutation.

(*iv*) Problems on formation of committees and distribution into groups will be of combination.

4.3.15 Greatest Value of ${}^nC_r$

To find for what value of r the number of combinations of n different objects taken r at a time is greatest.

We have,

$${}^nC_r = \frac{n!}{r!(n-r)!} = \frac{n-r+1}{r} \cdot \frac{n!}{(r-1)!(n-r)!(n-r+1)} = \frac{n-r+1}{r}\, {}^nC_{r-1}$$

$$\therefore \quad \frac{{}^nC_r}{{}^nC_{r-1}} = \frac{n-r+1}{r}$$

$$\therefore \quad {}^nC_r > {}^nC_{r-1} \text{ if } \frac{n-r+1}{r} > 1, \textit{ i.e.}, \text{ if } n - r + 1 > r, \textit{ i.e.}, \text{ if } n + 1 > 2r.$$

So, $\quad {}^nC_r > {}^nC_{r-1}$ if $r < \dfrac{n+1}{2}$ ...(1)

Similarly, $\quad {}^nC_r = {}^nC_{r-1}$ if $r = \dfrac{n+1}{2}$ ...(2)

and $\quad {}^nC_r < {}^nC_{r-1}$ if $r > \dfrac{n+1}{2}$ ...(3)

Case 1. Assume *n* be an even positive integer

Let $n = 2m$, where m is a positive integer. From (1), we get

$$ {}^nC_r > {}^nC_{r-1} \text{ if } r < \frac{2m+1}{2}, \text{ i.e., if } r < m + \frac{1}{2}$$

So, $\quad {}^nC_r > {}^nC_{r-1}$ when $r = 1, 2, 3, ..., m-1, m.$

$\therefore \quad {}^nC_1 > {}^nC_0, {}^nC_2 > {}^nC_1, ..., {}^nC_m > {}^nC_{m-1}$

Combining: $\quad {}^nC_0 < {}^nC_1 < {}^nC_2 < ... < {}^nC_{m-1} < {}^nC_m$ ...(4)

Also, from (3), we get

$$ {}^nC_r < {}^nC_{r-1} \text{ if } r > \frac{2m+1}{2}, \text{ i.e., if } r > m + \frac{1}{2}.$$

So, $\quad {}^nC_r < {}^nC_{r-1}$ when $r = m+1, m+2, m+3, ..., 2m-1, 2m$

$\therefore \quad {}^nC_{m+1} < {}^nC_m, {}^nC_{m+2} < {}^nC_{m+1}, ..., {}^nC_{2m} < {}^nC_{2m-1}$

Combining: $\quad {}^nC_m > {}^nC_{m+1} > {}^nC_{m+2} > ... > {}^nC_{2m-1} > {}^nC_{2m}$ ...(5)

From (4) and (5), we get

$$ {}^nC_0 < {}^nC_1 < {}^nC_2 < ... < {}^nC_m > {}^nC_{m+1} > {}^nC_{m+2} > ... > {}^nC_{2m}.$$

$\therefore$ We conclude that if $n = 2m$ then the greatest value of ${}^nC_r$ is ${}^nC_m$, *i.e.*, ${}^nC_r$ is greatest when $r = m = \dfrac{n}{2}$.

Hence, if n is an even positive integer then the greatest value of ${}^nC_r$ is

$$ {}^nC_{\frac{n}{2}} = \frac{n!}{\left(\frac{n}{2}\right)!\left(n-\frac{n}{2}\right)!} = \frac{n!}{\left\{\left(\frac{n}{2}\right)!\right\}^2}$$

Case 2. Assume n be an odd positive integer

Let $n = 2m + 1$, where m is a positive integer. From (1), we get

$$ {}^nC_r > {}^nC_{r-1} \text{ if } r < \frac{2m+1+1}{2}, \text{ i.e., if } r < m + 1$$

So, $\quad {}^nC_r > {}^nC_{r-1}$ when $r = 1, 2, 3, ..., m$

$\therefore \quad {}^nC_1 > {}^nC_0, {}^nC_2 > {}^nC_1, ..., {}^nC_m > {}^nC_{m-1}$

Combining: $\quad {}^nC_0 < {}^nC_1 < {}^nC_2 < ... < {}^nC_{m-1} < {}^nC_m$ ...(6)

From (2), we get $\quad {}^nC_r = {}^nC_{r-1}$ if $r = m + 1$, *i.e.*, ${}^nC_{m+1} = {}^nC_m$ ...(7)

Also, from (3), we get ${}^nC_r < {}^nC_{r-1}$ if $r > m + 1$.

So, $\quad {}^nC_r < {}^nC_{r-1}$ when $r = m+2, m+3, ..., 2m+1.$

$\therefore \quad {}^nC_{m+2} < {}^nC_{m+1}, {}^nC_{m+3} < {}^nC_{m+2}, ..., {}^nC_{2m+1} < {}^nC_{2m}.$

Combining : $\quad {}^nC_{m+1} > {}^nC_{m+2} > {}^nC_{m+3} > ... > {}^nC_{2m} > {}^nC_{2m+1}$ ...(8)

From (6) – (8), we get

$$^nC_0 < {}^nC_1 < {}^nC_2 < ... < {}^nC_m = {}^nC_{m+1} > {}^nC_{m+2} > ... > {}^nC_{2m+1}$$

∴ We conclude that if $n = 2m + 1$ then the greatest value of ^nC_r is $^nC_m = {}^nC_{m+1}$ where $n = 2m + 1$, *i.e.*, $m = \dfrac{n-1}{2}$.

Hence, if n is an odd positive integer then the greatest value of ^nC_r is

$$^nC_{\frac{n-1}{2}} = \frac{n!}{\left(\frac{n-1}{2}\right)!\left(\frac{n+1}{2}\right)!} = {}^nC_{\frac{n+1}{2}}$$

4.3.16 Summary

1. The number of combinations of n different objects taken r at a time $= {}^nC_r = \dfrac{n!}{r!(n-r)!}$.

2. $^nC_0 = {}^nC_n = 1, {}^nC_1 = {}^nC_{n-1} = n, {}^nC_2 = {}^nC_{n-2} = \frac{1}{2}n(n-1),$

$$^nC_3 = {}^nC_{n-3} = \frac{n(n-1)(n-2)}{3!}, {}^nC_r = {}^nC_{n-r}.$$

3. $^nC_r + {}^nC_{r-1} = {}^{n+1}C_r, {}^nC_r = \dfrac{n-r+1}{r}\, {}^nC_{r-1}, {}^nP_r = r!\, {}^nC_r$

4. Total number of combinations of $(p + q + r + ...)$ objects of which p are alike of one kind, q are alike of second kind, r are alike of third kind and so on taken any number at a time is $= \{(p + 1)(q + 1)(r + 1) ...\} - 1$.

5. Total number of combinations of $(m + p + q + r + ...)$ objects of which m are different from each other and from the rest and p are alike of one kind, q are alike of second kind, r are alike of third kind and so on taken any number at a time is $= \{2^m (p + 1)(q + 1)(r + 1) ...\} - 1$.

6. Number of ways in which $(m + n)$ different objects can be divided into two groups containing m and n objects respectively is

$$= {}^{m+n}C_m = \frac{(m+n)!}{m!n!} = {}^{m+n}C_n.$$

7. Number of ways in which $(m + n + p)$ different objects can be divided into three groups containing m, n and p objects respectively is

$$= {}^{m+n+p}C_m \times {}^{n+p}C_n = \frac{(m+n+p)!}{m!n!p!}$$

8. Also see Art. 4.3.14.

9. (*i*) If n is an even positive integer, then ^nC_r is greatest when $r = \dfrac{n}{2}$.

   (*ii*) If n is an odd positive integer, then ^nC_r is greatest when $r = \dfrac{n-1}{2}$, or $r = \dfrac{n+1}{2}$.

10. $^nC_{r_1} = {}^nC_{r_2} \Rightarrow r_1 = r_2$ or $r_1 + r_2 = n$.

ILLUSTRATIVE EXAMPLES—II

Example 1: *(i) If* $^nC_9 = {}^nC_{15}$, *find* $^{26}C_n$.

(ii) If $^nC_2 = 15$, *find n.*

(iii) If $^nP_r = {}^nP_{r+1}$ *and* $^nC_r = {}^nC_{r-1}$, *find n and r.*

(iv) If $^{10}P_r = 604800$ *and* $^{10}C_r = 120$, *find r.*

(v) If $^nC_6 : {}^{n-3}C_3 = 33 : 4$, *find n.*

Solution: (*i*) We know that $^nC_r = {}^nC_{n-r}$.

$\therefore \quad {}^nC_9 = {}^nC_{n-9}$. Given, $^nC_9 = {}^nC_{15}$.

$\therefore \quad {}^nC_{15} = {}^nC_{n-9}$. So, $15 = n - 9$, *i.e.*, $n = 15 + 9 = 24$.

$$\therefore \quad {}^{26}C_n = {}^{26}C_{24} = {}^{26}C_{26-24} = {}^{26}C_2 = \frac{1}{2} \times 26 \times (26 - 1) = 13 \times 25$$

$$= 325.$$

(*ii*) Given, $\quad {}^nC_2 = 15. \quad \therefore \quad \frac{1}{2}n(n-1) = 15, \quad$ or $\quad n^2 - n = 30,$

$\therefore \quad n^2 - n - 30 = 0, \quad$ or $\quad n^2 - 6n + 5n - 30 = 0, \quad$ or $\quad n(n-6) + 5(n-6) = 0$

$\therefore \quad (n-6)(n+5) = 0, \quad$ or $\quad n - 6 = 0 \qquad$ ($\because n + 5 \neq 0$, as n is a positive integer.)

$\therefore \quad n = 6.$

(*iii*) Given, $\quad {}^nP_r = {}^nP_{r+1} \quad \therefore \quad \frac{n!}{(n-r)!} = \frac{n!}{(n-r-1)!}, \quad$ or $\quad \frac{1}{n-r} = 1.$

$\therefore \quad n - r = 1 \qquad$ ...(1)

Also, given $\quad {}^nC_r = {}^nC_{r-1} \quad \therefore \quad r + r - 1 = n$

$\qquad [\because \; {}^nC_{r_1} = {}^nC_{r_2} \;\Rightarrow\; r_1 = r_2, \;$ or $\; r_1 + r_2 = n]$

or $\quad 2r - n = 1 \qquad$ ...(2)

Solving (1) and (2), we get $r = 2$ and $n = 3$.

(*iv*) Given, $\quad {}^{10}P_r = 604800$ and $^{10}C_r = 120 \qquad$ ...(1)

We know that $\quad {}^nP_r = r!\, {}^nC_r$.

$\therefore \quad {}^{10}P_r = r!\, {}^{10}C_r \quad$ or $\quad 604800 = r!\, 120$ [by (1)]

$\therefore \quad r! = 5040 = 24 \times 210 = 4! \times 5 \times 6 \times 7 = 7!$

$\therefore \quad r = 7.$

(*v*) Given, $\quad \frac{{}^nC_6}{{}^{n-3}C_3} = \frac{33}{4}, \quad$ or $\quad \frac{n!}{6!(n-6)!} \cdot \frac{3!(n-6)!}{(n-3)!} = \frac{33}{4}$

or $\quad \frac{n(n-1)(n-2)}{6 \times 5 \times 4} = \frac{33}{4}, \quad$ or $\quad n(n-1)(n-2) = 33 \times 30.$

or $\quad n(n-1)(n-2) = 9 \times 10 \times 11.$

$\therefore \quad n(n-1)(n-2) = 11 \times (11-1) \times (11-2)$

Hence, $\quad n = 11.$

Example 2: *How many committees of 6 members each can be formed from 8 official and 5 non-official members in the following cases:*

(*i*) *Each containing of 4 official and 2 non-official members.*

(*ii*) *Each consists of at least two non-official members.*

(*iii*) *A particular official member is always excluded.*

(*iv*) *A particular non-official member is always included.*

Solution: (*i*) 4 official members can be selected from 8 in

$$^8C_4 = \frac{8!}{4!(8-4)!} = \frac{8\times7\times6\times5}{4\times3\times2\times1} = 70 \text{ ways.}$$

2 non-officials out of 5 can be selected in

$$^5C_2 = \frac{5!}{2!3!} = \frac{5\times4}{2\times1} = 10 \text{ ways.}$$

So, by principle of counting, the number of ways in which the committee can be formed is $70 \times 10 = 700$.

(*ii*) At least two non-official members can be selected in the following ways:

1. No. of ways of forming a committee containing 2 non-officials and 4 officials is
$$= {}^5C_2 \times {}^8C_4 = 700.$$
2. No. of ways of forming a committee containing 3 non-officials and 3 officials is
$$= {}^5C_3 \times {}^8C_3 = \frac{5!}{3!2!} \times \frac{8!}{3!5!} = \frac{5\times4}{2\times1} \times \frac{8\times7\times6}{3\times2\times1} = 10 \times 56 = 560.$$
3. No. of ways of forming a committee containing 4 non-officials and 2 officials is
$$^5C_4 \times {}^8C_2 = 5 \times \frac{8!}{2!6!} = 5\times\frac{8\times7}{2\times1} = 140$$
4. No. of ways of forming a committee containing 5 non-officials and 1 official is
$$= {}^5C_5 \times {}^8C_1 = 1 \times 8 = 8.$$

∴ Total number of committees formed is

$$= 700 + 560 + 140 + 8 = 1408.$$

Alternative

No. of committees consisting of at least 2 non-officials

= No. of committees containing 6 members

– No. of committees containing 1 non-official

– No. of committee containing no non-official

$$= {}^{13}C_6 - {}^5C_1 \times {}^8C_5 - {}^8C_6 = \frac{13!}{6!7!} - 5\times\frac{8!}{5!3!} - \frac{8!}{6!2!}$$

$$= \frac{13\times12\times11\times10\times9\times8}{6\times5\times4\times3\times2\times1} - 5\times\frac{8\times7\times6}{3\times2\times1} - \frac{8\times7}{2\times1} = 13\times12\times11 - 280 - 28 = 1408.$$

(*iii*) A particular official member is always excluded.

Required number of ways

$$= {}^{13-1}C_6 = {}^{12}C_6 = \frac{12!}{6!(12-6)!} = \frac{12\times11\times10\times9\times8\times7}{6\times5\times4\times3\times2\times1}$$

$$= 11 \times 3 \times 4 \times 7 = 924$$

(*iv*) A particular non-official member is always included.

Required number of ways

$$= {}^{13-1}C_{6-1} = {}^{12}C_5 = \frac{12!}{5!7!} = \frac{12\times 11\times 10\times 9\times 8}{5\times 4\times 3\times 2\times 1}$$

$$= 11 \times 9 \times 8 = 792.$$

Example 3: (*i*) *Out of 14 footballers, two are goalkeepers, in how many ways a team of 11 footballers, containing only one goalkeeper, can be selected?*

(*ii*) *Of the 14 articles, 10 are of the same type and each of the remaining is of different type. Find the number of combinations if 10 articles are taken at a time.*

(*iii*) *An examinee is asked to answer 6 questions out of 12 questions. In two groups of the question paper, there are 6 questions in each group and the examinee can answer a maximum of 4 questions from a group. In how many ways can the examinee select 6 questions?*

(*iv*) *Prove that* ${}^{2n}C_n = \dfrac{1\cdot 3\cdot 5\ldots(2n-1)}{n!}\, 2^n$

(*v*) *Find the number of factors of the number 15400 (except 1 and 15400).*

(*vi*) *Find the number of divisors of 9600 including 1 and 9600.*

Solution: (*i*) 10 footballers (other than goalkeepers) can be selected from 12 in

$${}^{12}C_{10} = {}^{12}C_2 = \frac{1}{2} \times 12 \times (12-1) = 66 \text{ ways.}$$

One goalkeeper can be selected from 2 in ${}^{2}C_1 = 2$ ways.

So, by principle of counting, a team of 11 footballers containing only one goalkeeper can be selected in $66 \times 2 = 132$ ways.

(*ii*) r same type articles can be selected from 10 in ${}^{10}C_r$ ways and $(10-r)$ articles from others can be selected in ${}^{4}C_{10-r}$ ways.

∴ Required number of combinations

$$= \sum_{r=6}^{10} {}^{10}C_r \times {}^{4}C_{10-r}$$

$$= {}^{10}C_6 \times {}^{4}C_4 + {}^{10}C_7 \times {}^{4}C_3 + {}^{10}C_8 \times {}^{4}C_2 + {}^{10}C_9 \times {}^{4}C_1 + {}^{10}C_{10} \times {}^{4}C_0$$

$$= \frac{10!}{6!4!} + \frac{10!}{7!3!} \times 4 + \frac{1}{2} \times 10 \times 9 \times \frac{1}{2} \times 4 \times 3 + 10 \times 4 + 1$$

$$\left[\because\ {}^{n}C_r = {}^{n}C_{n-r},\ {}^{n}C_2 = \frac{1}{2}n(n-1)\right]$$

$$= \frac{10\times 9\times 8\times 7}{4\times 3\times 2\times 1} + \frac{10\times 9\times 8}{3\times 2\times 1} \times 4 + 45 \times 6 + 40 + 1$$

$$= 210 + 480 + 270 + 41 = 1001.$$

(*iii*) Let A and B be two groups each containing 6 questions. Now, the examinee can select 4 from A and 2 from B, or 3 from A and 3 from B, or 2 from A and 4 from B. Hence, the number of different ways in which the examinee can select 6 questions is

$$= {}^{6}C_4 \times {}^{6}C_2 + {}^{6}C_3 \times {}^{6}C_3 + {}^{6}C_2 \times {}^{6}C_4$$

$$= 2\,{}^6C_2 \times {}^6C_2 + {}^6C_3 \times {}^6C_3 \qquad [\because {}^nC_r = {}^nC_{n-r}]$$

$$= 2 \times \left\{\frac{1}{2} \times 6 \times 5\right\}^2 + \left\{\frac{6!}{3!3!}\right\}^2 \qquad \left[\because {}^nC_2 = \frac{1}{2}n(n-1)\right]$$

$$= 2 \times 15^2 + \left\{\frac{6 \times 5 \times 4}{3 \times 2 \times 1}\right\}^2 = 850.$$

Note: Where to multiply and where to add

When the occurrence of event B depends on the occurrence of event A then we multiply the corresponding number of ways by principle of counting to get the total number of ways. But when the events are independent we just add the corresponding number of ways.

(iv) $${}^{2n}C_n = \frac{(2n)!}{n!(2n-n)!} = \frac{2n(2n-1)(2n-2)(2n-3)\ldots 4\cdot 3\cdot 2\cdot 1}{n!n!}$$

$$= \frac{\{2n(2n-2)\ldots 4\cdot 2\}\{(2n-1)(2n-3)\ldots 3\cdot 1\}}{n!n!}$$

$$= \frac{2^n\{n(n-1)\ldots 2\cdot 1\}\{1\cdot 3\ldots(2n-3)(2n-1)\}}{n!n!}$$

$$= \frac{2^n n!\{1\cdot 3\cdot 5\ldots(2n-1)\}}{n!n!} = \frac{1\cdot 3\cdot 5\ldots(2n-1)\}}{n!}2^n.$$

(v) Clearly, $15400 = 2^3 \times 5^2 \times 7 \times 11$.

So, the prime numbers 2, 5, 7 and 11 appear 3 times, 2 times, 1 time and 1 time respectively in the number 15400.

$\therefore$ Number of different factors of 15400 (excluding 1 and including 15400)

$$= (3 + 1)(2 + 1)(1 + 1)(1 + 1) - 1 \quad \text{[see Art. 4.3.12]}$$
$$= 47.$$

$\therefore$ Number of factors of 15400 (except 1 and 15400)

$$= 47 - 1 = 46.$$

Note: An integer $p > 1$ is called a **prime number,** or simply a **prime** if its only positive divisors are 1 and p. Observe that the only even that is a prime is 2; rest all other primes are odd integers.

(vi) **Alternative approach**

Clearly, $9600 = 2^7 \times 3^1 \times 5^2$.

$\therefore$ $2^a3^b5^c$ are divisors of 9600 where integer a varies from 0 to 7, integer b varies from 0 to 1 and integer c varies from 0 to 2.

$\therefore$ By principle of counting, the total number of divisors of 9600 (including 1 and 9600) is $8 \times 2 \times 3 = 48$.

Example 4: *A box contains two white balls, three black balls and four red balls. In how many ways can three balls be drawn from the box if at least one black ball is to be included in the draw?*

Solution: Total number of balls = 2 + 3 + 4 = 9.

No. of black balls = 3.

Required number of ways = Total number of draw – No. of draw without black ball

$$= {}^9C_3 - {}^{9-3}C_3 = \frac{9!}{3!6!} - \frac{6!}{3!3!} = \frac{9 \times 8 \times 7}{3 \times 2 \times 1} - \frac{6 \times 5 \times 4}{3 \times 2 \times 1}$$

$$= 3 \times 4 \times 7 - 5 \times 4 = 64.$$

Example 5: *Prove that the product of any r consecutive positive integers is always divisible by r !*

Solution: We have to prove that the product $(n+1)(n+2)(n+3)\ldots(n+r)$ is divisible by $r!$ where $(n+1)$ is any positive integer.

Now, $$\frac{1}{r!}(n+1)(n+2)(n+3)\ldots(n+r) = \frac{1\cdot 2\cdot 3\ldots n(n+1)(n+2)(n+3)\ldots(n+r)}{r!\,1\cdot 2\cdot 3\ldots n}$$

$$= \frac{(n+r)!}{r!\,n!} = {}^{n+r}C_r = \text{a positive integer.}$$

Hence, the product of any r consecutive positive integers is always divisible by r ! **(Proved)**

Example 6: *A committee of 12 is to be formed from 9 women and 8 men. In how many ways this can be done if at least 5 women have to be included in a committee? In how many of these committees (i) the women are in majority (ii) the men are in majority?*

Solution: At least 5 women can be included in a committee of 12 in the following ways:

1. No. of ways of forming a committee containing 5 W and 7 M is
$$= {}^9C_5 \times {}^8C_7 = \frac{9!}{5!4!}\times 8 = \frac{9\times 8\times 7\times 6}{4\times 3\times 2\times 1}\times 8 = 9\times 8\times 7\times 2 = 1008.$$
2. No. of ways of forming a committee containing 6 W and 6 M is
$$= {}^9C_6 \times {}^8C_6 = \frac{9!}{6!3!}\times\frac{8!}{6!2!} = \frac{9\times 8\times 7}{3\times 2\times 1}\times\frac{8\times 7}{2\times 1} = 84\times 28 = 2352.$$
3. No. of ways of forming a committee containing 7 W and 5 M is
$$= {}^9C_7 \times {}^8C_5 = \frac{1}{2}\times 9\times(9-1)\times\frac{8!}{5!3!} = 36\times\frac{8\times 7\times 6}{3\times 2} = 2016$$
4. No. of ways of forming a committee containing 8 W and 4 M is
$$= {}^9C_8 \times {}^8C_4 = 9\times\frac{8!}{4!4!} = 9\times\frac{8\times 7\times 6\times 5}{4\times 3\times 2} = 630$$
5. No. of ways of forming a committee containing 9 W and 3 M is
$$= {}^9C_9 \times {}^8C_3 = 1\times\frac{8!}{3!5!} = 56.$$

∴ No. of ways of forming a committee consisting of at least 5 women
$$= 1008 + 2352 + 2016 + 630 + 56 = 6062$$

(*i*) Observe that in 3, 4, 5, the women are in majority.

∴ The total number of committees in which women are in majority
$$= 2016 + 630 + 56 = 2702.$$

(*ii*) In 1, the men are in majority.

∴ The total number of committees in which men are in majority = 1008.

Example 7: (*i*) *Find the number of different straight lines made by joining n different points on a plane, no three of which are collinear with the exception of m points which are collinear.*

(*ii*) *Determine also the number of triangles formed by these n points.*

Solution: (*i*) If no three points were lying on the same straight line, then the total number of straight lines would be

$${}^nC_2 = \frac{1}{2}n(n-1).$$

But m of the points lying on the same straight line, so

$$^mC_2 = \frac{1}{2}m(m-1)$$

straight lines are lost and in their place we get only one straight line on which the m points are lying.

$\therefore$ Required number of different straight lines

$$= \frac{1}{2}n(n-1) - \frac{1}{2}m(m-1) + 1$$

(*ii*) Since a triangle is formed by joining any three points not lying on the same straight line, so as before, the required number of triangles.

$$= {}^nC_3 - {}^mC_3 = \frac{1}{6}\, n(n-1)(n-2) - \frac{1}{6}\, m(m-1)(m-2)$$

Example 8: *The sides AB, BC, CA of a triangle ABC have 3, 4 and 5 interior points respectively on them. Find the number of triangles that can be constructed using these points as vertices.*

Solution: Here we have to use 3 + 4 + 5 = 12 points for forming the triangles. Since 3 points are required to form a triangle, the total number of triangles (which includes the triangles formed by collinear points and since these collinear points will not form triangles their number has to be subtracted from the total) is

$$^{12}C_3 = \frac{12\,!}{3\,!(12-3)\,!} = \frac{12 \times 11 \times 10}{3 \times 2 \times 1} = 220 \qquad ...(1)$$

But (1) includes

(*i*) the no. of triangles formed by 3 points on $AB = {}^3C_3 = 1$

(*ii*) the no. of triangles formed by 4 points on $BC = {}^4C_3 = 4$

(*iii*) the no. of triangles formed by 5 points on $CA = {}^5C_3 = {}^5C_{5-3}$

$$= \frac{1}{2} \times 5 \times (5-1) = 10$$

$\therefore$ Required number of triangles = 220 – (1 + 4 + 10) = 205.

Example 9: *Five points are given on one of two parallel straight lines and ten points are given on the other straight line. How many triangles can be formed by taking these points as vertices of a triangle?*

Solution: Let the two straight lines be marked as L_1, L_2 where five given points lie on L_1 and ten given points lie on L_2. Here we have to use 5 + 10 = 15 points for forming the triangles. Since 3 points are needed to form a triangle, the total number of triangles (including the triangles formed by collinear points and since these collinear points will not form triangles their number has to be subtracted from the total) is

$$^{15}C_3 = \frac{15\,!}{3\,!(15-3)\,!} = \frac{15 \times 14 \times 13}{3 \times 2} = 455 \qquad ...(1)$$

But (1) includes

(*i*) The no. of triangles formed by 5 points on L_1

$$= {}^5C_3 = {}^5C_2 = \frac{1}{2} \times 5 \times 4 = 10 \qquad \left[\because\ {}^nC_2 = \frac{1}{2}n(n-1)\right]$$

(*ii*) The no. of triangles formed by 10 points on L_2

$$= {}^{10}C_3 = \frac{10!}{3!7!} = \frac{10\times 9\times 8}{3\times 2} = 120.$$

$\therefore$ Required number of triangles formed by taking the points on L_1 and L_2 as vertices is

$$= 455 - (10 + 120) = 325.$$

Example 10: *Five balls of different colours are to be placed in three boxes of different sizes. Each box can hold all five balls. In how many different ways can we place the balls so that no box remains empty?*

Solution: Let the boxes be marked as *A*, *B*, *C*. As per problem, we have to ensure that no box remains empty and all five balls have to put in. There are following two cases:

I. Any two containing one each and the third one contains three

(*i*) $A \to 1, B \to 1, C \to 3$.

This can be done in ${}^5C_1 \times {}^4C_1 \times {}^3C_3 = 5 \times 4 \times 1 = 20$ ways

(*ii*) $A \to 1, B \to 3, C \to 1$.

This can be done in ${}^5C_1 \times {}^4C_3 \times {}^1C_1 = 5 \times 4 \times 1 = 20$ ways

(*iii*) $A \to 3, B \to 1, C \to 1$.

This can be done in ${}^5C_3 \times {}^2C_1 \times {}^1C_1 = \frac{1}{2} \times 5 \times 4 \times 2 \times 1 = 20$ ways

$\therefore$ In this case, no. of ways = 20 + 20 + 20 = 60.

II. Any two containing 2 each and the third one contains 1

(*i*) $A \to 2, B \to 2, C \to 1$.

This can be done in ${}^5C_2 \times {}^3C_2 \times {}^1C_1 = \frac{5!}{2!3!} \cdot \frac{3!}{2!1!} \cdot 1 = \frac{5!}{2!2!1!} = 30$ ways.

(*ii*) Similarly, $A \to 2, B \to 1, C \to 2$ can be done in 30 ways.

(*iii*) Also, $A \to 1, B \to 2, C \to 2$ can be done in 30 ways.

$\therefore$ In this case, no. of ways = 30 + 30 + 30 = 90.

Hence, required number of ways = 60 + 90 = 150.

Example 11: *m men and n women are to be seated in a row so that no two women sit together. If m > n, then show that the number of ways in which they can be seated is*

$$\frac{m!\,(m+1)!}{(m-n+1)!}$$

Solution: *m* men can be seated in *m* ! ways in a row. Now, there are $(m + 1)$ places in between them where *n* women can be seated so that no two women sit together. They can be seated in ${}^{m+1}P_n$ ways.

Hence, by using the principle of counting, given men and women can be seated (no two women sit together) in

$$m!\,{}^{m+1}P_n = \frac{m!\,(m+1)!}{(m+1-n)!} = \frac{m!\,(m+1)!}{(m-n+1)!} \text{ ways.}$$

4.4 THE PIGEONHOLE PRINCIPLE

A variety of combinatorial problems can be solved by using the **pigeonhole principle** which states that if there are more pigeons than pigeonholes, then there must be at least one pigeonhole occupied by more than one pigeon.

Notes: (*i*) This principle applies to other objects besides pigeons and pigeonholes.

(*ii*) This principle is also known as the **Dirichlet drawer principle**, after the German mathematician G. L. Dirichlet (1805–1859), who often used this principle in his work.

I. The first form of pigeonhole principle: If $(K + 1)$ or more objects are placed into K boxes, then at least one box must contain two or more of the objects.

Proof: Suppose that none of the K boxes contains more than one object and so the total number of objects would be at most K. This leads to a contradiction because there are at least $(K + 1)$ objects. Hence the result.

Illustration 1: Among any group of 367 people, there must be at least two people with the same birthday since there are at most 366 possible birthdays (because there are at most 366 days in a year, considering leap year).

Illustration 2: If any six integers from 1 to 10 be chosen, then at least two of them will have a sum 11.

Solution: Let $S = \{1, 2, 3, 4, 5, 6, 7, 8, 9, 10\}$. The different subsets of S each containing two integers with sum 11 are $S_1 = \{1, 10\}$, $S_2 = \{2, 9\}$, $S_3 = (3, 8)$, $S_4 = \{4, 7\}$, $S_5 = \{5, 6\}$.

Each of the six integers chosen from S must belong to any of these five subsets S_1, S_2, S_3, S_4 and S_5. These five subsets can be thought as pigeonholes (or boxes) and six chosen integers as pigeons (or objects). Since $5 < 6$, using the pigeonhole principle, we conclude that at least two of the chosen integers belong to the same set with a sum 11.

II. The second form of pigeonhole principle: If f is a function from a finite set X to a finite set Y and $|X| > |Y|$, where $|X|$ and $|Y|$ denote the number of elements in X and Y respectively, then $f(x_1) = f(x_2)$ for some $x_1, x_2 \in X$ with $x_1 \neq x_2$.

Proof: Let X be the set of objects (pigeons) and Y be the set of boxes (pigeonholes). Let us assign object x to box $f(x)$. Since $|X| > |Y|$, *i.e.*, the number of objects is greater than the number of boxes, by the first form of pigeonhole principle, at least two objects $x_1, x_2 \in X$ are assigned to the same box. Therefore, $f(x_1) = f(x_2)$ for some $x_1, x_2 \in X$ with $x_1 \neq x_2$. Hence the result.

Illustration 3: *If n processors are interconnected, then at least two processors are directly connected to the same number of processors.*

Solution: Let us designate the processors as $1, 2, 3, ..., n$ and let α_i be the number of processors to which processor i is directly connected, $i = 1, 2, ..., n$. So, the domain of the discrete function α_i is $X = \{1, 2, 3, ..., n\}$.

Observe that $\alpha_i \neq 0$ for some $i \in X$ and $\alpha_j \neq n - 1$ for some $j \in X$, where $i \neq j$, at the same time. Because if $\alpha_i = 0$ and $\alpha_j = n - 1$ for some $i, j \in X$ with $i \neq j$ happen at the same time, then we would have i^{th} processor not connected to any other processor while at the same time j^{th} processor is connected to all the other processors (including the ith processor). This is a contradiction. So, the range Y of the function α_i is a subset of either $\{0, 1, 2, ..., n - 2\}$ or $\{1, 2, 3, ..., n - 1\}$. In either case $|Y| < n = |X|$. Using the second form of pigeonhole principle, we have $\alpha_i = \alpha_j$ for some $i \neq j$ implying that at least two processors are directly connected to the same number of processors.

Illustration 4: *Let A be some fixed 12-element subset of {1, 2, 3, ..., 99, 100}. Show that A possesses two distinct 6-element subset, the sums of whose elements are equal.*

Solution: Let X be the family of 6-element subsets of A and let for each $B \in X, f(B)$ be the sum of the numbers in B. Observe that $f(B) \geq 1 + 2 + 3 + 4 + 5 + 6 = 21$ and $f(B) \leq 100 + 99 + 98 + 97 + 96 + 95 = 585$. So, $f: X \to Y$ where $Y = \{21, 22, 23, ..., 585\}$. Since $|Y| = 585 - 20 = 565$ and $|X| = C(12, 6) = 924$, *i.e.*, $|X| > |Y|$, using pigeonhole principle , X possesses at least two distinct 6-element subsets with the same image under f. Hence the result.

III. The generalized pigeonhole principle: If N objects (pigeons) are placed into K boxes (pigeonholes), then there is at least one box containing at least $\lceil N/K \rceil$ objects where the ceiling function $\lceil N/K \rceil$ denotes the smallest integer that is greater than or equal to N/K.

Proof: Suppose that none of the boxes contains more than $\lceil N/K \rceil - 1$ objects and so the K boxes contain a total number of objects less than or equal to

$$K(\lceil N/K \rceil - 1) < K\left\{\left(\frac{N}{K}+1\right)-1\right\} = N \qquad \left(\because \lceil N/K \rceil < \frac{N}{K}+1\right).$$

It leads to a contradiction since there are a total of N objects. Hence the result.

Illustration 5: *Find the minimum number of students required in a discrete mathematics class to be sure that at least seven will receive the same grade, if there are four possible grades : A, B, C, D.*

Solution: Using generalized pigeonhole principle, the minimum number of students required to guarantee that at least seven students receive the same grade is the smallest integer N such that $\lceil N/4 \rceil = 7$. The smallest integer is $N = 4 \times 6 + 1 = 25$. Therefore, 25 is the minimum number of students required to be sure that at least 7 students will receive the same grade.

Illustration 6: *In a group of 6 people, each pair of individuals consists of 2 friends or 2 enemies. Prove that there are either 3 mutual friends or 3 mutual enemies in the group.*

Solution: Let A be any one of the 6 people of the group. Using generalized pigeonhole principle, there are at least $\lceil 5/2 \rceil = 3$ among the other 5 people who are friends of A or enemies of A. Let B, C, D are friends of A and so if any 2 of these individuals are friends then these 2 and A form a group of 3 mutual friends. Otherwise, B, C, D form a set of 3 mutual enemies. In the second case, let us suppose B, C, D are enemies of A. If any 2 of these individuals are enemies then these 2 and A form a group of 3 mutual enemies, otherwise B, C, D form a group of 3 mutual friends.

Illustration 7: *Find the least number of area codes required to ensure that the 25 million phones in a state have distinct 10-digit telephone numbers. Assume that telephone numbers are of the form NXX-NXX-XXXX, where the first three digits form the area code, N denotes a digit from 2 to 9 inclusive and X represents any digit.*

Solution: Using product rule, there are $8 \times 10 \times 10 = 800$ area codes with format *NXX* and also there are $10 \times 10 \times 10 \times 10 = 10,000$ station codes with format *XXXX*. Thus there are $800 \times 10,000 = 8,000,000$ different phone numbers with format *NXX – XXXX*. Using generalized pigeonhole principle, among 25 million telephones, at least $\lceil 25,000,000/8,000,000 \rceil = 4$ of them must have identical phone numbers. Hence, at least four area codes are needed to ensure that all 10-digit numbers are different.

Illustration 8: *Prove that among any (n +1) positive integers not exceeding 2n there must be an integer that divides one of the other integers.*

Solution: Let us write each of the $(n + 1)$ integers $a_1, a_2, ..., a_{n+1}$ as a power of 2 times an odd integer. So, let $a_j = 2^{K_j} q_j$ for $j = 1, 2, ..., n + 1$, where K_j is a nonnegative integer and q_j is odd. Each of $q_1, q_2, ..., q_{n+1}$ is an odd positive integer less than $2n$. Since there are only n odd positive integers less than $2n$, it follows from the pigeonhole principle that two of the integers $q_1, q_2, ..., q_{n+1}$ must be identical. Hence there are integers i and j such that $q_i = q_j = q$ (say) and so $a_i = 2^{K_i} q$, $a_j = 2^{K_j} q$. Therefore, if $K_i < K_j$, then a_i divides a_j; while if $K_i > K_j$, then a_j divides a_i. Hence the result.

Illustration 9: *A patient is given a prescription of 45 pills with the instruction to take at least one pill per day for 30 days. Show that there must be a period of consecutive days during which the patient takes a total of exactly 14 pills.*

Solution: Let a_i be the number of pills the patient has taken through the end of the ith day. We have $1 \le a_1 < a_2 < ... < a_{30} \le 45$ since the patient takes at least one pill per day and at most 45 pills in 30 days. So, $a_1 + 14 < a_2 + 14 < ... < a_{30} + 14 \le 45 + 14 = 59$ and hence all 60 integers: $a_1, a_2, ..., a_{30}, a_1 + 14, a_2 + 14, ..., a_{30} + 14$ lie in the range between 1 and 59. Thus, we have 60 pigeons in 59 pigeonholes which implies that there must be 2 of these integers are identical (by pigeonhole principle). As $a_1, a_2, ..., a_{30}$ are all different and also $a_1 + 14, a_2 + 14, ..., a_{30} + 14$ are all different, there exist i and j such that $a_j = a_i + 14$. Thus, between days i and j, the patient takes exactly 14 pills. Hence the result.

4.5 THE PRINCIPLE OF INCLUSION AND EXCLUSION

The number of elements in the union of finite disjoint sets can be counted by the sum rule. However, if the sets are not disjoint in general, the sum rule must be modified. The technique of counting the number of elements in the union of finite sets is called the **principle of inclusion-exclusion.**

Illustration 1: *Out of 15 persons 10 can speak Hindi and 8 can speak English. How many persons can speak Hindi only?*

Solution: Let A and B represents the set of persons who can speak Hindi and English respectively.

Given: $n(A) = 10, n(B) = 8$

and $n(A \cup B) = 15.$

$\therefore$ $n(A) + n(B) - n(A \cap B) = 15$

or $10 + 8 - n(A \cap B) = 15$

$\therefore$ $n(A \cap B) = 3.$

Hence $n(A) - n(A \cap B) = 10 - 3 = 7$ persons can speak Hindi only.

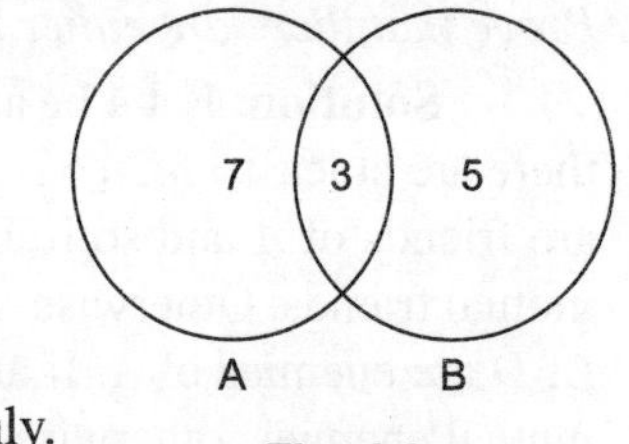

Fig. 4.3

Alternative Statement

When two tasks can be done at the same time, we cannot use the ordinary sum rule in general to count the number of ways to do one of the two tasks. Adding the number of ways to do each task leads to an over count because the ways to do both tasks are counted twice. To correctly count the number of ways to do one of the two tasks, we have to add the number of ways to do each of the two tasks and then subtract the number of ways to do both tasks. This counting technique is called the **principle of inclusion and exclusion.**

We can phrase this counting principle in terms of finite sets. Let A_1 and A_2 be two sets having $|A_1|$ and $|A_2|$ elements respectively. Let T_i be the task of choosing an element from A_i $(i = 1, 2)$ and so $|A_1|$ ways to do T_1 and $|A_2|$ ways to do T_2. The number of ways to do either T_1 or T_2 is the sum of the number of ways to do T_1 and the number of ways to do T_2 minus the number of ways to do both T_1 and T_2. There are $|A_1 \cup A_2|$ ways to do either T_1 or T_2 and $|A_1 \cap A_2|$ ways to do both T_1 and T_2.

$$\therefore \quad |A_1 \cup A_2| = |A_1| + |A_2| - |A_1 \cap A_2|.$$

Illustration 2: *A discrete mathematics class contains 52 students majoring in Computer Science, 17 students majoring in Mathematics and 9 joint Mathematics and Computer Science majors. Find the number of students in this class if every student is majoring in Computer Science only or Mathematics only or both Computer Science and Mathematics.*

Solution: Let A denotes the set of students in the class majoring in Computer Science only and B denotes the set of students in the class majoring in Mathematics only. Then $A \cap B$ represents the set of students in the class that are joint Computer Science and Mathematics majors. Since every student in the class is majoring in either Computer Science or Mathematics or both, it follows that the number of students in the class is $|A \cup B|$.

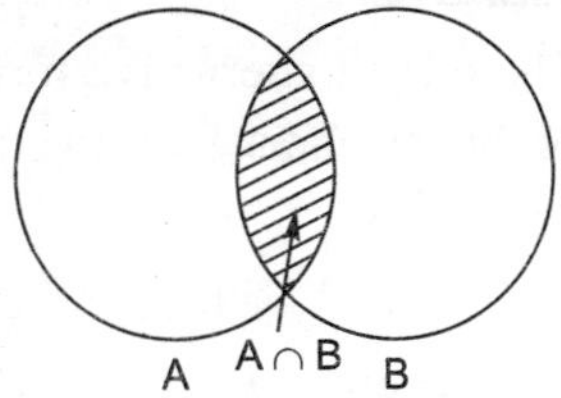

Fig. 4.4

Now, $|A \cup B| = |A| + |B| - |A \cap B| = 52 + 17 - 9 = 60.$

Therefore, there are 60 students in the class.

Theorem 1: (The Principle of Inclusion and Exclusion)

If $A_1, A_2, ..., A_n$ be finite sets, then

$$|A_1 \cup A_2 \cup ... \cup A_n| = \sum_{1 \le i \le n} |A_i| - \sum_{1 \le i < j \le n} |A_i \cap A_j| + \sum_{1 \le i < j < k \le n} |A_i \cap A_j \cap A_k| - ... + (-1)^{n+1} |A_1 \cap A_2 \cap ... \cap A_n|$$

where $|A_i|$ represents the number of elements in A_i $(i = 1, 2, ..., n)$.

Proof: Let us prove this formula by showing that an element in the union is counted exactly once by the right side of the equation. Suppose α be a member of exactly r of the sets $A_1, A_2, ..., A_n$; $1 \le r \le n$. This element is counted $C(r, 1)$ times by $\Sigma |A_i|$, $C(r, 2)$ times by $\Sigma |A_i \cap A_j|$ and in general $C(r, m)$ times by the summation involving m of the sets A_i. Therefore, this element is counted exactly

$$C(r, 1) - C(r, 2) + C(r, 3) - ... + (-1)^{r+1} C(r, r) \qquad ...(1)$$

times by the expression on the right side of this equation.

Using Binomial theorem, we have

$$(1 + x)^r = C(r, 0) + x\, C(r, 1) + x^2 C(r, 2) + x^3 C(r, 3) + ... + x^r C(r, r).$$

This is an identity and it is true for all values of x. Putting $x = -1$, we get

$$(1 - 1)^r = C(r, 0) - C(r, 1) + C(r, 2) - C(r, 3) + ... + (-1)^r C(r, r)$$

$$\Rightarrow \quad C(r, 1) - C(r, 2) + C(r, 3) - .... + (-1)^{r+1} C(r, r) = C(r, 0) = 1 \qquad ...(2)$$

From (1) and (2) we conclude that each element in the union is counted exactly once by the expression on the right side of the equation. This completes the proof.

Notes: (*i*) The inclusion-exclusion principle gives a formula for the number of elements in the union of n finite sets for every positive integer n.

(*ii*) For any three finite sets A_1, A_2, A_3:

$$|A_1 \cup A_2 \cup A_3| = |A_1| + |A_2| + |A_3| - |A_1 \cap A_2| - |A_2 \cap A_3| - |A_3 \cap A_1| + |A_1 \cap A_2 \cap A_3|.$$

(*iii*) For any four finite sets A_1, A_2, A_3, A_4;

$$|A_1 \cup A_2 \cup A_3 \cup A_4| = |A_1| + |A_2| + |A_3| + |A_4| - |A_1 \cap A_2| - |A_1 \cap A_3| - |A_1 \cap A_4| - |A_2 \cap A_3| - |A_2 \cap A_4| - |A_3 \cap A_4| + |A_1 \cap A_2 \cap A_3| + |A_1 \cap A_2 \cap A_4| + |A_1 \cap A_3 \cap A_4| + |A_2 \cap A_3 \cap A_4| - |A_1 \cap A_2 \cap A_3 \cap A_4|.$$

(*iv*) If $A_1, A_2, ..., A_n$ are mutually disjoint finite sets, then $|A_1 \cup A_2 \cup ... \cup A_n| = \sum_{i=1}^{n} |A_i|$.

4.6 AN ALTERNATIVE FORM OF THE PRINCIPLE OF INCLUSION AND EXCLUSION

There is an alternative form of the principle of inclusion-exclusion which is useful to solve counting problems that ask for the number of elements in a set having none of n properties $P_1, P_2, ..., P_n$.

Statement

Let A_i be the subset consisting of the elements with property P_i. The number of elements with all the properties $P_{i_1}, P_{i_2}, ..., P_{i_k}$ is denoted by $N(P_{i_1} P_{i_2} ... P_{i_k})$.

$$\therefore \quad |A_{i_1} \cap A_{i_2} \cap ... \cap A_{i_k}| = N(P_{i_1} P_{i_2} ... P_{i_k})$$

If the total number of elements in the underlying set is denoted by N and the number of elements with none of the properties $P_1, P_2, ..., P_n$ is denoted by $N(P_1' P_2' ... P_n')$, it follows that (using the principle of inclusion-exclusion):

$$N(P_1' P_2' ... P_n') = N - |A_1 \cap A_2 \cap ... \cap A_n|$$

$$= N - \sum_{1 \le i \le n} N(P_i) + \sum_{1 \le i < j \le n} N(P_i P_j) - \sum_{1 \le i < j < k \le n} N(P_i P_j P_k)$$

$$+ ... + (-1)^n N(P_1 P_2 ... P_n).$$

Let us state and prove an useful theorem.

Theorem 2: *The number of r-combinations from a set with n elements is C(n + r – 1, r) when repetition of elements is allowed.*

Proof: Let us represent each r-combination from a set with n elements by a list of $(n - 1)$ bars and r stars when repetition of elements is allowed. Here the $(n - 1)$ bars are used to mark off n distinct cells, with the ith cell containing a star for each time the ith element of the set occurs in the combination. For example, a 7-combination of a set with four elements is represented with three bars and seven stars. Here

** | **** || *

represents the combination containing exactly 2 of the first element, four of the second element, none of the third element and one of the fourth element of the set.

As we have seen, each different list containing $(n - 1)$ bars and r stars corresponds to an r-combination from the set with n elements (when repetition is allowed).

The number of such lists is $C(n - 1 + r, r)$, because each list corresponds to a choice of the r positions to place the r stars from the $(n - 1 + r)$ positions that contain r stars and $(n - 1)$ bars.

Illustration 1: *A cookie shop has 5 different kinds of cookies. How many different ways can 8 cookies be chosen? Assume that only the type of cookie matters (not the individual cookies or the order in which they are chosen).*

Solution: The number of ways to choose 8 cookies is the number of 8-combinations from a set with 5 elements. Using Theorem 2, this occurs in $C(5 + 8 - 1, 8) = C(12, 8) = \frac{12!}{8!4!} = \frac{12 \times 11 \times 10 \times 9}{4 \times 3 \times 2 \times 1}$ $= 495$ ways. Thus, there are 495 different ways to choose 8 cookies.

Illustration 2: *How many solutions does the equation $x_1 + x_2 + x_3 = 9$ have, where x_1, x_2 and x_3 are nonnegative integers with $x_1 \le 2$, $x_2 \le 3$ and $x_3 \le 5$?*

Solution: To apply the principle of inclusion-exclusion, let a solution have property P_1 is $x_1 > 2$, property P_2 is $x_2 > 3$ and property P_3 is $x_3 > 5$. The number of solutions satisfying the inequalities $x_1 \le 2, x_2 \le 3$ and $x_3 \le 5$ is

$$N(P_1' P_2' P_3') = N - N(P_1) - N(P_2) - N(P_3) + N(P_1P_2) + N(P_1P_3) + N(P_2P_3) - N(P_1P_2P_3).$$

To count the number of solutions, we note that a solution corresponds to a way of selecting 9 items from a set with 3 elements, so that x_1 items of type one, x_2 items of type two and x_3 items of type three are chosen. Therefore, the number of solutions is equal to the number of 9-combinations (repetition allowed) from a set of 3 elements. Using Theorem 2, it follows that

$$N = \text{Total number of solutions} = C(3 + 9 - 1, 9) = C(11, 9) = C(11, 2) = \frac{1}{2} \times 11 \times 10 = 55.$$

A solution to this equation subject to $x_1 \ge 3, x_2 \ge 0, x_3 \ge 0$ corresponds to a selection of 9 items with x_1 items of type one, x_2 items of type two and x_3 items of type three, where, in addition, there is at least three items of type one. So, choose three items of type one and then select six additional items. Using Theorem-2, it follows that

$$N(P_1) = (\text{number of solutions with } x_1 \ge 3)$$
$$= C(3 + 6 - 1, 6) = C(8, 6) = C(8, 2) = \frac{1}{2} \times 8 \times 7 = 28.$$
$$N(P_2) = (\text{number of solutions with } x_2 \ge 4)$$
$$= C(3 + 5 - 1, 5) = C(7, 5) = C(7, 2) = \frac{1}{2} \times 7 \times 6 = 21.$$
$$N(P_3) = (\text{number of solutions with } x_3 \ge 6)$$
$$= C(3 + 3 - 1, 3) = C(5, 3) = C(5, 2) = \frac{1}{2} \times 5 \times 4 = 10.$$
$$N(P_1P_2) = (\text{number of solutions with } x_1 \ge 3 \text{ and } x_2 \ge 4)$$
$$= C(3 + 2 - 1, 2) = C(4, 2) = \frac{1}{2} \times 4 \times 3 = 6.$$
$$N(P_1P_3) = (\text{number of solutions with } x_1 \ge 3 \text{ and } x_3 \ge 6)$$
$$= C(3 + 0 - 1, 0) = 1$$
$$N(P_2P_3) = (\text{number of solutions with } x_2 \ge 4 \text{ and } x_3 \ge 6)$$
$$= 0$$
$$N(P_1P_2P_3) = (\text{number of solutions with } x_1 \ge 3, x_2 \ge 4, x_3 \ge 6) = 0$$

Inserting these quantities into the formula for $N(P_1' P_2' P_3')$ shows that the number of solutions with $x_1 \le 2, x_2 \le 3$ and $x_3 \le 5$ equals

$$N(P_1' P_2' P_3') = 55 - 28 - 21 - 10 + 6 + 1 + 0 - 0 = 3.$$

4.7 THE NUMBER OF ONTO FUNCTIONS

The principle of inclusion-exclusion can also be used to find the number of onto functions from a set with m elements to a set with n elements. Let us first consider the following example:

Example 1: (Counting Functions) *Determine the number of functions from a set with m elements to one with n elements.*

Solution: A function corresponds to a choice of one of the n elements in the co-domain for each of the m elements in the domain and so by the product rule there are $n \cdot n \dots n = n^m$ functions from a set with m elements to one with n elements.

Example 2: *Find the number of onto functions from a set with 6 elements to a set with 3 elements.*

Solution: Suppose that the elements in the co-domain are b_1, b_2, b_3 and P_1, P_2, P_3 be the properties that b_1, b_2, b_3 are not in the range of the function respectively. So, a function is onto if and only if it has none of the properties P_1, P_2, P_3. By the principle of inclusion-exclusion, it follows that the number of onto functions from a set with 6 elements to a set with 3 elements is

$$N(P_1' P_2' P_3') = N - \{N(P_1) + N(P_2) + N(P_3)\} + \{N(P_1P_2) + N(P_1P_3) + N(P_2P_3)\} - N(P_1P_2P_3),$$

where N is the total number of functions from a set with 6 elements to one with 3 elements and so $N = 3^6$ (see Example 1).

(*i*) Observe that $N(P_i)$ denotes the total number of functions that do not have b_i in their range and so there are two choices for the value of the function at each element of the domain. Thus, $N(P_i) = 2^6$ and there are $C(3, 1)$ terms of this kind.

(*ii*) Note that $N(P_iP_j)$ is the number of functions that do not have b_i, b_j in their range and so there is only one choice for the value of the function at each element of the domain. Therefore, $N(P_iP_j) = 1^6 = 1$ and there are $C(3, 2)$ terms of this kind.

(*iii*) Also, note that $N(P_1P_2P_3) = 0$, because $N(P_1P_2P_3)$ is the number of functions that have none of b_1, b_2, b_3 in their range. Clearly, there are no such functions.

Therefore, the number of onto functions from a set with 6 elements to one with 3 elements is

$$3^6 - C(3, 1)2^6 + C(3, 2)1^6 - 0 = 729 - 192 + 3 = 540.$$

The general result that states how many onto functions there are from a set with m elements to one with n element is stated in the following theorem. The proof is left as an exercise.

Theorem 3: If m and n be positive integers with $m \geq n$, then there are $n^m - C(n, 1)(n-1)^m + C(n, 2)(n-2)^m - .... + (-1)^{n-1} C(n, n-1)1^m$ onto functions from a set with m elements to a set with n elements.

Example 3: *In how many ways 6 different jobs can be assigned to 4 different employees if every employee is assigned at least one job?*

Solution: Consider the assignment of jobs as a function from the set of 6 jobs to the set of 4 employees. An assignment where each employee gets at least one job is the same as an onto function from the set of jobs to the set of employees. Hence, by Theorem 3 (where $m = 6$ and $n = 4$), it follows that there are

$$4^6 - C(4, 1)(4-1)^6 + C(4, 2)(4-2)^6 - C(4, 3)(4-3)^6$$
$$= 4096 - 2916 + 384 - 4 = 1560$$

ways to assign the jobs so that every employee is assigned at least one job.

4.8 DERANGEMENTS

Definition: A permutation of objects that leaves no object in its original position is called a **derangement.**

Illustration: The permutation 231 is a derangement of 123 because no number is left in its original position. However, 321 is not a derangement of 123 since this permutation leaves 2 fixed.

Let D_n represent the number of derangements of n objects. For example, $D_3 = 2$, because the derangements of 123 are 231 and 312 only. In the following theorem the value of D_n, for all positive integers n, is stated.

Theorem 4: *The number of derangements of a set with n members is*

$$D_n = n!\left\{1 - \frac{1}{1!} + \frac{1}{2!} - \frac{1}{3!} + \ldots + (-1)^n \frac{1}{n!}\right\}.$$

Proof: Suppose a permutation have property P_i if it fixes element i in its original position. The number of derangements is the number of permutations having none of the properties P_i for $i = 1, 2, ..,$ n, and so $D_n = N(P_1' P_2' \ldots P_n')$. Using the principle of inclusion-exclusion, we get

$$D_n = N - \sum_{i} N(P_i) + \sum_{i<j} N(P_i P_j) - \sum_{i<j<k} N(P_i P_j P_k)$$

$$+ \ldots. + (-1)^n N(P_1 P_2 \ldots P_n),$$

where N is the total number of permutations of n elements and so $N = n!$

(*i*) $N(P_i) = (n-1)!$ since it is the number of permutations that fix the ith element in its original position. It follows from the product rule because each of the remaining $(n-1)$ positions can be filled arbitrarily.

$$\therefore \qquad \sum_{1 \le i \le n} N(P_i) = C(n, 1)\,(n-1)!,$$

as there are $C(n, 1)$ ways to choose 1 element from n element.

(*ii*) Similarly, $$\sum_{1 \le i < j \le n} N(P_i P_j) = C(n, 2)\,(n-2)!$$

and in general $$\sum_{1 \le i_1 < i_2 < \ldots < i_m \le n} N(P_{i_1} P_{i_2} \ldots P_{i_m}) = C(n, m)\,(n-m)!$$

as there are $C(n, m)$ ways to choose m elements from n elements $(m \le n)$.

Consequently, inserting these quantities into the formula for D_n, we get

$$D_n = n! - C(n, 1)\,(n-1)! + C(n, 2)(n-2)! - \ldots + (-1)^n C(n, n)\,(n-n)!$$

$$= n! - \frac{n!}{1!(n-1)!} \cdot (n-1)! + \frac{n!}{2!(n-2)!} \cdot (n-2)! - \ldots + (-1)^n \frac{n!}{n!0!} \cdot 0!$$

$$= n!\left\{1 - \frac{1}{1!} + \frac{1}{2!} - \ldots + (-1)^n \frac{1}{n!}\right\}.$$

Example 1: *At a party there are n men and n women. Find the number of ways n women choose male partners*

(*i*) *for the first dance,*

(*ii*) *for the second dance provided everyone has to change partner.*

Solution: (*i*) For the first dance, the number of ways to choose male partners is $n!$.

(*ii*) For the second dance, every woman has to choose as a partner a man other than the one with whom she first danced. So, for this case the number of ways is the nth derangement number

$$D_n = n!\left\{1 - \frac{1}{1!} + \frac{1}{2!} - \frac{1}{3!} + \ldots + (-1)^n \frac{1}{n!}\right\}.$$

Example 2: *Find the total number of derangements of 1234.*

Solution: The possible derangements of 1234 are

2143	2341	2413
3142	3421	3412
4123	4321	4312

Thus there are 9 possible derangements of 1234 and this can be verified by the formula:

$$D_n = n\,!\left\{1 - \frac{1}{1!} + \frac{1}{2!} - \frac{1}{3!} + \ldots + (-1)^n \frac{1}{n!}\right\}$$

Putting $n = 4$, we have

$$D_4 = 4\,!\left\{1 - \frac{1}{1!} + \frac{1}{2!} - \frac{1}{3!} + \frac{1}{4!}\right\}$$

$$= \frac{4\,!}{2\,!} - \frac{4\,!}{3\,!} + \frac{4\,!}{4\,!} = 12 - 4 + 1 = 9.$$

Example 3: *Find the number of ways of putting 5 letters in the addressed envelope in such a way that*

(*i*) *no letter is being put in the right envelope,*

(*ii*) *at least two of them in the wrong envelope.*

Solution: (*i*) Here $n = 5$ and so the required number of putting the letters is given by

$$D_5 = 5\,!\left\{1 - \frac{1}{1!} + \frac{1}{2!} - \frac{1}{3!} + \frac{1}{4!} - \frac{1}{5!}\right\}$$

$$= \frac{5\,!}{2\,!} - \frac{5\,!}{3\,!} + \frac{5\,!}{4\,!} - \frac{5\,!}{5\,!} = 60 - 20 + 5 - 1 = 44.$$

(*ii*) The number of ways in which at least 2 of them are in the wrong envelope is

$$= \sum_{r=2}^{5} C(5, r)\, D_r = C(5, 2)\, 2\,!\left\{1 - \frac{1}{1!} + \frac{1}{2!}\right\} + C(5, 3)\, 3!\left\{1 - \frac{1}{1!} + \frac{1}{2!} - \frac{1}{3!}\right\}$$

$$+ C(5, 4)\, 4\,!\left\{1 - \frac{1}{1!} + \frac{1}{2!} - \frac{1}{3!} + \frac{1}{4!}\right\}$$

$$+ C(5, 5)\, 5\,!\left\{1 - \frac{1}{1!} + \frac{1}{2!} - \frac{1}{3!} + \frac{1}{4!} - \frac{1}{5!}\right\}$$

$$= \frac{1}{2} \times 5 \times 4 + \frac{1}{2} \times 5 \times 4 \times (3 - 1) + 5\left\{\frac{4!}{2!} - \frac{4!}{3!} + \frac{4!}{4!}\right\} + \frac{5!}{2\,!} - \frac{5!}{3\,!} + \frac{5!}{4\,!} - \frac{5!}{5\,!}$$

$$= 10 + 20 + 45 + (60 - 20 + 5 - 1) = 119.$$

Example 4: (Hatcheck Problem)

A new employee checks the hats of n people at a restaurant, forgetting to put claim check numbers on the hats. When customers return for their hats, the checker gives them back hats chosen at random from the remaining hats. What is the probability that no one receives the correct hat?

Solution: The answer is the number of ways the hats can be arranged so that there is no hat in its original position divided by n ! (the number of permutations of n hats).

$$\therefore \quad \text{Required probability} = \frac{D_n}{n!} = 1 - \frac{1}{1!} + \frac{1}{2!} - \ldots + (-1)^n \frac{1}{n!}.$$

The values of this probability are displayed in the following table:

n	2	3	4	5	6	7
$\frac{D_n}{n!}$	0.50000	0.33333	0.37500	0.36667	0.36806	0.36786

Note: We know that

$$e^{-1} = 1 - \frac{1}{1!} + \frac{1}{2!} - \ldots + (-1)^n \frac{1}{n!} + \ldots \approx 0.368.$$

Since this is an alternating series with $t_n \to 0$ as $n \to \infty$, the probability that no one receives the correct hat converges to $e^{-1} \approx 0.368$.

ILLUSTRATIVE EXAMPLES—III

Example 1: *If 11 natural numbers (positive integers) between 1 to 380 (both inclusive), show that there exist at least 2 among these 11 numbers whose difference is at most 38.*

Solution: Arrange the numbers 1, 2, 3, ..., 380 in 10 boxes in such a way that the first box contains 1, 2, ..., 38, the second contains 39, 40, ..., 76, etc. There are 11 numbers to be chosen. Select these numbers from these 10 boxes. By the pigeonhole principle, at least one box will contain two of these eleven numbers and so these two numbers differ by 38 or less.

Example 2: *If five distinct numbers are chosen from {1, 2, ..., 8}, prove that two of them are consecutive.*

Solution: Let the chosen distinct numbers be a_1, a_2, a_3, a_4, a_5 and so $a_1 + 1, a_2 + 1, a_3 + 1, a_4 + 1, a_5 + 1$ be their successors.

If one of $a_1, a_2, \ldots, a_5$ is 8, say, $x_i = 8$, then $x_i + 1 = 9 \notin \{1, 2, \ldots, 8\}$. So, 9 of these 10 numbers ($a_1, a_2, \ldots a_5, a_1 + 1, \ldots, a_5 + 1$) lie between 1 and 8. On the other hand, if none of the $a_1, a_2, \ldots a_5$ is 8, then all the 10 numbers lie between 1 and 8. So, at least nine of these ten numbers lie between 1 and 8. By pigeonhole principle, two of these nine numbers are equal. Let them be a_i and $a_j + 1$ form some i and j (since $a_1, a_2, \ldots, a_5$ are distinct). Then a_j and $a_i = a_j + 1$ are consecutive. **(Proved)**

Example 3: *Let S denotes a square with each side has length 2 unit. Determine the minimum number of points to be chosen arbitrarily from the interior of S such that distance between two of the points will be less than $\sqrt{2}$ unit.*

Solution: Join middle points of the opposite sides of S by line segments which partitions S into four subsquares S_1, S_2, S_3 and S_4. So, each side of every S_r ($r = 1, 2, 3, 4$) has a length 1 unit. Therefore, the length of the diagonal of every S_r is $\sqrt{2}$ unit.

If any two points belong to any of S_1, S_2, S_3 and S_4, then their distance will be less than $\sqrt{2}$ unit.

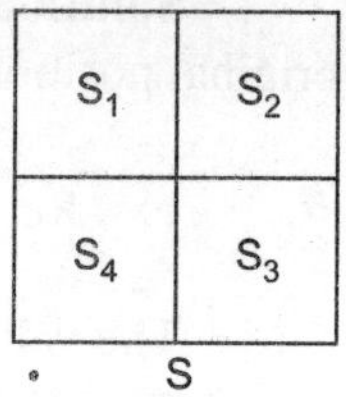

Fig. 4.5

Let us consider S_1, S_2, S_3 and S_4 as 4 pigeonholes. Using pigeonhole principle, if at least 5 points (pigeons) of the interior of S are chosen arbitrarily, two will belong to any of S_1, S_2, S_3 and S_4 and so the distance between two of the 5 arbitrarily selected points will be less than $\sqrt{2}$ unit.

Hence, the minimum number of points is 5.

Example 4: *If seven points are chosen within a regular hexagon with length of each side is 1, show that there are two points such that the distance between them is at most 1 unit.*

Solution: Let us divide the regular hexagon into six equilateral triangles by drawing three diagonals (as shown in the adjacent figure). The side of each equilateral triangle is of length 1 unit and so by pigeonhole principle, 2 points (out of 7 chosen points) must belong to the same triangle. So, the distance between these two points is at most 1 unit. Because the farthest points in an equilateral triangle are the vertices and the distance between the vertices is 1 unit.

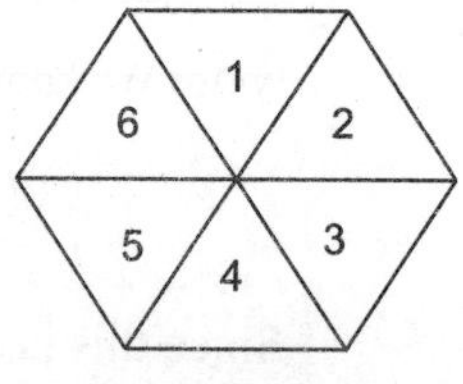

Fig. 4.6

Example 5: *If (x_i, y_i, z_i), $i = 1(1)\,9$, be a set of nine distinct points with integer co-ordinates in xyz-space, then show that the midpoint of at least one pair of these points has integer co-ordinates.*

Solution: The co-ordinates of the midpoints of the line segment joining the points (a, b, c) and (d, e, f) are $\{(a + d)/2, (b + e)/2, (c + f)/2\}$.

This midpoint has integer co-ordinates if and only if a and d have the same parity, b and e have the same parity, c and f have the same parity. Since there are $2^3 = 8$ possible triples of parity [such as (even, odd, odd) etc.], by the pigeonhole principle at least 2 of the 9 points have the same triple of parities. The midpoint of the line segment joining two such points has integer co-ordinates.

Example 6: *A computer network consists of 6 computers and each computer is directly connected to zero or more of the other computers. Show that there are at least two computers in the network that are directly connected to the same number of other computers.*

Solution: Since there are six computers, the number of other computers a computer is connected to is an integer between 0 and 5 (both inclusive). However, 0 and 5 cannot both occur. Because if some computer is connected to no others, then no computer is connected to all five others and if some computer is connected to all five others, then no computer is connected to no others. Since there are at most five possibilities for the number of computers a computer is connected to, using pigeonhole principle there are at least two computers in the set of six connected to the same number of others.

Example 7: *Among 100 students, 32 study mathematics, 20 study physics, 45 study chemistry, 7 study mathematics and physics, 15 study mathematics and chemistry, 10 study physics and chemistry, 30 do not study any of the three subjects. Determine the number of students studying exactly one of the three subjects.*

Solution: Let A, B, C be the sets of students studying mathematics, physics, chemistry respectively. Given, $|A| = 32$, $|B| = 20$, $|C| = 45$, $|A \cap B| = 7$, $|A \cap C| = 15$, $|B \cap C| = 10$ and $|(A \cup B \cup C)'| = 30$.

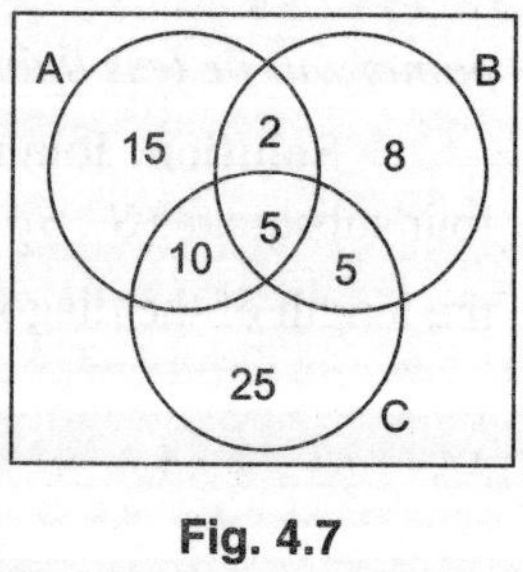

Fig. 4.7

So, $|A \cup B \cup C| = 100 - |(A \cup B \cup C)'| = 100 - 30 = 70$.

We know that

$$|A \cup B \cup C| = |A| + |B| + |C| - |A \cap B| - |A \cap C| - |B \cap C| + |A \cap B \cap C|$$

$\Rightarrow \quad 70 = 32 + 20 + 45 - 7 - 15 - 10 + |A \cap B \cap C|$

$\Rightarrow \quad |A \cap B \cap C| = 5.$

$\therefore$ Number of students studying mathematics only

$= 32 - (10 + 5 + 2) = 15.$

Number of students studying physics only

$= 20 - (5 + 5 + 2) = 8.$

Number of students studying chemistry only

$= 45 - (10 + 5 + 5) = 25.$

$\therefore$ Number of students studying exactly one subject = 15 + 8 + 25 = 48.

Example 8: *Find how many integers between 1 and 897 are divisible by 5 but not by 9?*

Solution: Total number of integers between 1 and 897 divisible by 5 is [897/5] = 179.

Total number of integers between 1 and 897 divisible by both 5 and 9 is [897/5 × 9] = 19.

Hence, the number of integers between 1 and 897 divisible by 5 but not by 9 is 179 – 19 = 160.

Example 9: *How many words of length 7 can be formed from the letters {a, b, c, d} if the first letter must be 'd' or the last letter be 'c'.*

Solution: There are 4^6 words that begin with '*d*', there are 4^6 words that end with '*c*' and there are 4^5 words that both begin with '*d*' and end with '*c*'. By the principle of inclusion-exclusion, there are $4^6 + 4^6 - 4^5 = 4^5 (8 - 1) = 7168$ words that begin with '*d*' or end with '*c*'.

Example 10: *Find the number of r-digit quintary sequences (made up with 0, 1, 2, 3) that contain at least one '0', one '1' and one '2' by using the principle of inclusion-exclusion.*

Solution: Let N be the number of r-digit quintary sequences and so $N = 4^r$. Let P_1, P_2 and P_3 respectively represent the property that r-digit quintary sequences contain no 0's, no 1's and no 2's. Then the number of r-digit quintary sequences that contain at least one 0, one 1 and one 2 is

$$N(P_1'P_2'P_3') = N - N(P_1) - N(P_2) - N(P_3) + N(P_1P_2) + N(P_1P_3) + N(P_2P_3) - N(P_1P_2P_3)$$

$$= 4^r - 3^r - 3^r - 3^r + 2^r + 2^r + 2^r - 1$$

$$= 4^r - 3 \times 3^r + 3 \times 2^r - 1.$$

Example 11: *How many positive integers not exceeding 500 are divisible by 7 or 11?*

Solution: Let A denotes the set of positive integers not exceeding 500 that are divisible by 7 and let B be the set of positive integers not exceeding 500 that are divisible by 11. So, $A \cup B$ is the set of positive integers not exceeding 500 that are divisible by 7 or 11 and $A \cap B$ is the set of positive integers not exceeding 500 that are divisible by both 7 and 11.

$$\therefore \quad |A \cup B| = |A| + |B| - |A \cap B|$$

$$= [500/7] + [500/11] - [500/(7 \times 11)]$$

$$= 71 + 45 - 6 = 110.$$

Therefore, there are 110 positive integers, not exceeding 500, that are divisible by 7 or 11.

Note: $[x]$ means the greatest integer not exceeding x.

Example 12: *Find the total number of integers n, $1 \le n \le 1000$, that are divisible by at least one of 2, 3 and 7.*

Solution: We know that the total number of integers from 1 to m that are divisible by k is $[m/k]$, where $[x]$ denotes the largest integer less than or equal to x.

Let A, B, C be the sets of integers between 1 and 1000 that are divisible by 2, 3, 7 respectively.

$\therefore$ $|A|$ = Number of integers between 1 and 1000 that are divisible by 2.

$= [1000/2] = 500.$

Similarly, $|B| = [1000/3] = 333$ and

$|C| = [1000/7] = 142.$

Now, AB, AC, BC and ABC are the sets of integers between 1 and 1000 that are divisible by 2 and 3; 2 and 7; 3 and 7; 2 and 3 and 7 respectively.

$\therefore$ $|AB| = [1000/6] = 166, |AC| = [1000/14] = 71,$

$|BC| = [1000/21] = 47$ and $|ABC| = [1000/42] = 23,$

since l.c.m. $\{2, 3\} = 6$, l.c.m. $\{2, 7\} = 14$, l.c.m. $\{3, 7\} = 21$ and l.c.m. $\{2, 3, 7\} = 42$.

$\therefore$ Required number of integers $= |A \cup B \cup C|$

$= |A| + |B| + |C| - |AB| - |AC| - |BC| + |ABC|$

$= 500 + 333 + 142 - 166 - 71 - 47 + 23 = 714.$

Example 13: *Find the total number of integers n, $1 \le n \le 1000$, that are not divisible by 5, 6 and 8.*

Solution: Let N be the total number of integers between 1 and 1000 (both inclusive) and so $N =$ 1000. Let A, B, C be the sets of integers between 1 and 1000 that are divisible by 5, 6, 8 respectively.

$\therefore$ $|A|$ = Number of integers between 1 and 1000 that are divisible by 5.

$= [1000/5] = 200.$

Similarly, $|B| = [1000/6] = 166$ and $|C| = [1000/8] = 125$. Now, AB, AC, BC and ABC are the sets of integers between 1 and 1000 that are divisible by 5 and 6; 5 and 8; 6 and 8; 5, 6 and 8 respectively.

$\therefore$ $|AB| = [1000/30] = 33, |AC| = [1000/40] = 25,$

$|BC| = [1000/24] = 41$ and $|ABC| = [1000/120] = 8,$

since l.c.m. $\{5, 6\} = 30$, l.c.m. $\{5, 8\} = 40$, l.c.m. $\{6, 8\} = 24$ and l.c.m. $\{5, 6, 8\} = 120$.

$\therefore$ Required number of integers $= |A'B'C'|$

$= N - |A| - |B| - |C| + |AB| + |AC| + |BC| - |ABC|$

$= 1000 - 200 - 166 - 125 + 33 + 25 + 41 - 8 = 600.$

Example 14: *Find the number of bit strings of length eight that either start with a 1 bit or end with the two bits 00.*

Solution: First task: Construction of a bit string of length eight starting with a 1 bit can be done in $2^7 = 128$ ways. It follows by the product rule, because the first bit can be chosen in only one way and each of the other seven bits can be chosen in two ways.

Second task: Construction of a bit string of length eight ending with the two bits 00 can be done in $2^6 = 64$ ways. It follows by the product rule, because each of the first six bits can be chosen in two ways and the last two bits can be chosen in only one way.

Both tasks: Construction of a bit string of length eight that starts with a 1 and ends with 00 can be done in $2^5 = 32$ ways. It follows by the product rule, because the first bit can be chosen in only one way, each of the second through the sixth bits can be chosen in two ways and the last two bits can be chosen in one way.

Conclusion: Therefore, the number of bit strings of length eight that start with a 1 or end with 00, which equals the number of ways to do either the first task or the second task, equals $128 + 64 - 32 = 160$.

Example 15: *Find the number of ways in which 10 books on Discrete Mathematics, 15 books on Java and 20 books on C^{++} can be arranged so that neither all the books on C^{++}, nor all the books on Java, nor still all the books on Discrete Mathematics are together.*

Solution: Let N be the total number of permutations of all 45 books, then

$$N = \frac{45!}{10!15!20!} \quad \text{(see Art. 4.3.5).}$$

Let P_1, P_2 and P_3 be the property that all the books on Discrete Mathematics, all the books on Java and all the books on C^{++} are together respectively. Denote a book on Discrete Mathematics, Java and C^{++} by D, J and C respectively. Then

$$N(P_1) = \frac{36!}{15!20!},\ N(P_2) = \frac{31!}{20!10!} \text{ and } N(P_3) = \frac{26!}{10!15!}$$

$N(P_1P_2)$ is the number of arrangements of the block of D's, the block of J's and 20 C's.

$$\therefore \qquad N(P_1P_2) = \frac{22!}{20!}$$

Similarly, $\qquad N(P_1P_3) = \frac{17!}{15!}$ and $N(P_2P_3) = \frac{12!}{10!}$

Now, $\qquad N(P_1P_2P_3) = 3!$

$$\therefore \quad N(P_1'P_2'P_3') = \frac{45!}{10!15!20!} - \left\{\frac{36!}{15!20!} + \frac{31!}{20!10!} + \frac{26!}{10!15!}\right\} + \frac{22!}{20!} + \frac{17!}{15!} + \frac{12!}{10!} - 3!$$

This is the required number of ways.

Example 16: *Find the number of primes not exceeding 100.*

Solution: We know that a composite integer is divisible by a prime not exceeding its square root. So, to find the number of primes not exceeding 100, first observe that composite integers not exceeding 100 must have a prime factor not exceeding 10. The only primes less than 10 are 2, 3, 5, 7 and so the primes not exceeding 100 are these 4 primes and those positive integer (> 1) and not exceeding 100 that are divisible by none of 2, 3, 5, 7. To apply the principle of inclusion and exclusion, let us assume the followings:

(*i*) P_1 be the property that an integer is divisible by 2.

(*ii*) P_2 be the property that an integer is divisible by 3.

(*iii*) P_3 be the property that an integer is divisible by 5.

(*iv*) P_4 be the property that an integer is divisible by 7.

Therefore, the number of primes not exceeding 100 is $4 + N(P_1'P_2'P_3'P_4')$.

As there are 99 positive integers greater than 1 and not exceeding 100, using the principle of inclusion-exclusion, we have

$$\begin{aligned} &N(P_1'P_2'P_3'P_4') \\ &= 99 - N(P_1) - N(P_2) - N(P_3) - N(P_4) + N(P_1P_2) + N(P_1P_3) + N(P_1P_4) + N(P_2P_3) + N(P_2P_4) \\ &\quad + N(P_3P_4) - N(P_1P_2P_3) - N(P_1P_2P_4) - N(P_1P_3P_4) - N(P_2P_3P_4) + N(P_1P_2P_3P_4). \end{aligned}$$

The number of integers (> 1) and not exceeding 100 that are divisible by all the primes in a subset of {2, 3, 5, 7} is [100/N], where N is the product of the primes in this subset (since any two of these primes have no common factor except 1). Consequently,

$N(P_1'P_2'P_3'P_4')$

$$= 99 - [100/2] - [100/3] - [100/5] - [100/7] + [100/2 \times 3]$$
$$+ [100/2 \times 5] + [100/2 \times 7] + [100/3 \times 5] + [100/3 \times 7]$$
$$+ [100/5 \times 7] - [100/2 \times 3 \times 5] - [100/2 \times 3 \times 7] - [100/2 \times 5 \times 7]$$
$$- [100/3 \times 5 \times 7] + [100/2 \times 3 \times 5 \times 7]$$
$$= 99 - 50 - 33 - 20 - 14 + 16 + 10 + 7 + 6 + 4 + 2 - 3 - 2 - 1 - 0 + 0 = 21.$$

Therefore, there are 4 + 21 = 25 primes not exceeding 100.

Note: An integer $p > 1$ is said to be a **prime number,** or simply a **prime,** if its only positive divisors are 1 and p.

An integer (> 1) which is not prime is said to be **composite.**

EXERCISE—4A

Section–1: Fill in the Blanks

1. The value of $^{10}C_8$ is _______ .
2. If $^nP_r = {^nC_r}$, then $r =$ _______ .
3. $^nC_r + {^nC_{r-1}} =$ _______ .
4. If $^nC_{r_1} = {^nC_{r_2}}$, then $r_1 = r_2$ or $r_1 + r_2 =$ _______ .
5. If $^nP_r = m\,{^nC_r}$, then $m =$ _______ .
6. In a football championship, there were played 153 matches. Every two teams played on match with each other. The number of teams, participating is _______ .
7. The faces of a cube can be painted by 6 different colours in _______ ways.
8. The total number of ways in which six '+' and four '–' signs can be arranged in a line such that no two '–' signs occur together is _______ .
9. The number of parallelograms that can be formed from a set of four parallel lines intersecting another set of three parallel lines is _______ .
10. The minimum number of students in a class to be sure that 4 out of them are born in the same month is _______ .
11. The total number of different bit strings having length 7 is _______ .
12. The minimum number of students required in a Discrete Mathematics class to be sure that at least six will receive the same grade, if there are five possible grades, is _______ .
13. If seven integers from 1 to 12 are chosen, then two of them will add upto _______ .
14. If 9 colours are used to paint 100 houses, then at least _______ houses will be of the same colour.
15. Among 50 students in a class, 26 passed in first semester and 21 passed in second semester examination. If 17 did not pass in either semester, then _______ passed in both semesters.

 [**Hint:** Given : $|(A \cup B)'| = 17 \;\Rightarrow\; |A \cup B| = 50 - 17 = 33.$

 $\Rightarrow \quad 33 = |A \cup B| = |A| + |B| - |A \cap B| = 26 + 21 - |A \cap B|$

 $\Rightarrow \quad |A \cap B| = 14.$]

Section–2: True/False

1. The number of ways in which 6 different books can be distributed among 5 students is 5^6.

   (*a*) True (*b*) False

2. The letters of the word PENCIL are arranged in all possible ways. The number of ways in which N always occur next to E is ${}^5P_5$.

   (*a*) True (*b*) False

3. Number of permutations of n different things taken r at a time when m particular things always occur is ${}^{n-m}P_{r-m}\, r\,!$

   (*a*) True (*b*) False

4. Number of arrangements of n distinct objects taken r at a time in which m particular objects never occur is ${}^{n-m}P_r$, provided $n - m \geq r$.

   (*a*) True (*b*) False

5. The total number of combinations of n different objects taken any number at a time is $2^n + 1$.

   (*a*) True (*b*) False

6. Number of combinations of n different things taken r at a time when m particular objects always occur is ${}^{n-m}C_{r-m}$.

   (*a*) True (*b*) False

7. If $n_1, n_2, ..., n_p$ are p positive integers whose sum is an even integer, then the number of odd integers among them is odd.

   (*a*) True (*b*) False

8. The number of solutions of $x + y + z = 17$ subject to $x \geq 1, y \geq 2$ and $z \geq 3$ is 78 (x, y, z are integers).

   (*a*) True (*b*) False

   [**Hint :** Put $x = u + 1, y = v + 2, z = w + 3$ and so the given equation becomes $u + v + w = 11$ with u, v, w are non-negative integers. The number of solutions is therefore $C(3 + 11 - 1, 11) = C(13, 11) = C(13, 2) = \frac{1}{2} \times 13 \times 12 = 78$ (see Art. 4.6).]

9. There are 220 positive integers not exceeding 1000 that are divisible by either 7 or 11.

   (*a*) True (*b*) False

   [**Hint :** $n(A \cup B) = n(A) + n(B) - n(A \cap B)$
   $= [1000/7] + [1000/11] - [1000/77]$
   $= 142 + 90 - 12 = 220$]

10. The number of onto functions from {L, A, B, O, U, R} to {A, R, T} is 540.

    (*a*) True (*b*) False

11. The number of ways in which 6 different dolls can be distributed to 3 girls, so that each girl gets at least one doll, is $C(8, 5)$.

    (*a*) True (*b*) False

12. If ten people are chosen in any way, then at least two of them will have been born on the same day.

    (*a*) True (*b*) False

MULTIPLE CHOICE QUESTIONS

Section–3

1. Which one of the following is the value of n when ${}^nC_4 = {}^nC_3$?
   (a) 0 (b) 2 (c) 5 (d) 7.
2. 0 ! is always taken as
   (a) 1 (b) 0 (c) ∞ (d) none of these.
3. If $(n-1)! \times 210 = (n+2)!$ then $n =$
   (a) 7 (b) 6 (c) 5 (d) 4.
4. If ${}^{13}C_{3r} = {}^{13}C_{r+1}$, then $r =$
   (a) 2 (b) 3 (c) 4 (d) $\frac{1}{2}$.
5. The number of ways in which n distinct objects can be put into four different boxes is
   (a) ${}^nP_4$ (b) $4n$ (c) n^4 (d) 4^n.
6. There are 20 true-false questions. The number of ways they can be answered is
   (a) 20 ! (b) 20 (c) 2^{20} (d) 20^2.
7. The letters of the word ENGLISH are arranged in all possible ways. The number of ways in which G always occur next to N is
   (a) ${}^6P_6$ (b) ${}^6P_6 \times 2!$ (c) ${}^7P_2 \times {}^5P_5$ (d) ${}^7P_7$.
8. There are 3 letter boxes in a post office. In how many ways can a man post 5 distinct letters?
   (a) 3×5 (b) 3^5 (c) 5^3 (d) ${}^5P_3$.
9. The number of all 7-digit numbers, which have all the digits distinct, is
   (a) ${}^9C_7$ (b) 9^7 (c) $9 \times {}^9C_6$ (d) none of these.
10. The number of arrangements of n different objects taken r at a time which include a particular object is
    (a) $n\,{}^{n-1}P_r$ (b) $n\,{}^{n-1}P_{r-1}$ (c) $r\,{}^{n-1}P_r$ (d) $r\,{}^{n-1}P_{r-1}$.
11. The number of arrangements of n different objects taken r at a time which exclude a particular object is
    (a) ${}^{n-1}P_r$ (b) $r\,{}^{n-1}P_r$ (c) $r\,{}^{n-1}P_{r-1}$ (d) $n\,{}^{n-1}P_{r-1}$.
12. A coin is tossed n times, the number of all the possible outcomes is
    (a) $2n$ (b) 2^n (c) ${}^nC_2$ (d) ${}^nP_2$.
13. Number of diagonals of an n-sided polygon is
    (a) $\frac{1}{2}n(n-3)$ (b) ${}^nC_2$ (c) ${}^nC_2 - 1$ (d) none of these.
14. The number of selections of n distinct objects taken r at a time which include a particular object is
    (a) ${}^{n-1}C_r$ (b) $r\,{}^{n-1}C_{r-1}$ (c) ${}^{n-1}C_{r-1}$ (d) $n\,{}^{n-1}C_{r-1}$.
15. How many ways are there to travel in xyz space from the origin (0, 0, 0) to the point (4, 3, 5) by taking unit steps in positive x, y, z directions only?
    (a) 4 ! 3 ! 5 ! (b) 60 (c) 12 !/(5 ! 4 ! 3 !) (d) 3^{12}.

    [WBUT 2012]
16. The number of integers between 1 and 567 which are divisible by 3 but not by 5 is
    (a) 189 (b) 37 (c) 226 (d) 152.
17. In a survey conducted on 200 people, it was found that 140 drink tea, 80 drink coffee and 40 drink both. The number of people drinking at least one beverage is
    (a) 220 (b) 180 (c) 200 (d) none of these.

18. Six gentlemen attend a party, where before joining the party, they leave their over-coats in a clock room. After the party, the overcoats get mixed up and are returned to the gentlemen in a random manner. The probability that none receives his own coat is

(*a*) e^{-1} (*b*) e^{-2} (*c*) e^{-3} (*d*) e^{-4}.

19. A cookie shop has 4 different kinds of cookies. Total number of different ways of choosing 6 cookies is

(*a*) 4^6 (*b*) 24 (*c*) 84 (*d*) 6^4.

20. If n be the minimum number of integers from {1, 2, ..., 100} so that the sum of the two of these n integers is even, then n =

(*a*) 3 (*b*) 4 (*c*) 5 (*d*) 6.

21. If n be the minimum number of integers from {1, 2, ..., 100} so that the difference of two of these n integers is a multiple of 5, then n =

(*a*) 3 (*b*) 4 (*c*) 5 (*d*) 6.

22. The number of integers from 100 to 1000 which are either even or end in a 7 is

(*a*) [900/7] (*b*) 541 (*c*) 361 (*d*) 900 – [900/7] .

23. The number of 3-digit decimal strings that have exactly two 8's is

(*a*) 9 (*b*) 18 (*c*) 27 (*d*) 30.

24. The number of 3-digit numbers (from 100 to 999) that have exactly two 8's is

(*a*) 10 (*b*) 20 (*c*) 29 (*d*) 30.

25. If n pigeonholes are occupied by $(n + 1)$ or more pigeons, then the number of pigeonholes occupied by more than one pigeon is

(*a*) at least one (*b*) exactly one (*c*) at most one (*d*) none of these.

26. In a department with 13 professors, the number of professors born in the same month

(*a*) is 2 (*b*) is 3 (*c*) is 4 (*d*) cannot be decided.

27. A bag contains balls of four different colours. The smallest number of balls to be drawn so that two among the drawn balls are of the same colour is

(*a*) 3 (*b*) 4 (*c*) 5 (*d*) 6.

28. Total number of binary strings of length n is

(*a*) n (*b*) 2^n (*c*) $2^{n+1} - 1$ (*d*) n^2.

29. Total number of onto functions from $\{a, b, c, d, e\}$ to $\{a, b, c\}$ is

(*a*) 3 (*b*) 5 (*c*) 15 (*d*) 150.

30. The total number of positive integers having at most 3 digits and divisible by 6 or 13 is

(*a*) 230 (*b*) 226 (*c*) 224 (*d*) 220.

31. The number of ways of assigning 3 rooms to 20 students in a dormitary is

(*a*) $C(20, 3)$ (*b*) 3^{20} (*c*) $3^{20} - 3 \times 2^{20} + 3$ (*d*) $20^3 - 20 \times 2^3 + 3$.

32. The number of non-negative integer solutions of $x_1 + x_2 + x_3 = 5$ is

(*a*) $C(6, 5)$ (*b*) $C(7, 5)$ (*c*) $C(8, 5)$ (*d*) $C(9, 5)$.

33. The number of integer solutions of $x + y + z = 5$; $x, y, z \geq 1$ is

(*a*) $C(7, 5)$ (*b*) $C(7, 2)$ (*c*) $C(4, 2)$ (*d*) $C(7, 3)$.

34. Let A and B be sets with cardinalities (number of elements) m and n respectively. The number of one-one mappings (injections) from A to B, when $m < n$, is

(*a*) m^n (*b*) ${}^nP_m$ (*c*) ${}^mC_n$ (*d*) ${}^nC_m$

(*e*) ${}^mP_n$. [*GATE 1992*]

35. How many 4-digit even numbers have all 4 digits distinct?

(*a*) 2240 (*b*) 2296 (*c*) 2620 (*d*) 4536. [*GATE 2001*]

36. If ${}^nC_{r-1} = 36$, ${}^nC_r = 84$ and ${}^nC_{r+1} = 126$, then n and r are respectively.

(*a*) 9, 6 (*b*) 9, 3

(*c*) 6, 3 (*d*) 6, 2. [*AIEEE 2002*]

ANSWERS

Section–1

1. 45	**2.** 0 or 1	**3.** ${}^{n+1}C_r$	**4.** n
5. $r\,!$	**6.** 18	**7.** 1	**8.** ${}^7C_4$
9. ${}^4C_2 \times {}^3C_2 = 18$	**10.** 37	**11.** 128	**12.** 26
13. 13	**14.** 12	**15.** 14	

Section–2

1. (*a*)	**2.** (*a*)	**3.** (*b*)	**4.** (*a*)
5. (*b*)	**6.** (*a*)	**7.** (*b*)	**8.** (*a*)
9. (*a*)	**10.** (*a*)	**11.** (*b*)	**12.** (*a*)

Section–3

1. (*d*)	**2.** (*a*)	**3.** (*c*)	**4.** (*b*)
5. (*d*)	**6.** (*c*)	**7.** (*a*)	**8.** (*b*)
9. (*c*)	**10.** (*d*)	**11.** (*a*)	**12.** (*b*)
13. (*a*)	**14.** (*c*)	**15.** (*c*)	**16.** (*d*)
17. (*b*)	**18.** (*a*)	**19.** (*c*)	**20.** (*a*)
21. (*d*)	**22.** (*b*)	**23.** (*d*)	**24.** (*c*)
25. (*a*)	**26.** (*d*)	**27.** (*c*)	**28.** (*b*)
29. (*d*)	**30.** (*a*)	**31.** (*c*)	**32.** (*b*)
33. (*c*)	**34.** (*c*)	**35.** (*d*)	**36.** (*b*).

[Hint of 20 : Let $S_1 = \{1, 3, 5, ..., 99\}$ and $S_2 = \{2, 4, 6, ..., 100\}$ (pigeonholes). Then given three integers from $\{1, 2, ..., 100\}$, two of them belong to the same set S_1 or S_2 and their sum is even. So, $n = 3$.

Hint of 21: Let $S_1 = \{1, 6, ..., 96\}$, $S_2 = \{2, 7, ..., 97\}$, $S_3 = \{3, 8, ..., 98\}$, $S_4 = \{4, 9, ..., 99\}$ and $S_5 = \{5, 10, ..., 100\}$ (pigeonholes). Then given six integers from $\{1, 2, ..., 100\}$, two of them belong to the same subset S_i and their difference is a multiple of 5. So, $n = 6$]

EXERCISE—4B

1. Explain the concepts of permutation and combination.

2. Find n in each of the following cases:

(*i*) ${}^nP_5 : {}^nP_3 = 2 : 1$ (*ii*) ${}^{n-1}P_3 : {}^{n+1}P_3 = 5 : 12$

(*iii*) ${}^{n+2}P_8 : {}^{n-2}P_4 = 57 : 16$

3. Find r in each of the following cases:

(*i*) ${}^{15}C_{3r} = {}^{15}C_{r+3}$ (*ii*) ${}^{56}P_{r+6} : {}^{54}P_{r+3} = 30800 : 1$

(*iii*) ${}^{2n}C_r = {}^{2n}C_{r+2}$ (*iv*) ${}^nP_r : {}^nC_{n-r} = 120 : 1$

4. Find n and r in each of the following cases:

(*i*) ${}^{n+r}P_2 = 56$ and ${}^{n-r}P_2 = 12$. (*ii*) ${}^nP_r = {}^nP_{r+1}$ and ${}^nC_r = {}^nC_{r-1}$.

5. Prove that

(i) ${}^nC_r = \dfrac{n-r+1}{r}\,{}^nC_{r-1}$ (ii) ${}^nC_r + {}^{n-1}C_{r-1} + {}^{n-1}C_{r-2} = {}^{n+1}C_r$.

(iii) ${}^nP_r = n\,{}^{n-1}P_{r-1} = (n-r+1)\,{}^nP_{r-1}$ (iv) ${}^1P_1 + 2\,{}^2P_2 + 3\,{}^3P_3 + \ldots + n\,{}^nP_n = {}^{n+1}P_{n+1} - 1$.

6. Find the number of words formed out of the letters of the words ARTICLE such that vowels occupy even places.

7. In how many other ways can the letters of the word GANESHPURI be rearranged?

8. Find the number of different words that can be formed with the letters of the word ORDINATE such that

(i) Beginning with O (ii) Beginning with O and ending with E

(iii) The vowels occupy odd places.

9. Find how many words can be formed of the letters in the word FAILURE, the four vowels always coming together.

10. Find the number of ways of arranging the letters

AAAAA *BBB* *CCC* *D* *EE* *F*

in a row if the letters B are separated from one another.

11. (i) How many different words can be constructed by taking 4 letters at a time out of the letters of the word MATHEMATICS?

(ii) A five letter word is to be formed such that the letters appearing in the odd positions are taken from the unrepeated letters of the word MATHEMATICS whereas the letters which occupy even places are taken from among the repeated letters.

12. How many words each containing five different letters can be formed by taking 2 consonants from 9 different consonants and 3 vowels from 5 different vowels?

13. Find the number of different words that can be formed from 12 consonants and 5 vowels by taking 4 consonants and 3 vowels in each word.

14. How many different numbers between 4000 and 5000 can be formed using the digits 2, 3, 4, 5, 6, 7?

15. How many odd numbers of 5 significant digits can be formed using the digits 3, 2, 7, 4, 0?

16. Find the number of different factors of 3528 which are greater than 1 and less than 3528.

17. How many numbers greater than 23000 can be formed from the digits 1, 2, 3, 4, 5?

18. Out of 18 points in a plane, no three are in the same straight line except five points which are collinear. How many

(i) straight lines (ii) triangles can be formed by joining them?

19. In a plane there are 10 points out of which no three are collinear except the four which lie on a straight line. By joining these 10 points how many (a) straight lines, (b) triangles can be obtained?

20. If 20 straight lines be drawn on a plane, no two of them being parallel and no three of them being concurrent, how many points of intersection will be there?

21. Find the number of committees of 5 members that can be formed from 6 gentlemen and 4 ladies if each committee has at least one lady and two gentlemen.

22. A committee of 5 is to be formed from six ladies and four gentlemen. In how many ways this can be done so that the committee contains (i) exactly two ladies (ii) at least two ladies (iii) at most two ladies?

23. A candidate is required to answer 6 out of 10 questions which are divided into 2 groups each containing 5 questions, and he/she is not permitted to attempt more than 4 from any group. In how many different ways can he/she make up his choice?

24. How many different triangles can be formed by joining the angular points of a decagon? Find also the number of the diagonals of the decagon.

25. At an election there are 5 candidates and 3 members are to be elected and a voter is entitled to vote for any number to be elected. In how many ways may a voter choose a vote?

26. From 6 gentlemen and 4 ladies, a committee of 5 is formed. In how many ways can this be done, so as to include at least one lady ?

27. In a group of 15 boys there are 7 boy-scouts. In how many ways can 12 boys be elected, so as to include (*i*) exactly 6 boy-scouts; (*iii*) at least 6 boy-scouts?

28. A cricket team consisting of 11 players is to be selected from 2 groups consisting of 6 and 8 players respectively. In how many ways can the selection be made on the supposition that the group of six shall contribute no fewer than 4 players?

29. Out of 17 consonants and 5 vowels, how many different words can be formed, each consisting of 3 consonants and 2 vowels?

30. If P_r denotes the number of permutations of n different things taken r at a time, then show that

$$\frac{P_1}{1!}+\frac{P_2}{2!}+\frac{P_3}{3!}+\ldots+\frac{P_n}{n!}=2^n-1.$$

$$\left[\textbf{Hint: } \frac{{}^nP_r}{r!}={}^nC_r\right]$$

31. Find the number of numbers less than 1000 and divisible by 5 which can be formed using the digits 0, 1, 2, 3, 4, 5, 6, 7, 8, 9, each digit occurring not more than once in each number.

32. In how many ways can a pack of 52 cards be divided equally among 4 players in order?
[**Hint :** 13 cards can be given to first in ${}^{52}C_{13}$ ways. Similarly to second in ${}^{39}C_{13}$ ways etc.
$\therefore$ Total number of ways = ${}^{52}C_{13}\times{}^{39}C_{13}\times{}^{26}C_{13}\times{}^{13}C_{13}$].

33. In how many ways can you divide 52 cards in 4 sets, three of them having 17 cards each and the fourth one just 1 card?
[**Hint:** ${}^{52}C_{17}\times{}^{35}C_{17}\times{}^{18}C_{17}\times{}^{1}C_{1}$]

34. To fill 12 vacancies, there are 25 candidates of which 5 are from scheduled castes. If 3 vacancies are reserved for scheduled caste candidates while the rest are open to all, find the number of ways in which the selections can be made.

35. From 4 officers and 8 Jawans in how many ways can 6 be chosen.
(*i*) to include exactly one officer. (*ii*) to include at least one officer.

36. A debating team consists of r boys and s girls. Find the number n of ways they can sit in a row if (*i*) the boys and girls are each to sit together, (*ii*) just the girls are to sit together.

37. If any five numbers from 2 to 9 are chosen, then show that two of them will add to 11.

38. What is the minimum number of elements that one needs to take from the set $S = \{1, 2, ..., 9\}$ to be sure that two of the numbers add up to 10?

39. (*i*) How many cards must be drawn from a standard deck of 52 cards to be sure that at least three cards of the same suit are chosen?
(*ii*) How many must be drawn to guarantee that at least three hearts are chosen?

40. Prove that if any eight positive integers are selected, two of them will have the same remainder when divided by 7.

41. If eleven integers are chosen from the set $\{1, 2, 3, ..., 20\}$, then show that one of them must be a multiple of another.

42. During a month with 30 days a baseball team plays at least one game per day, but not more than 45 games. Prove that there must be a period of some number of consecutive days during which the team must play exactly 14 games.

43. A drawer contains a dozen blue socks and a dozen black socks, all unmatched. A man takes socks out at random in the dark.

(*i*) How many socks must be taken out to be sure that he has at least two socks of same colour?

(*ii*) How many socks must be taken out to be sure that he has at least two black socks?

44. A computer network consists of 6 computers. Each computer is directly connected to at least one of the other computers. Show that there are at least 2 computers in the network that are directly connected to the same number of other computers.

45. State the principle of inclusion-exclusion. Find the number of positive integers n, $1 \le n \le 2000$ that are not divisible by 2, 3 or 5 but are divisible by 7.

46. Survey is taken on methods of computer travel. Each respondent is asked to check bus, train or automobile, as a major method of travelling to work. More than one answer is permitted. The results reported were as follows:

(*i*) 30 people checked bus
(*ii*) 35 people checked train
(*iii*) 100 people checked automobile
(*iv*) 15 people checked bus and train
(*v*) 15 people checked bus and automobile
(*vi*) 20 people checked train and automobile
(*vii*) 5 people checked all three methods

How many respondents completed their surveys?

47. Survey of 500 television watchers produced the following information:

285 watch football games; 195 watch hockey games; 115 watch basket ball games; 50 watch hockey and basket ball games; 50 do not watch any of the three games; 45 watch football and basket ball games; 70 watch football and hockey games.

(*i*) How many people in the survey watch all three games?

(*ii*) How many people watch exactly one of the three games?

48. How many solutions are there for the equation $x_1 + x_2 + x_3 + x_4 = 20$?

(*i*) If all x_i must be non-negative integers.

(*ii*) If all x_i must be non-negative integers and x_4 is atmost 10.

49. Find the number of integral solutions of the equation

$$x_1 + x_2 + x_3 + x_4 = 20$$

subject to $x_1 \ge 3, x_2 \ge 1, x_3 \ge 0, x_4 \ge 5.$

50. What is the number of integral solutions of the equation

$$x_1 + x_2 + x_3 = 11$$

subject to $x_1 \le 3, x_2 \le 4$ and $x_3 \le 6.$

51. After a test, 10 students collect their books (which are identical) which were left outside the examination hall before the test. None of them receive their own book. Find the number of ways in which this can happen. Also, find the number of ways of just one student getting his own book.

ANSWERS

2. (*i*) 5 (*ii*) 8 (*iii*) 19

3. (*i*) 3 (*ii*) 41 (*iii*) $n - 1$ (*iv*) 5

4. (*i*) $n = 6, r = 2$ (*ii*) $n = 3, r = 2$

6. 144 **7.** 10 ! – 1

8. (*i*) 7 ! (*ii*) 6 ! (*iii*) 576

9. 576 **10.** 95135040

11. (*i*) 2454 (*ii*) 540
12. 43200 **13.** $4950 \times 7!$ **14.** 60 **15.** 36
16. 34 **17.** 90
18. (*i*) 144 (*ii*) 806
19. (*a*) 40 (*b*) 116 **20.** 190 **21.** 240
22. (*i*) 60 (*ii*) 246 (*iii*) 66
23. 200 **24.** 120, 35 **25.** 25 **26.** 246
27. (*i*) 196 (*ii*) 252 **28.** 344 **29.** 816000
31. 154 **32.** $\dfrac{52\,!}{(13\,!)^4}$ **33.** $\dfrac{52\,!}{(17\,!)^3}$ **34.** 1679600
35. (*i*) 224 (*ii*) 896 **38.** 6
39. (*i*) 9 (*ii*) 42 **43.** (*i*) 3 (*ii*) 14
46. 120 **47.** (*i*) 20 (*ii*) 325
48. (*i*) $C(23, 20)$ (*ii*) $C(23, 20) - 66$ **49.** 364 **50.** 6
51. $10\,!\left\{1-\dfrac{1}{1!}+\dfrac{1}{2\,!}-...+\dfrac{1}{10\,!}\right\}$;

$$10\,!\left\{1-\frac{1}{1!}+\frac{1}{2\,!}-...-\frac{1}{9\,!}\right\}.$$

Module III

CHAPTER 5

Recurrence Relations

5.1 INTRODUCTION

Many counting problems cannot be solved easily using the methods studied in the previous chapter. As for example, how many bit strings of length n do not have two consecutive ones? To solve this problem, let a_n be the number of bit strings of length n that do not contain the pattern 11. An argument can be given that shows $a_n = a_{n-1} + a_{n-2}$, $n \geq 3$ with initial conditions $a_1 = 2$ and $a_2 = 3$ (see Example 9). This equation is called a **recurrence relation** (or **difference equation**).

As computers mostly deal with discrete quantities, the theory of recurrence relations (dealing with discrete quantities) has assumed importance in applications to computer science in general and analysis of algorithms in particular. In this chapter, we will study formulation of different counting problems in terms of recurrence relations and discuss solution of linear recurrence relations with constant coefficients (upto second order) by (*i*) iterative method (*ii*) characteristic roots method and (*iii*) generating functions method.

5.2 RECURRENCE RELATION AND ITS SOLUTION

Definition 1: A **recurrence relation** for the sequence (function of positive integers) $\{a_n\}$ is an equation that relates a_n in terms of one or more of the preceeding terms of the sequence, namely, $a_0, a_1, \ldots, a_{n-2}, a_{n-1}$, for all integers n with $n \geq n_0$, where n_0 is a non-negative integer.

A sequence is called a **solution** of a recurrence relation if the terms of the sequence satisfy the recurrence relation.

Example 1: If a_n be the nth term of the sequence {3, 5, 7, 9, 11, 13, ...}, then the recurrence relation corresponding to this sequence is $a_n = a_{n-1} + 2$, $n \geq 2$, with initial condition $a_1 = 3$.

Example 2: *Show that the sequence* $\{2, 3, 4, 5, \ldots, n+1, \ldots\}$ *for* $n \geq 1$ *satisfies the recurrence relation* $a_n = 2a_{n-1} - a_{n-2}$, $n \geq 3$.

Solution: If a_n denotes the nth term of the given sequence, then $a_n = n + 1$, $n \geq 1$.

$\therefore$ $\quad a_{n-1} = (n-1) + 1 = n$ and $a_{n-2} = (n-2) + 1 = n - 1$.

Now, $\quad 2a_{n-1} - a_{n-2} = 2n - (n-1) = n + 1 = a_n$.

$\therefore$ $\quad a_n = 2a_{n-1} - a_{n-2}$, $n \geq 3$. Hence the result.

Example 3: *Prove that the sequence* $\{a_n\}$ *defined by* $a_n = 4.2^n + 7.3^n$ *is a solution of the recurrence relation* $a_n - 5a_{n-1} + 6a_{n-2} = 0$.

Solution: Given: $a_n = 4.2^n + 7.3^n$

$\Rightarrow$ $a_{n-1} = 4.2^{n-1} + 7.3^{n-1}$; $a_{n-2} = 4.2^{n-2} + 7.3^{n-2}$

$\therefore$ $a_n - 5a_{n-1} + 6a_{n-2}$

$$= 4 \,.\, 2^n + 7.3^n - 5(4.2^{n-1} + 7.3^{n-1}) + 6\,(4.\, 2^{n-2} + 7.3^{n-2})$$

$$= 4(2^n - 5 \,.\, 2^{n-1} + 6 \,.\, 2^{n-2}) + 7(3^n - 5 \,.\, 3^{n-1} + 6 \,.\, 3^{n-2})$$

$$= 4 \,.\, 2^{n-2}\,(4 - 10 + 6) + 7 \,.\, 3^{n-2}\,(9 - 15 + 6) = 0.$$

Hence the result.

Observations: $a_n = A \,.\, 2^n + B \,.\, 3^n$, $b_n = A \,.\, 2^n$ and $c_n = B \,.\, 3^n$ are also solutions of the recurrence relation $a_n - 5a_{n-1} + 6a_{n-2} = 0$.

Example 4: *Determine a recurrence relation for the sequence* $\{a_n\}$ *given by* $a_n = A \,.\, (-2)^n + B \,.\, 3^n$.

Solution: Given, $a_n = A \,.\, (-2)^n + B \,.\, 3^n$ ...(1)

Replacing n by $n - 1$, we get

$$a_{n-1} = A \,.\, (-2)^{n-1} + B \,.\, 3^{n-1} \quad ...(2)$$

From (1) and (2), we get (eliminating B) :

$$a_n - 3a_{n-1} = A\{(-2)^n - 3 \,.\, (-2)^{n-1}\} = A \,.\, (-5)\,(-2)^{n-1} \quad ...(3)$$

Replacing n by $n - 1$ in (3), we get

$$a_{n-1} - 3a_{n-2} = A\,(-5)\,(-2)^{n-2}$$

$\Rightarrow$ $-2\,(a_{n-1} - 3a_{n-2}) = A\,(-5)\,(-2)^{n-1}$

$\Rightarrow$ $-2\,(a_{n-1} - 3a_{n-2}) = a_n - 3a_{n-1}$ [by (3)]

$\Rightarrow$ $a_n - a_{n-1} - 6a_{n-2} = 0$

This is the required recurrence relation.

5.3 FORMULATION (OR MODELLING) OF DIFFERENT COUNTING PROBLEMS IN TERMS OF RECURRENCE RELATIONS

Recurrence relations can be used to model a wide variety of counting problems, such as finding compound interest, counting rabbits on an island and counting bit strings with certain properties. If a recurrence relation is solved, then we can understand that the counting problem is solved.

Example 5: *Soham deposits ₹ 90,000 in a savings account at a bank. The bank gives 9.5% interest compounded annually. Formulate the recurrence relation to compute the amount Soham will have in his account at the end of* n^{th} *year.*

Solution: Let P_n denote the amount in the account after n years.

$\therefore$ P_o = The initial amount = ₹ 90,000

Now, P_n = The amount at the end of $(n-1)^{th}$ year + the interest (on it) for the nth year

$$= P_{n-1} + \frac{9.5}{100}\,P_{n-1} = 1.095\,P_{n-1}$$

Hence the required recurrence relation is $P_n = 1.095\,P_{n-1}$, with P_o = ₹ 90,000.

Example 6: *There are guests in a gathering and each guest shakes hand with everyone else exactly once. Construct a recurrence relation representing the number of handshakes occurred in the gathering.*

Solution: Let the number of guests present in the gathering be n and let a_n denote the number of handshakes occurred in the gathering. Obviously $a_1 = 0$.

$\therefore$ a_n = Number of handshakes occurred among the gathering of $(n - 1)$ guests + number of handshakes done by the nth guest with the $(n - 1)$ number of guests

$= a_{n-1} + (n - 1)$.

Hence the required recurrence relation is

$a_n = a_{n-1} + (n - 1)$, with $a_1 = 0$.

Example 7: *A person climbs a staircase by climbing either (i) only one step in a single stride or (ii) two steps in a single stride. Construct a recurrence relation for the number of ways of climbing n stairs.*

Solution: Let a_n represent the number of ways of climbing n stairs. In arriving n steps, the person can climb either one step or two steps in his last stride. For these two choices, the number of ways are a_{n-1} and a_{n-2}. Therefore, $a_n = a_{n-1} + a_{n-2}$, this is the required recurrence relation.

Example 8: *Lieonardo Fibonacci, an Italian mathematician, proposed the following problem around the year 1202 :*

If (i) two new born rabbits, one male and one female, are placed on an island;

(ii) when they are 2 months old, each mixed pair gives birth to a mixed pair and in each month afterwards,

(iii) no rabbit dies.

Construct a recurrence relation for the number of pairs of rabbits on the island at the beginning of the nth month.

Solution: Let us denote by a_n the number of pairs of rabbits at the beginning of the nth month. Since a pair can give births to a baby pair only after 2 months, $a_1 = a_2 = 1$. One baby pair is born at the beginning of the third month and so $a_3 = 1 + 1 = 2$. The calculations for nine months is given in the following table:

Number of pairs	*Months*								
	1	2	3	4	5	6	7	8	9
Adult	0	1	1	2	3	5	8	13	21
Baby	1	0	1	1	2	3	5	8	13
Total	1	1	2	3	5	8	13	21	34

From the table, it is clear that $a_n = a_{n-1} + a_{n-2}$. This is the required recurrence relation. The sequence $\{a_n\}$ satisfies the recurrence relation

$$a_n = a_{n-1} + a_{n-2}, \quad \text{for } n \geq 3$$

together with the initial conditions $a_1 = 1$ and $a_2 = 1$. This sequence is called **Fibonacci sequence.** This recurrence relation and the initial conditions uniquely determine the sequence $\{a_n\}$. The number of pairs of rabbits on the island at the beginning of nth month is given by a_n called nth **Fibonacci number.**

Example 9: *Determine a recurrence relation and give initial conditions for the number of bit strings of length n that do not have two consecutive 1. How many such bit strings are there of length six ?*

Solution: Let b_n denote the number of bit strings of length n that do not have two consecutive 1 and so can be counted as

(*i*) that begin with 0,

(*ii*) that begin with 10.

As the sets of strings of types (*i*) and (*ii*) are disjoint, using sum rule, b_n will be equal to the sum of the numbers of bit strings of types (*i*) and (*ii*). Suppose that an n bit string starts with 0 and does not contain the pattern 11. Observe that $(n-1)$ bit string not containing 11 can follow the initial 0, there are b_{n-1} such bit string of type (*i*). In an n bit string starts with 10 and does not contain pattern 11, the $(n-2)$ bit string following the initial 10 can not contain the pattern 11 and so there are b_{n-2} bit strings of type (*ii*).

Thus, $$b_n = b_{n-1} + b_{n-2}, \quad n \geq 3.$$

The initial conditions are $b_1 = 2$ (as 0 and 1 are the only strings of unit length, not containing the pattern 11) and $b_2 = 3$, since the valid bit strings (not containing the pattern 11) of length 2 are 00, 01, 10.

To obtain b_6, we use the recurrence relation four times to find that

$$b_3 = b_2 + b_1 = 3 + 2 = 5,$$
$$b_4 = b_3 + b_2 = 5 + 3 = 8,$$
$$b_5 = b_4 + b_3 = 8 + 5 = 13,$$
$$b_6 = b_5 + b_4 = 13 + 8 = 21.$$

Remark: Note that $\{b_n\}$ satisfies the same recurrence relation as the Fibonacci sequence. Since $b_1 = a_3$ and $b_2 = a_4$ (see Ex. 8), it follows that $b_n = a_{n+2}$.

Example 10: ***(Codeword enumeration):*** *A computer system considers a string of decimal digits as a valid codeword if it has an even number of 0 digits. Construct a recurrence relation for the number of valid n-digit codewords.*

Solution: Let a_n denote the number of valid n-digit codewords. So, $a_1 = 9$ as there are 10 one-digit strings and only one (the string 0) is not valid. A recurrence relation can be obtained for this sequence by considering how a valid n-digit string can be made from strings of $(n-1)$ digits. There are two ways to make a valid string with n digits from a string with $(n-1)$ digits:

(*i*) A valid string of n digits can be derived by appending a valid string of $(n-1)$ digits with a digit other than 0. This process can be done in 9 ways and so a valid string with n digits can be formed in this manner in $9a_{n-1}$ ways.

(*ii*) A valid string of n digits can be made by appending a 0 to a string of length $(n-1)$ that is invalid (it yields a string with an even number of 0 digits as an invalid string of length $n-1$ has an odd number of 0 digits). The number of ways it can be done equals the number of invalid $(n-1)$-digit strings. Since there are 10^{n-1} decimal strings of length $(n-1)$ and a_{n-1} are valid, there are $10^{n-1} - a_{n-1}$ valid n–digit strings obtained by appending an invalid string of length $(n-1)$ with a 0.

Since all valid strings of length n are produced in one of these two ways, using sum rule, it follows that there are

$$a_n = 9a_{n-1} + (10^{n-1} - a_{n-1})$$
$$= 8a_{n-1} + 10^{n-1}$$

valid strings of length n. So, the required recurrence relation is

$$a_n = 8a_{n-1} + 10^{n-1}, n \geq 2,$$

with initial condition $a_1 = 9$.

5.4 LINEAR RECURRENCE RELATION WITH CONSTANT COEFFICIENTS

Definition 2: A linear homogeneous recurrence relation with constant coefficients, for the sequence $\{a_r\}$, is a recurrence relation of the form:

$$a_n = c_1 a_{n-1} + c_2 a_{n-2} + \ldots\ldots + c_k a_{n-k} \qquad ...(1)$$

where $c_1, c_2, \ldots\ldots, c_k$ are real numbers with $c_k \neq 0$. The recurrence relation (1) is

(*i*) *linear* since the right-hand side is a sum of multiples of the previous terms (occurs to the first power) of a_n in the sequence $\{a_r\}$,

(*ii*) *homogeneous* since no terms occur that are not multiples of the a_is,

(*iii*) of *order k* because a_n is expressed in terms of the previous k terms of the sequence.

The coefficients of the terms of the recurrence relation (1) are all *constants*, rather than functions that depend on n.

Illustration: (*i*) The recurrence relation

$a_n = 2a_{n-1} + a_{n-3}^2$ is a 3rd order ($k = 3$) non-linear recurrence relation as a_{n-3}^2 is not of first degree (does not occur to the first power).

(*ii*) The recurrence relation $a_n = a_{n-1} + 1$ is a 1st order linear non-homogeneous (as the term 1 is present) recurrence relation with constant coefficients. The recurrence relations occurred in Examples 6 and 10 are also 1st order linear non-homogeneous recurrence relations with constant coefficients.

(*iii*) $b_n = nb_{n-1} + b_{n-2}$ is a 2nd order linear homogeneous recurrence relation with variable coefficient (as the coefficient n is not constant).

(*iv*) $f_n = f_{n-1}f_{n-2} + n$ is a 2nd order non-linear (as the term $f_{n-1}f_{n-2}$ is not of 1st degree) non-homogeneous (as the term n is present) recurrence relation with constant coefficient.

Note: Linear homogeneous recurrence relations are studied for two reasons: (*i*) they often occur in modelling (formulation) of problems, (*ii*) they can be systematically solved.

There are no general method that will enable one to solve all recurrence relations. The three methods of solving recurrence relations are discussed in the following articles.

5.5 SOLUTION OF LINEAR RECURRENCE RELATIONS WITH CONSTANT COEFFICIENTS BY ITERATIVE METHOD

In this method the recurrence relation for a_n is used repeatedly to solve for a general expression for a_n in terms of n which is described through the following examples.

ILLUSTRATIVE EXAMPLES—I

Example 1: *Solve the recurrence relation*

$$a_n = 2a_{n-1} + 1, \quad n \geq 2$$

subject to the initial condition $a_1 = 1$.

Solution: $a_n = 2a_{n-1} + 1$ (1st stage) ...(1)

Replacing n by $n-1, n-2, n-3, \ldots\ldots, 4, 3, 2$ in (1), we get

$$a_{n-1} = 2a_{n-2} + 1$$
$$a_{n-2} = 2a_{n-3} + 1$$
....
....
$$a_4 = 2a_3 + 1$$
$$a_3 = 2a_2 + 1$$
$$a_2 = 2a_1 + 1$$

$\therefore$ $a_n = 2(2a_{n-2} + 1) + 1 = 2^2 a_{n-2} + 2 + 1$ (2nd stage)

$= 2^2 (2a_{n-3} + 1) + 2 + 1 = 2^3 a_{n-3} + 2^2 + 2 + 1$ (3rd stage)

....

$= 2^k a_{n-k} + 2^{k-1} + 2^{k-2} + ... + 2^2 + 2 + 1$ (k^{th} stage)

$= ... = 2^{n-1} a_1 + 2^{n-2} + 2^{n-3} + ... + 2^2 + 2 + 1$ (stage $k = n - 1$)

$= 2^{n-1} + 2^{n-2} + 2^{n-3} + ... + 2^2 + 2 + 1$ ($\because$ $a_1 = 1$)

$\therefore$ $a_n = 2^n - 1, n \geq 2$, which is the required solution.

Example 2: *Solve the recurrence relation 'handshake' problem (see Ex. 6, Art. 5.3):*

$$a_n = a_{n-1} + (n - 1), n \geq 2$$

subject to the initial condition $a_1 = 0$.

Solution: $a_n = a_{n-1} + (n - 1)$ (1st stage) ...(1)

Replacing n by $n - 1, n - 2, n - 3, ..., 4, 3, 2$ in (1), we get

$$a_{n-1} = a_{n-2} + n - 1 - 1 = a_{n-2} + n - 2$$
$$a_{n-2} = a_{n-3} + n - 3$$
$$a_{n-3} = a_{n-4} + n - 4$$
....
$$a_4 = a_3 + 3$$
$$a_3 = a_2 + 2$$
$$a_2 = a_1 + 1$$

$\therefore$ $a_n = a_{n-2} + (n - 2) + (n - 1)$ (2nd stage)

$= a_{n-3} + (n - 3) + (n - 2) + (n - 1)$ (3rd stage)

....

$= a_{n-k} + (n - k) + (n - k + 1) + (n - k + 2) + ... + (n - 1)$ (k^{th} stage)

....

$= a_1 + 1 + 2 + 3 + ... + (n - 3) + (n - 2) + (n - 1)$ (stage $k = n - 1$)

$= 0 + \dfrac{(n-1)(n-1+1)}{2} = \dfrac{1}{2} n (n - 1)$ [$\because$ $a_1 = 0$]

$\therefore$ $a_n = \dfrac{1}{2} n(n - 1), n \geq 2,$

which is the required solution.

Example 3: *Solve the following recurrence relation using iterative method:*

$$a_n = 2a_{n-1} + 3^n, n \geq 2$$

subject to the initial condition $a_1 = 3$.

Solution: $$a_n = 2a_{n-1} + 3^n \quad \text{(1st stage)} \qquad ...(1)$$

Replacing n by $n-1, n-2, n-3, ..., 4, 3, 2$ in (1), we get

$$a_{n-1} = 2a_{n-2} + 3^{n-1}$$
$$a_{n-2} = 2a_{n-3} + 3^{n-2}$$
$$a_{n-3} = 2a_{n-4} + 3^{n-3}$$
$$\cdots\cdots\cdots$$
$$a_4 = 2a_3 + 3^4$$
$$a_3 = 2a_2 + 3^3$$
$$a_2 = 2a_1 + 3^2$$

$\therefore$ $$a_n = 2(2a_{n-2} + 3^{n-1}) + 3^n = 2^2a_{n-2} + 2.3^{n-1} + 3^n \quad (2^{\text{nd}} \text{ stage})$$
$$= 2^2(2a_{n-3} + 3^{n-2}) + 2.3^{n-1} + 3^n = 2^3a_{n-3} + 2^2\,3^{n-2} + 2.\,3^{n-1} + 3^n \quad \text{(3rd stage)}$$
$$\cdots\cdots\cdots$$
$$= 2^ka_{n-k} + 2^{k-1}.\,3^{n-k+1} + 2^{k-2}.\,3^{n-k+2} + ... + 2^2.3^{n-2} + 2.3^{n-1} + 3^n \quad (k^{\text{th}} \text{ stage})$$
$$\cdots\cdots\cdots$$
$$= 2^{n-1}a_1 + 2^{n-2}.\,3^2 + 2^{n-3}.\,3^3 + ... + 2^2.\,3^{n-2} + 2.\,3^{n-1} + 3^n \quad (\text{stage } k = n-1)$$
$$= 2^n.\frac{3}{2} + 2^n.\left(\frac{3}{2}\right)^2 + 2^n\left(\frac{3}{2}\right)^3 + ... + 2^n.\left(\frac{3}{2}\right)^{n-2} + 2^n.\left(\frac{3}{2}\right)^{n-1} + 2^n\left(\frac{3}{2}\right)^n$$
$$= 2^n\left\{\frac{3}{2} + \left(\frac{3}{2}\right)^2 + \left(\frac{3}{2}\right)^3 + ... + \left(\frac{3}{2}\right)^{n-2} + \left(\frac{3}{2}\right)^{n-1} + \left(\frac{3}{2}\right)^n\right\}$$
$$= 2^n\frac{\frac{3}{2}\left\{\left(\frac{3}{2}\right)^n - 1\right\}}{\frac{3}{2} - 1} = 2^n.3\left\{\left(\frac{3}{2}\right)^n - 1\right\} = 3(3^n - 2^n)$$

$\therefore$ $a_n = 3(3^n - 2^n), n \geq 2$, which is the required solution.

Example 4: *Solve the following recurrence relation using iterative method:*

$$t_n = 4(t_{n-1} - t_{n-2}),\ n \geq 2,\ for\ t_1 = 2,\ t_0 = 1.$$

Solution: $$t_n = 4(t_{n-1} - t_{n-2}) \quad \text{(1st stage)} \qquad ...(1)$$

Replacing n by $n-1, n-2, n-3, ..., 4, 3, 2$ in (1), we get

$$t_{n-1} = 4(t_{n-2} - t_{n-3})$$
$$t_{n-2} = 4(t_{n-3} - t_{n-4})$$
$$t_{n-3} = 4(t_{n-4} - t_{n-5})$$
$$\cdots\cdots\cdots$$
$$t_4 = 4(t_3 - t_2)$$
$$t_3 = 4(t_2 - t_1)$$
$$t_2 = 4(t_1 - t_0)$$

$$\begin{aligned}
\therefore \quad t_n &= 4\{4(t_{n-2} - t_{n-3}) - 4(t_{n-3} - t_{n-4})\} \\
&= 4\{4t_{n-2} - 2\,.\,4t_{n-3} + 4t_{n-4}\} \\
&= 4\{4\,.\,4(t_{n-3} - t_{n-4}) - 2\,.\,4t_{n-3} + 4t_{n-4}\} \\
&= 4^2(2t_{n-3} - 3t_{n-4}) \quad \text{(2nd stage)} \\
&= 4^2\{2\,.4(t_{n-4} - t_{n-5}) - 3\,.\,4(t_{n-5} - t_{n-6})\} \\
&= 4^3(2\,.\,t_{n-4} - 5t_{n-5} + 3t_{n-6}) \\
&= 4^3\{2\,.\,4(t_{n-5} - t_{n-6}) - 5t_{n-5} + 3t_{n-6}\} \\
&= 4^3(3t_{n-5} - 5t_{n-6}) \quad \text{(3rd stage)} \\
&\ldots\ldots\ldots\ldots\ldots\ldots\ldots\ldots\ldots\ldots \\
&= 4^k\{kt_{n-(2k-1)} - (2k-1)t_{n-2k}\} \quad (k^{\text{th}} \text{ stage}) \\
&\ldots\ldots\ldots\ldots\ldots\ldots\ldots\ldots\ldots\ldots \\
&= 4^{n/2}\left\{\frac{n}{2}t_1 - (n-1)t_0\right\} \quad \left[\text{At this stage } k = \frac{n}{2} \text{ as } n \text{ is even}\right] \\
&= 2^n\left\{\frac{n}{2}.2 - (n-1).1\right\} \quad [\because \; t_1 = 2 \text{ and } t_0 = 1]
\end{aligned}$$

$\therefore$ $t_n = 2^n$, $n \geq 2$, which is the required solution.

Example 5: *Solve, by iterative method, the recurrence relation (see Ex. 10, Art. 5.3):*

$$a_n = 8a_{n-1} + 10^{n-1}, \; n \geq 2,$$

with initial condition $a_1 = 9$.

Solution: $a_n = 8a_{n-1} + 10^{n-1}$ (1st stage) ...(1)

Replacing n by $n - 1, n - 2, n - 3, \ldots, 3, 2$ in (1), we get

$$\begin{aligned}
a_{n-1} &= 8a_{n-2} + 10^{n-2} \\
a_{n-2} &= 8a_{n-3} + 10^{n-3} \\
a_{n-3} &= 8a_{n-4} + 10^{n-4} \\
&\ldots\ldots\ldots\ldots\ldots\ldots \\
a_3 &= 8a_2 + 10^2 \\
a_2 &= 8a_1 + 10
\end{aligned}$$

From above, we get

$$\begin{aligned}
a_n &= 8(8a_{n-2} + 10^{n-2}) + 10^{n-1} = 8^2a_{n-2} + 8\,.\,10^{n-2} + 10^{n-1} \quad \text{(2nd stage)} \\
&= 8^2(8a_{n-3} + 10^{n-3}) + 8\,.\,10^{n-2} + 10^{n-1} \\
&= 8^3a_{n-3} + 8^2 10^{n-3} + 8\,.\,10^{n-2} + 10^{n-1} \quad \text{(3rd stage)} \\
&\ldots\ldots\ldots\ldots\ldots\ldots\ldots\ldots\ldots\ldots \\
&= 8^k a_{n-k} + 8^{k-1}10^{n-k} + 8^{k-2}.10^{n-k+1} + \ldots + 8\,.\,10^{n-2} + 10^{n-1} \quad (k^{\text{th}} \text{ stage}) \\
&\ldots\ldots\ldots\ldots\ldots\ldots\ldots\ldots\ldots\ldots \\
&= 8^{n-1}a_1 + 8^{n-2}\,10^1 + 8^{n-3}\,10^2 + \ldots + 8\,.\,10^{n-2} + 10^{n-1} \quad (\text{stage } k = n-1) \\
&= 9\,.\,8^{n-1} + 8^{n-1}\left\{\frac{10}{8} + \left(\frac{10}{8}\right)^2 + \ldots + \left(\frac{10}{8}\right)^{n-2} + \left(\frac{10}{8}\right)^{n-1}\right\} \quad (\because \; a_1 = 9)
\end{aligned}$$

$$= 9 \,.\, 8^{n-1} + 8^{n-1} \frac{\frac{10}{8}\left\{\left(\frac{10}{8}\right)^{n-1} - 1\right\}}{\frac{10}{8} - 1} = 9.8^{n-1} + 5(10^{n-1} - 8^{n-1})$$

$\therefore \qquad a_n = 4.8^{n-1} + 5.10^{n-1}, n \geq 2,$

which is the required solution.

Note: $\qquad a_n = \frac{1}{2}(8^n + 10^n).$

Example 6: *In an air-traffic control system the desired altitude of an aircraft h_n is computed by a computer every second and is compared with the actual altitude of the aircraft, a_{n-1}, determined by tracking radar 1 second earlier. Depending on whether h_n is larger or smaller than a_{n-1}, the altitude of the aircraft will be changed accordingly. Specifically, the change in altitude at the nth second, $a_n - a_{n-1}$, is proportional to the difference $h_n - a_{n-1}$. Determine a_n, given that*

$$h_n = \begin{cases} 1000\,(3/2)^n, & 0 \leq n \leq 10 \\ 1000\,(3/2)^{10}, & n > 10 \end{cases}$$

the proportional constant $k = 3$, $a_0 = 0$.

Solution: Given, $a_n - a_{n-1} = 3(h_n - a_{n-1})$

$\therefore \qquad a_n = 3h_n - 2a_{n-1}$

So, $\qquad a_1 = 3h_1 - 2a_0 = 3h_1$, since $a_0 = 0$.

$$a_2 = 3h_2 - 2a_1 = 3h_2 - 2(3h_1) = 3(h_2 - 2h_1)$$

$$a_3 = 3h_3 - 2\,.\,3(h_2 - 2h_1) = 3(h_3 - 2h_2 + 2^2h_1)$$

....

$$a_{10} = 3(h_{10} - 2h_9 + 2^2h_8 - \ldots\ldots. - 2^9h_1)$$

For n ≥ 11: $\qquad a_n = 3h_n - 2a_{n-1} = 3h_n - 2(3h_{n-1} - 2a_{n-2})$

$$= 3(h_n - 2h_{n-1}) + 2^2a_{n-2}$$

$$= 3(h_n - 2h_{n-1}) + 2^2\{3(h_{n-2} - 2h_{n-3}) + 2^2a_{n-4}\}$$

$$= 3(h_n - 2h_{n-1} + 2^2\,h_{n-2} - 2^3\,h_{n-3}) + 2^4a_{n-4}$$

....

$$= 3\{h_n - 2h_{n-1} + 2^2h_{n-2} - \ldots + (-1)^{n-10}\,h_{10}\} + 2^{n-9}a_9.$$

5.6 SOLUTION OF LINEAR RECURRENCE RELATIONS WITH CONSTANT COEFFICIENTS BY CHARACTERISTIC ROOTS METHOD

A linear k^{th} order recurrence relation (or difference equation) with constant coefficients is of the form:

$$a_n = c_1a_{n-1} + c_2a_{n-2} + \ldots + c_ka_{n-k} + f(n) \qquad \ldots(1)$$

where $c_1, c_2, \ldots, c_k$ are real numbers (with $c_k \neq 0$) and $f(n)$ is a function depending only on n in general.

In characteristic roots method, the solution of (1) is obtained as the sum of two parts, the homogeneous solution which satisfies (1) when $f(n)$ is set to 0, and the particular solution which satisfies (1) with $f(n)$ not identically zero.

5.6.1 Linear Homogeneous Recurrence Relations with Constant Coefficients

The basic approach for solving **linear homogeneous recurrence relations** [$f(n) = 0$ in (1)] is to search for solutions of the form $a_n = r^n$, where r is a constant. Note that $a_n = r^n$ is a solution of the recurrence relation

$$a_n = c_1a_{n-1} + c_2a_{n-2} + ... + c_ka_{n-k} \qquad ...(2)$$

iff

$$r^n = c_1r^{n-1} + c_2r^{n-2} + ... + c_kr^{n-k}$$

$$\Rightarrow \quad r^k = c_1r^{k-1} + c_2r^{k-2} + ... + c_k \qquad \text{(Dividing both sides by } r^{n-k})$$

$$\Rightarrow \quad r^k - c_1r^{k-1} - c_2r^{k-2} - ..... - c_k = 0 \qquad ...(3)$$

Consequently, the sequence $\{a_n\}$ with $a_n = r^n$ is a solution of (2) if and only if r is a solution of (3) called the **characteristic equation** of the recurrence relation (2). The solutions of (3) are called the **characteristics roots** of the recurrence relation (2). These characteristic roots are used to give an explicit formula for all the solutions of the recurrence relation (2).

Here we will develop results that deal with linear second order homogeneous recurrence relations with constant coefficients. First, consider the case when there are two distinct roots of (3) for $k = 2$.

Theorem 1: *If $r^2 - c_1r - c_2 = 0$ has two distinct roots r_1, r_2, then the sequence $\{a_n\}$ is a solution of the recurrence relation $a_n = c_1a_{n-1} + c_2a_{n-2}$ iff $a_n = A\,r_1^n + B\,r_2^n$ where A and B are constants.*

Proof: Since r_1, r_2 are roots of $r^2 - c_1r - c_2 = 0$, it follows that

$$r_1^2 = c_1r_1 + c_2;\ r_2^2 = c_1r_2 + c_2 \qquad ...(4)$$

Now,

$$\begin{aligned} c_1a_{n-1} + c_2a_{n-2} &= c_1(A\,r_1^{n-1} + B\,r_2^{n-1}) + c_2(A\,r_1^{n-2} + B\,r_2^{n-2}) \\ &= A\,r_1^{n-2}(c_1r_1 + c_2) + B\,r_2^{n-2}(c_1r_2 + c_2) \\ &= A\,r_1^{n-2}\,.\,r_1^2 + B\,r_2^{n-2}\,.\,r_2^2 \qquad \text{[Using (4)]} \\ &= A\,r_1^n + B\,r_2^n = a_n. \end{aligned}$$

This shows that the sequence $\{a_n\}$ with $a_n = A\,r_1^n + B\,r_2^n$ is a solution of the recurrence relation $a_n = c_1a_{n-1} + c_2a_{n-2}$.

Theorem 1 does not apply when the characteristic equation has identical roots. This case can be handled using Theorem 2 stated below:

Theorem 2: *If $r^2 - c_1r - c_2 = 0$ has only one root r_0, then a sequence $\{a_n\}$ is a solution of the recurrence relation $a_n = c_1a_{n-1} + c_2a_{n-2}$ iff $a_n = A\,r_0^n + B\,nr_0^n$ where A and B are constants.*

Proof: Since r_0 is the only root of $r^2 - c_1r - c_2 = 0$, it follows that $r_0^2 = c_1r_0 + c_2$ ...(5)

Also, $r^2 - c_1r - c_2 \equiv (r - r_0)^2$, is an identity

$$\Rightarrow \quad r^2 - c_1r - c_2 \equiv r^2 - 2r_0r + r_0^2$$

$$\Rightarrow \quad c_1 = 2r_0 \text{ and } c_2 = -\,r_0^2 \qquad ...(6)$$

Now,

$$\begin{aligned} c_1a_{n-1} + c_2a_{n-2} &= c_1\{A\,r_0^{n-1} + B(n-1)r_0^{n-1}\} + c_2\{A\,r_0^{n-2} + B(n-2)r_0^{n-2}\} \\ &= A\,r_0^{n-2}(c_1r_0 + c_2) + B\,r_0^{n-2}\{c_1(n-1)r_0 + c_2(n-2)\} \\ &= A\,r_0^{n-2}\,.\,r_0^2 + B\,r_0^{n-2}\{2(n-1)r_0^2 - (n-2)r_0^2\} \qquad \text{[By (5) and (6)]} \\ &= A\,r_0^n + B\,nr_0^n = a_n. \end{aligned}$$

Thus the sequence $\{a_n\}$ with $a_n = A\,r_0^n + B\,nr_0^n = (A + Bn)r_0^n$ is a solution of the recurrence relation $a_n = c_1a_{n-1} + c_2a_{n-2}$.

ILLUSTRATIVE EXAMPLES—II

Example 1: *Solve, by characteristic roots method, the recurrence relation*

$$a_n = a_{n-1} + 2a_{n-2} \text{ with } a_0 = 0,\ a_1 = 6. \text{ Hence find } a_8.$$

Solution: Putting $a_n = r^n$ in the given recurrence relation, we have

$$r^n = r^{n-1} + 2r^{n-2} \Rightarrow r^2 - r - 2 = 0 \Rightarrow r^2 - 2r + r - 2 = 0 \Rightarrow (r-2)(r+1) = 0$$

This is the characteristic equation with roots 2 and –1.

So, the general solution of the given recurrence relation is

$$a_n = A\,2^n + B(-1)^n, \text{ where } A \text{ and } B \text{ are constants.}$$

Using the given conditions $a_0 = 0$ and $a_1 = 6$, we have

$$A + B = 0 \text{ and } 2A - B = 6.$$

Solving, we get $A = 2$ and $B = -2$.

Hence the required solution is $a_n = 2 \,.\, 2^n - 2 \,.\, (-1)^n$, or $a_n = 2^{n+1} + 2(-1)^{n+1}$

$$\therefore \quad a_8 = 2^{8+1} + 2(-1)^{8+1} = 512 - 2 = 510.$$

Example 2: *Solve the recurrence relation of the Fibonacci sequence of numbers (see Ex. 8, Art. 5.3):*

$$a_n = a_{n-1} + a_{n-2}, \text{ for } n \geq 3$$

subject to the initial conditions $a_1 = 1$ and $a_2 = 1$.

Solution: Given, $a_n = a_{n-1} + a_{n-2}\,;\ n \geq 3$ ...(1)

Putting $a_n = r^n$ in (1), we have $r^n = r^{n-1} + r^{n-2}$

$\therefore$ $r^2 - r - 1 = 0$. This is the characteristic equation with roots $(1 \pm \sqrt{5})/2$. So, the general solution of (1) is $a_n = A\left(\frac{1+\sqrt{5}}{2}\right)^n + B\left(\frac{1-\sqrt{5}}{2}\right)^n$, where A and B are constants.

Using the given conditions $a_1 = 1$ and $a_2 = 1$, we have

$$A\left(\frac{1+\sqrt{5}}{2}\right) + B\left(\frac{1-\sqrt{5}}{2}\right) = 1;\ A\left(\frac{1+\sqrt{5}}{2}\right)^2 + B\left(\frac{1-\sqrt{5}}{2}\right)^2 = 1$$

i.e., $$A(1+\sqrt{5}) + B(1-\sqrt{5}) = 2 \text{ and } A(3+\sqrt{5}) + B(3-\sqrt{5}) = 2.$$

Solving, we get $$A = \frac{1}{\sqrt{5}} \text{ and } B = -\frac{1}{\sqrt{5}}.$$

Thus the nth Fibonacci number is given explicitly by $a_n = \frac{1}{\sqrt{5}}\left\{\left(\frac{1+\sqrt{5}}{2}\right)^n - \left(\frac{1-\sqrt{5}}{2}\right)^n\right\}$

Example 3: *Solve, by characteristic roots method, the recurrence relation:*

$$a_n = 6a_{n-1} - 9a_{n-2}$$

subject to the initial conditions $a_0 = 1$ and $a_1 = 6$.

Solution: Given $a_n = 6a_{n-1} - 9a_{n-2}$ ...(1)

Putting $a_n = r^n$ in (1), we get $r^n = 6r^{n-1} - 9r^{n-2}$

$$\therefore \quad r^2 - 6r + 9 = 0, \text{ or } (r-3)^2 = 0.$$

This is the characteristic equation with equal roots 3, 3. So, the general solution of (1) is $a_n = (A + Bn)3^n$, where A and B are arbitrary constants.

Using the initial conditions $a_0 = 1$ and $a_1 = 6$, we get

$$A = 1 \text{ and } (A + B)3 = 6, \text{ i.e., } A = B = 1.$$

Therefore, the required solution is $a_n = (1 + n)3^n$.

5.6.2 Linear Nonhomogeneous Recurrence Relations with Constant Coefficients

A linear k^{th} order nonhomogeneous recurrence relation with constant coefficients is of the form:

$$a_n = c_1 a_{n-1} + c_2 a_{n-2} + \dots + c_k a_{n-k} + f(n) \qquad \dots(7)$$

where $c_1, c_2, \dots, c_k$ are real numbers (with $c_k \neq 0$) and $f(n)$ is a function not identically zero depending only on n.

The recurrence relation

$$a_n = c_1 a_{n-1} + c_2 a_{n-2} + \dots + c_k a_{n-k} \qquad \dots(8)$$

is called the **associated homogeneous recurrence relation** and it plays an important role in the solution of the nonhomogeneous recurrence relation (7).

The main feature about linear nonhomogeneous recurrence relation with constant coefficients is that every solution is the sum of a particular solution of it and a solution of the associated linear homogeneous recurrence relation as stated in the following theorem:

Theorem 3: *If $\{a_n^{(p)}\}$ is a particular solution of the linear nonhomogeneous recurrence relation with constant coefficients (7), then every solution of (7) is of the form $\{a_n^{(p)} + a_n^{(h)}\}$, where $\{a_n^{(h)}\}$ is a solution of the associated homogeneous recurrence relation (8).*

Proof: Since $\{a_n^{(p)}\}$ is a particular solution of the nonhomogeneous recurrence relation (7), we have

$$a_n^{(p)} = c_1 a_{n-1}^{(p)} + c_2 a_{n-2}^{(p)} + \dots + c_k a_{n-k}^{(p)} + f(n).$$

Now, let $\{b_n\}$ is a second solution of (7), so we have

$$b_n = c_1 b_{n-1} + c_2 b_{n-2} + \dots + c_k b_{n-k} + f(n).$$

Subtracting the first of these two equations from the second, we get

$$b_n - a_n^{(p)} = c_1(b_{n-1} - a_{n-1}^{(p)}) + c_2(b_{n-2} - a_{n-2}^{(p)}) + \dots + c_k(b_{n-k} - a_{n-k}^{(p)}).$$

It follows that $\{b_n - a_n^{(p)}\}$ is a solution of the associated linear homogeneous recurrence relation (8), say $\{a_n^{(h)}\}$. Consequently, $b_n = a_n^{(p)} + a_n^{(h)}$, for all n.

Theorem 3 shows that the key to solve linear nonhomogeneous recurrence relations with constant coefficients is to find a particular solution. Then every solution is a sum of this solution and a solution of the associated linear homogeneous recurrence relation. Although there is no general method for finding such a solution that works for every function $f(n)$, there are techniques (sometimes called the **trial sequence method**) that works for certain types of functions $f(n)$, such as polynomials and powers of constants.

The following table shows a list of trial solutions $T(n)$ against specified function $f(n)$, where A, $A_0, A_1, A_2, \ldots$ represent the unknown constant coefficients to be determined.

	$f(n)$	*Trial solution* $T(n)$
1.	b^n (where b is not a root of the characteristic equation)	Ab^n
2.	b^n (where b is a root of the characteristic equation with multiplicity s)	An^sb^n
3.	Polynomial $P(n)$ of degree m.	$A_0 + A_1n + A_2n^2 + \ldots + A_mn^m$
4.	$c^nP(n)$, where $P(n)$ is a polynomial of degree m and c is not a root of the characteristic equation.	$c^n(A_0 + A_1n + A_2n^2 + \ldots + A_mn^m)$
5.	$c^nP(n)$, where $P(n)$ is a polynomial of degree m and c is a root of the characteristic equation with multiplicity s.	$n^sc^n(A_0 + A_1n + A_2n^2 + \ldots + A_mn^m)$

Notes: (*i*) If $f(n)$ is a constant (*i.e.*, a polynomial of degree zero) then $T(n) = A$.

(*ii*) If $f(n)$ is a linear combination of the functions given in the table, then $T(n)$ is to be taken as the linear combination of the corresponding trial solutions given in the table with different unknown constant coefficients to be determined.

ILLUSTRATIVE EXAMPLES—III

Example 1: *Solve the recurrence relation*

$$a_n - 5a_{n-1} + 6a_{n-2} = 2$$

with initial conditions $a_0 = 1$ *and* $a_1 = -1$.

Solution: The associated homogeneous recurrence relation is

$$a_n - 5a_{n-1} + 6a_{n-2} = 0 \qquad \ldots(1)$$

Putting $a_n = r^n$ in (1), we have

$r^n - 5r^{n-1} + 6r^{n-2} = 0 \Rightarrow r^2 - 5r + 6 = 0 \Rightarrow r^2 - 3r - 2r + 6 = 0 \Rightarrow (r-3)(r-2) = 0.$

$\therefore \quad r = 3, 2$. So, the solution of (1) is $a_n^{(h)} = C_1 3^n + C_2 2^n$

To find a particular solution of the given equation, let $a_n^{(p)} = A$ and substituting it in the given equation, we get

$$A - 5A + 6A = 2 \Rightarrow A = 1.$$

Thus, $a_n^{(p)} = 1$ is a particular solution and so the general solution is

$$a_n = a_n^{(p)} + a_n^{(h)} = 1 + C_1 3^n + C_2 2^n \qquad \ldots(2)$$

To find C_1 and C_2, putting $n = 0$ and $n = 1$ in (2), we get

$$a_0 = 1 + C_1 + C_2, \text{ or } 1 = 1 + C_1 + C_2 \ (\because \ a_0 = 1) \Rightarrow C_1 + C_2 = 0 \qquad \ldots(3)$$

$$a_1 = 1 + 3C_1 + 2C_2, \text{ or } -1 = 1 + 3C_1 + 2C_2 \ (\because \ a_1 = -1) \Rightarrow 3C_1 + 2C_2 = -2 \qquad \ldots(4)$$

Solving (3) and (4), we get $C_1 = -2$ and $C_2 = 2$.

Putting these values of C_1 and C_2 in (2), the required solution is $a_n = 1 - 2.\ 3^n + 2^{n+1}$.

Example 2: *Solve the recurrence relation*

$$a_n - a_{n-1} - 2a_{n-2} = 2n^2.$$

Solution: The associated homogeneous recurrence relation is

$$a_n - a_{n-1} - 2a_{n-2} = 0 \qquad \ldots(1)$$

Putting $a_n = r^n$ in (1), we have

$$r^n - r^{n-1} - 2r^{n-2} = 0 \Rightarrow r^2 - r - 2 = 0 \Rightarrow r^2 - 2r + r - 2 = 0$$

$\therefore$ $(r-2)(r+1) = 0 \Rightarrow r = 2, -1.$

So, the solution of (1) is $a_n^{(h)} = C_1 2^n + C_2(-1)^n$

Let a particular solution of the given equation is $a_n^{(p)} = A_0 + A_1 n + A_2 n^2$ ($\because$ $f(n) = 2n^2$ is a polynomial of degree 2). Substituting in the given equation, we get

$$A_0 + A_1 n + A_2 n^2 - \{A_0 + A_1(n-1) + A_2(n-1)^2\} - 2\{A_0 + A_1(n-2) + A_2(n-2)^2\} = 2n^2$$

or $$(-2A_0 + 5A_1 - 9A_2) + (-2A_1 + 10A_2)n - 2A_2 n^2 = 2n^2$$

This is an identity in n and so the constant term and the coefficient of n, n^2 on both sides will be equal.

$$\therefore \quad -2A_0 + 5A_1 - 9A_2 = 0$$
$$-2A_1 + 10A_2 = 0$$
$$-2A_2 = 2$$

Solving, we get $A_0 = -8, A_1 = -5$ and $A_2 = -1$.

$\therefore$ $a_n^{(p)} = 8\ 5n - n^2$.

Hence the general solution of the given recurrence relation is

$$a_n = C_1 2^n + C_2(-1)^n - 8 - 5n - n^2,$$

where C_1, C_2 are arbitrary constants.

Example 3: *Solve the recurrence relation*

$$a_{n+2} + 5a_{n+1} + 6a_n = n^2 - 2n + 1.$$

Solution: Replacing $n + 2$ by n (*i.e.*, n by $n - 2$), the given equation becomes

$$a_n + 5a_{n-1} + 6a_{n-2} = (n-2-1)^2$$

or $$a_n + 5a_{n-1} + 6a_{n-2} = n^2 - 6n + 9 \quad \text{...(1)}$$

The associated homogeneous recurrence relation is

$$a_n + 5a_{n-1} + 6a_{n-2} = 0 \quad \text{...(2)}$$

Putting $a_n = r^n$ in (2), we have

$$r^n + 5r^{n-1} + 6r^{n-2} = 0 \Rightarrow r^2 + 5r + 6 = 0 \Rightarrow r^2 + 3r + 2r + 6 = 0$$

$\therefore$ $(r+3)(r+2) = 0 \Rightarrow r = -3, -2$

So, the solution of (2) is $a_n^{(h)} = C_1(-3)^n + C_2(-2)^n$.

Let a particular solution of (1) is $a_n^{(p)} = A_0 + A_1 n + A_2 n^2$. Substituting in (1), we get

$$A_0 + A_1 n + A_2 n^2 + 5\{A_0 + A_1(n-1) + A_2(n-1)^2\} + 6\{A_0 + A_1(n-2) + A_2(n-2)^2\}$$
$$= n^2 - 6n + 9$$

or $$12A_2 n^2 + (12A_1 - 34A_2)n + (12A_0 - 17A_1 + 29A_2) = n^2 - 6n + 9\cdot$$

This is an identity in n. Equating the coefficients of like powers of n and constant terms from both sides, we get

$$12A_2 = 1;\ 12A_1 - 34A_2 = -6;\ 12A_0 - 17A_1 + 29A_2 = 9$$

Solving, we get $A_0 = \dfrac{151}{864}$; $A_1 = -\dfrac{19}{72}$; $A_2 = \dfrac{1}{12}$.

$$\therefore \quad a_n^{(p)} = \frac{151}{864} - \frac{19}{72}n + \frac{1}{12}n^2.$$

Hence the general solution of the given recurrence relation is

$$a_n = C_1(-3)^n + C_2(-2)^n + \frac{151}{864} - \frac{19}{72}n + \frac{1}{12}n^2,$$

where C_1, C_2 are arbitrary constants.

Example 4: *Solve the recurrence relation*

$$a_n - 6a_{n-1} + 8a_{n-2} = 7^n.$$

Solution: The associated homogeneous recurrence relation is

$$a_n - 6a_{n-1} + 8a_{n-2} = 0 \qquad ...(1)$$

Putting $a_n = r^n$ in (1), we have

$$r^n - 6r^{n-1} + 8r^{n-2} = 0 \Rightarrow r^2 - 6r + 8 = 0 \Rightarrow r^2 - 4r - 2r + 8 = 0$$

$$\therefore \quad (r-4)(r-2) = 0 \Rightarrow r = 4, 2.$$

So, the solution of (1) is $a_n^{(h)} = C_1 4^n + C_2 2^n$.

Let a particular solution of the given equation is $a_n^{(p)} = A7^n$ ($\because$ 7 is not a root of the characteristic equation).

Substituting in the given equation, we get

$$A7^n - 6A7^{n-1} + 8A7^{n-2} = 7^n$$

or

$$\left(A - \frac{6}{7}A + \frac{8}{7^2}A\right)7^n = 7^n, \quad \text{or} \quad \frac{49 - 42 + 8}{49}A = 1$$

$$\therefore \quad A = \frac{49}{15}. \quad \therefore \quad a_n^{(p)} = \frac{49}{15}7^n.$$

Hence the general solution of the given recurrence relation is

$$a_n = C_1 4^n + C_2 2^n + \frac{49}{15}7^n,$$

where C_1, C_2 are arbitrary constants.

Example 5: *Solve the recurrence relation*

$$a_n - 4a_{n-1} + 4a_{n-2} = 2^n.$$

Solution: The associated homogeneous recurrence relation is

$$a_n - 4a_{n-1} + 4a_{n-2} = 0 \qquad ...(1)$$

Putting $a_n = r^n$ in (1), we have

$$r^n - 4r^{n-1} + 4r^{n-2} = 0 \Rightarrow r^2 - 4r + 4 = 0 \Rightarrow (r-2)^2 = 0 \Rightarrow r = 2, 2.$$

So, the characteristic equation has a single root 2 with multiplicity two and hence the solution of (1) is $a_n^{(h)} = (C_1 + C_2 n)\,2^n$.

Let a particular solution of the given equation is $a_n^{(p)} = An^2 2^n$, since 2 is a root of the associated characteristic equation with multiplicity two.

Substituting in the given equation, we get

$$An^2 . 2^n - 4A(n-1)^2 2^{n-1} + 4A(n-2)^2 2^{n-2} = 2^n$$

or

$$A\{n^2 - 2(n-1)^2 + (n-2)^2\}\,2^n = 2^n$$

or

$$2A = 1, \text{ or, } A = \frac{1}{2}.$$

$$\therefore \quad a_n^{(p)} = \frac{1}{2}n^2 2^n.$$

Hence the general solution of the given recurrence relation is

$$a_n = (C_1 + C_2 n)\, 2^n + \frac{1}{2} n^2\, 2^n,$$

where C_1, C_2 are arbitrary constants.

Example 6: *Solve the recurrence relation*

$$t_n - t_{n-1} - 12t_{n-2} = 4^n + n^2.$$

Solution: The associated homogeneous recurrence relation is

$$t_n - t_{n-1} - 12t_{n-2} = 0 \qquad ...(1)$$

Putting $t_n = r^n$ in (1), we have

$$r^n - r^{n-1} - 12r^{n-2} = 0 \;\Rightarrow\; r^2 - r - 12 = 0 \;\Rightarrow\; r^2 - 4r + 3r - 12 = 0.$$

$$\therefore \qquad (r-4)(r+3) = 0 \;\Rightarrow\; r = 4, -3.$$

So, the solution of (1) is $t_n^{(h)} = C_1 4^n + C_2(-3)^n$

Particular solution corresponding to 4^n

$$t_n - t_{n-1} - 12t_{n-2} = 4^n \qquad ...(2)$$

Since 4 is a simple root (multiplicity 1) of the characteristic equation, choose $t_n^{(p)} = An4^n$ and substitute in (2).

$$\therefore \qquad An4^n - A(n-1)\,4^{n-1} - 12A(n-2)\,4^{n-2} = 4^n$$

or
$$A\left\{n - \frac{1}{4}(n-1) - \frac{3}{4}(n-2)\right\} 4^n = 4^n$$

or
$$\frac{7}{4}A = 1, \text{ or } A = \frac{4}{7}. \quad \therefore \quad t_n^{(p)} = \frac{4}{7}\, n4^n.$$

Particular solution corresponding to n^2

$$t_n - t_{n-1} - 12t_{n-2} = n^2 \qquad ...(3)$$

Choose $t_n^{(p)} = A_0 + A_1 n + A_2 n^2$ and substitute in (3).

$$\therefore \quad A_0 + A_1 n + A_2 n^2 - \{A_0 + A_1(n-1) + A_2(n-1)^2\} - 12\{A_0 + A_1(n-2) + A_2(n-2)^2\} = n^2$$

or
$$-12A_2 n^2 + (-12A_1 + 50A_2)n + (-12A_0 + 25A_1 - 49A_2) = n^2.$$

This is an identity in n and so equating the coefficients of like powers of n and constant terms from both sides, we get

$$-12A_2 = 1;\; -12A_1 + 50A_2 = 0;\; -12A_0 + 25A_1 - 49A_2 = 0.$$

Solving, we get
$$A_0 = -\frac{331}{864},\; A_1 = -\frac{25}{72};\; A_2 = -\frac{1}{12}.$$

Hence the general solution of the given recurrence relation is

$$t_n = C_1 4^n + C_2(-3)^n + \frac{4}{7} n\, 4^n - \frac{331}{864} - \frac{25}{72}\, n - \frac{1}{12}\, n^2,$$

where C_1, C_2 are arbitrary constants.

Example 7: *Solve the recurrence relation*

$$a_n - 6a_{n-1} + 9a_{n-2} = (n+1)\, 3^n$$

with initial conditions $a_0 = 1$ and $a_1 = \frac{7}{2}$. Hence find a_4.

Solution: The associated homogeneous recurrence relation is

$$a_n - 6a_{n-1} + 9a_{n-2} = 0 \quad ...(1)$$

Putting $a_n = r^n$ in (1), we have

$$r^n - 6r^{n-1} + 9r^{n-2} = 0 \Rightarrow r^2 - 6r + 9 = 0 \Rightarrow (r-3)^2 = 0 \Rightarrow r = 3, 3.$$

So, the characteristic equation has a single root 3 with multiplicity two and hence the solution of (1) is $a_n^{(h)} = (C_1 + C_2 n)3^n$.

Since here $f(n) = (n+1)\,3^n$ and 3 is a root of the associated characteristic equation with multiplicity two, so we choose a particular solution as $a_n^{(p)} = n^2 3^n(A_0 + A_1 n)$. Substituting it in the given equation, we get

$$n^2 3^n(A_0 + A_1 n) - 6[(n-1)^2\,3^{n-1}\{A_0 + A_1(n-1)\}] + 9\,[(n-2)^2\,3^{n-2}\{A_0 + A_1(n-2)\}] = (n+1)\,3^n$$

or $$n^2\,3^n(A_0 + A_1 n) - 2(n-1)^2\,3^n\{A_0 + A_1(n-1)\} + (n-2)^2\,3^n\{A_0 + A_1(n-2)\} = (n+1)\,3^n$$

or $$n^2(A_0 + A_1 n) - 2(n-1)^2\{A_0 + A_1(n-1)\} + (n-2)^2\{A_0 + A_1(n-2)\} = n+1$$

$\therefore \quad 6A_1 n + (2A_0 - 6A_1) = n + 1$

This is an identity in n and so

$$6A_1 = 1 \quad \text{and} \quad 2A_0 - 6A_1 = 1.$$

Solving, $$A_0 = 1 \text{ and } A_1 = \frac{1}{6}.$$

$\therefore$ $$a_n^{(p)} = n^2 3^n\left(1 + \frac{n}{6}\right).$$

Hence the general solution of the given recurrence relation is

$$a_n = (C_1 + C_2 n)\,3^n + n^2\,3^n\left(1 + \frac{n}{6}\right), \quad ...(2)$$

where C_1, C_2 are arbitrary constants.

Putting $n = 0$ and $n = 1$ respectively in (2) and using the given conditions, we get

$$a_0 = C_1 = 1 \text{ and } a_1 = (1 + C_2)3 + 3 \times \frac{7}{6} = \frac{7}{2} \quad (\because \; C_1 = 1)$$

$\therefore$ $$C_2 = -1$$

$\therefore$ Required solution is

$$a_n = (1-n)3^n + n^2\,3^n\left(1 + \frac{n}{6}\right)$$

$\therefore$ $$a_4 = (1-4)3^4 + 4^2\,3^4\left(1 + \frac{4}{6}\right) = -243 + 16 \times 27 \times 5 = 1917.$$

Example 8: *It is given that the solution of the recurrence relation* $Aa_n + Ba_{n-1} + Ca_{n-2} = 6$ *is* $C_1 3^n + C_2(-2)^n + 3$. *Find A, B and C.*

Solution: In the given solution, $C_1 3^n + C_2(-2)^n$ is the solution of the associated homogeneous recurrence relation $Aa_n + Ba_{n-1} + Ca_{n-2} = 0$.

So, 3, –2 are the roots of the characteristic equation $Ar^2 + Br + C = 0$.

$\therefore \quad A3^2 + B3 + C = 0, \quad \text{or} \quad 9A + 3B + C = 0 \quad ...(1)$

and $\quad A(-2)^2 + B(-2) + C = 0, \quad \text{or} \quad 4A - 2B + C = 0 \quad ...(2)$

As per the given solution, 3 is a particular solution of the recurrence relation

$$Aa_n + Ba_{n-1} + Ca_{n-2} = 6$$

So, $a_n = 3$ must satisfy this relation.

$$\therefore \quad 3A + 3B + 3C = 6, \quad \text{or} \quad A + B + C = 2 \qquad \text{...(3)}$$

Subtracting (2) from (1), we get

$$5A + 5B = 0 \quad \text{or} \quad B = -A.$$

Putting $B = -A$ in (3), we get $C = 2$

Putting $B = -A$ and $C = 2$ in (1), we get $6A + 2 = 0$, *i.e.*,

$$A = -\frac{1}{3}. \text{ So, } B = -A = \frac{1}{3}.$$

$$\therefore \quad A = -\frac{1}{3}, B = \frac{1}{3} \text{ and } C = 2.$$

5.7 SOLUTION OF LINEAR RECURRENCE RELATIONS WITH CONSTANT COEFFICIENTS BY GENERATING FUNCTIONS METHOD

Definition 3: The **generating function** for a sequence $a_0, a_1, a_2, ..., a_k, ...$ of real numbers is denoted and defined by the following infinite series:

$$G(x) = a_0 + a_1x + a_2x^2 + ... + a_kx^k + ... \text{ to } \infty = \sum_{k=0}^{\infty} a_k x^k.$$

Note: The generating function for the sequence $\{a_k\}$ defined here is sometimes called the *ordinary generating function* of $\{a_k\}$ to distinguish it from other types of generating function for this sequence.

Illustrations: (*i*) The generating function of the sequence

$$\{a_n = 1^n\} = \{1, 1, 1, ... \text{ to } \infty\} \text{ is}$$

$$G(x) = 1^0x^0 + 1^1x^1 + 1^2x^2 + ... + 1^nx^n + ... \text{ to } \infty$$

$$= 1 + x + x^2 + ... + x^n + ... \text{ to } \infty$$

$$= (1 - x)^{-1} = \frac{1}{1-x}, \text{ for } |x| < 1.$$

(*ii*) The generating function of the sequence $\{a_n = 2^n\} = \{2^0, 2^1, 2^2, ...\}$ is

$$G(x) = 2^0x^0 + 2^1x^1 + 2^2x^2 + ... + 2^nx^n + ... + \text{ to } \infty.$$

$$= (1 - 2x)^{-1} = \frac{1}{1-2x}, \text{ for } |2x| < 1 \Rightarrow |x| < \frac{1}{2}.$$

The generating functions can also be defined for finite sequences of real numbers by extending a finite sequence $a_0, a_1, a_2, ..., a_n$ into an infinite sequence by setting $a_{n+r} = 0$, for $r = 1, 2, 3, ...$ So, the generating function $G(x)$ of this infinite sequence $\{a_n\}$ is a polynomial of degree n since $a_rx^r = 0, \forall r > n$, that is, $G(x) = a_0 + a_1x + a_2x^2 + ... + a_nx^n$.

The following theorem states results on how to add and how to multiply two generating functions.

Theorem 4: If $f(x) = \sum_{n=0}^{\infty} a_n x^n$ and $g(x) = \sum_{n=0}^{\infty} b_n x^n$, then $f(x) + g(x) = \sum_{n=0}^{\infty} (a_n + b_n)\, x^n$. ...(1)

and
$$f(x)\, g(x) = a_0 b_0 + (a_0 b_1 + a_1 b_0)\, x + (a_0 b_2 + a_1 b_1 + a_2 b_0)\, x^2$$
$$+ (a_0 b_3 + a_1 b_2 + a_2 b_1 + a_3 b_0)\, x^3 + \ldots + \left\{ \sum_{i=0}^{n} a_i\, b_{n-i} \right\} x^n + \ldots = \sum_{n=0}^{\infty} \left\{ \sum_{i=0}^{n} a_i\, b_{n-i} \right\} x^n \quad \ldots(2)$$

Proofs are omitted.

Note: This special kind of multiplication is called **convolution.**

Illustration: We know that $(1 - x)^{-1} = 1 + x + x^2 + x^3 + x^4 + \ldots$ to ∞. So, using Theorem 4, we have

$$(1 - x)^{-2} = (1 - x)^{-1}\, (1 - x)^{-1} = \sum_{n=0}^{\infty} \left\{ \sum_{i=0}^{n} 1 \right\} x^n = \sum_{n=0}^{\infty} (n + 1)\, x^n$$

$$(1 - x)^{-3} = (1 - x)^{-2}\, (1 - x)^{-1} = \sum_{n=0}^{\infty} \left\{ \sum_{i=0}^{n} a_i\, b_{n-i} \right\} x^n = \sum_{n=0}^{\infty} \left\{ \sum_{i=0}^{n} (i + 1)\,.\,1 \right\} x^n$$

$$= \sum_{n=0}^{\infty} \frac{1}{2}(n + 1)\,(n + 2)\, x^n.$$

Note: When generating functions are used to solve counting problems, they are usually considered to be formal power series and questions regarding the convergence of these series are ignored.

Some results

(*i*) $(1 + x)^{-1} = 1 - x + x^2 - x^3 + \ldots = \sum_{n=0}^{\infty} (-1)^n x^n$ [Provided $|x| < 1$]

(*ii*) $(1 - x)^{-1} = 1 + x + x^2 + x^3 + \ldots = \sum_{n=0}^{\infty} x^n$ [Provided $|x| < 1$]

(*iii*) $(1 + x)^{-2} = 1 - 2x + 3x^2 - 4x^3 + \ldots = \sum_{n=0}^{\infty} (-1)^n (n+1) x^n$ [Provided $|x| < 1$]

(*iv*) $(1 - x)^{-2} = 1 + 2x + 3x^2 + 4x^3 + \ldots = \sum_{n=0}^{\infty} (n+1) x^n$ [Provided $|x| < 1$]

(*v*) $(1 + x)^{-3} = 1 - 3x + 6x^2 - 10x^3 + \ldots = \sum_{n=0}^{\infty} (-1)^n \frac{1}{2}(n+1)(n+2) x^n$ [Provided $|x| < 1$]

$$(vi)\ (1-x)^{-3} = 1 + 3x + 6x^2 + 10x^3 + \ldots = \sum_{n=0}^{\infty} \frac{1}{2}(n+1)(n+2)x^n \qquad \text{[Provided } |x| < 1]$$

$$(vii)\ (1+x)^{-n} = 1 - nx + \frac{n(n+1)}{2!}x^2 - \frac{n(n+1)(n+2)}{3!}x^3 + \ldots$$

$$+ (-1)^r \frac{1}{r!} n(n+1)(n+2) \ldots (n+r-1)x^r + \ldots$$

$$= 1 + \sum_{r=1}^{\infty} (-1)^r \cdot \frac{1}{r!} n(n+1)(n+2) \ldots (n+r-1)x^r \qquad \text{[Provided } |x| < 1]$$

$$(viii)\ (1-x)^{-n} = 1 + nx + \frac{n(n+1)}{2!}x^2 + \frac{n(n+1)(n+2)}{3!}x^3 + \ldots + \frac{1}{r!} n(n+1)(n+2) \ldots (n+r-1)x^r + \ldots$$

$$= 1 + \sum_{r=1}^{\infty} \frac{1}{r!} n(n+1)(n+2) \ldots (n+r-1)x^r \qquad \text{[Provided } |x| < 1]$$

ILLUSTRATIVE EXAMPLES—IV

Example 1: *Solve the recurrence relation*

$$a_n = 2a_{n-1}, \text{ for } n = 1, 2, 3, \ldots$$

with initial condition $a_0 = 3$. Use generating function method. [WBUT–2004, 2007]

Solution: Let $G(x)$ be the generating function for the sequence $\{a_n\}$, that is, $G(x) = \sum_{n=0}^{\infty} a_n x^n$.

$$\therefore \quad xG(x) = \sum_{n=0}^{\infty} a_n x^{n+1} = \sum_{k=1}^{\infty} a_{k-1} x^k \quad (\text{Put } k = n+1 \Rightarrow n = k-1)$$

$$= \sum_{n=1}^{\infty} a_{n-1} x^n$$

$$\therefore \quad G(x) - 2xG(x) = \sum_{n=0}^{\infty} a_n x^n - 2\sum_{n=1}^{\infty} a_{n-1} x^n$$

$$= a_0 + \sum_{n=1}^{\infty} (a_n - 2a_{n-1})\, x^n$$

$$= 3 \qquad [\because\ a_0 = 3 \text{ and } a_n = 2a_{n-1}, \text{ for } n = 1, 2, 3, \ldots]$$

$$\therefore \quad (1-2x)\,G(x) = 3 \Rightarrow G(x) = \frac{3}{1-2x} = 3(1-2x)^{-1} = 3\sum_{n=0}^{\infty} 2^n x^n$$

$$\therefore \quad G(x) = \sum_{n=0}^{\infty} 3 \cdot 2^n x^n$$

Consequently, $a_n = 3.2^n;\ n = 1, 2, 3, \ldots$

Example 2: *Using generating functions method, solve the recurrence relation*

$$a_n - a_{n-1} - 6a_{n-2} = 0, \text{ for } n = 2, 3, 4, \ldots$$

with initial conditions $a_0 = 2$ *and* $a_1 = 1$.

Solution: Let $G(x)$ be the generating function for the sequence $\{a_n\}$, that is, $G(x) = \sum_{n=0}^{\infty} a_n x^n$.

Multiplying both sides of the given recurrence relation by x^n and summing from $n = 2$ to ∞, we get

$$0 = \sum_{n=2}^{\infty} a_n x^n - \sum_{n=2}^{\infty} a_{n-1} x^n - 6\sum_{n=2}^{\infty} a_{n-2} x^n$$

$$= \sum_{n=2}^{\infty} a_n x^n - x\sum_{n=2}^{\infty} a_{n-1} x^{n-1} - 6x^2 \sum_{n=2}^{\infty} a_{n-2} x^{n-2}$$

$$= \sum_{n=2}^{\infty} a_n x^n - x\sum_{k=1}^{\infty} a_k x^k - 6x^2 \sum_{r=0}^{\infty} a_r x^r \quad (\text{put } k = n-1, r = n-2)$$

$$= \{G(x) - a_0 - a_1 x\} - x\{G(x) - a_0\} - 6x^2 G(x)$$

$$= \{G(x) - 2 - x\} - x\{G(x) - 2\} - 6x^2 G(x) \qquad [\because \; a_0 = 2, a_1 = 1]$$

$$= (1 - x - 6x^2)G(x) - 2 + x$$

$$\therefore \qquad G(x) = \frac{2-x}{1-x-6x^2} = \frac{2-x}{(1-3x)(1+2x)} \qquad \ldots(1)$$

Let $$\frac{2-x}{(1-3x)(1+2x)} = \frac{A}{1-3x} + \frac{B}{1+2x}$$

$$\therefore \qquad 2 - x = A(1+2x) + B(1-3x) \qquad \ldots(2)$$

This is an identity in x and hence it is true for all values of x.

Putting $x = -\frac{1}{2}$ in (2), we get $2 + \frac{1}{2} = B\left(1+\frac{3}{2}\right) \Rightarrow B = 1.$

Putting $x = \frac{1}{3}$ in (2), we get $2 - \frac{1}{3} = A\left(1+\frac{2}{3}\right) \Rightarrow A = 1.$

So, from (1): $$G(x) = (1-3x)^{-1} + (1+2x)^{-1} = \sum_{n=0}^{\infty} 3^n x^n + \sum_{n=0}^{\infty} (-1)^n 2^n x^n$$

$$= \sum_{n=0}^{\infty} \{3^n + (-1)^n 2^n\} x^n$$

Consequently, $a_n = 3^n + (-1)^n 2^n$; $n = 0, 1, 2, 3, 4, \ldots$

Example 3: *Using generating function, solve the recurrence relation:*

$$a_n - 7a_{n-1} + 10a_{n-2} = 0, \text{ for all } n > 1 \text{ with } a_0 = a_1 = 3.$$ *[WBUT 2005]*

Solution: Let $G(x)$ be the generating function for the sequence $\{a_n\}$, that is, $G(x) = \sum_{n=0}^{\infty} a_n x^n$.

Multiplying both sides of the given recurrence relation by x^n and summing from $n = 2$ to ∞, we get

$$0 = \sum_{n=2}^{\infty} a_n x^n - 7\sum_{n=2}^{\infty} a_{n-1}x^n + 10\sum_{n=2}^{\infty} a_{n-2}x^n$$

$$= \sum_{n=2}^{\infty} a_n x^n - 7x\sum_{n=2}^{\infty} a_{n-1}x^{n-1} + 10x^2\sum_{n=2}^{\infty} a_{n-2}x^{n-2}$$

$$= \sum_{n=2}^{\infty} a_n x^n - 7x\sum_{k=1}^{\infty} a_k x^k + 10x^2\sum_{r=0}^{\infty} a_r x^r \quad (\text{put } k = n-1,\ r = n-2)$$

$$= \{G(x) - a_0 - a_1 x\} - 7x\{G(x) - a_0\} + 10x^2 G(x)$$

$$= \{G(x) - 3 - 3x\} - 7x\{G(x) - 3\} + 10x^2 G(x) \qquad [\because\ a_0 = a_1 = 3]$$

$$= (10x^2 - 7x + 1)G(x) - 3 + 18x$$

$$\therefore \quad G(x) = \frac{3-18x}{10x^2 - 7x + 1} = \frac{3-18x}{(2x-1)(5x-1)} \qquad \ldots(1)$$

Let $\dfrac{3-18x}{(2x-1)(5x-1)} = \dfrac{A}{2x-1} + \dfrac{B}{5x-1}$

$$\therefore \quad 3 - 18x = A(5x-1) + B(2x-1) \qquad \ldots(2)$$

This is an identity in x and hence it is true for all values of x.

Putting $x = \dfrac{1}{2}$ in (2), we get $3 - \dfrac{18}{2} = A\left(\dfrac{5}{2} - 1\right) \Rightarrow A = -4$.

Putting $x = \dfrac{1}{5}$ in (2), we get $3 - \dfrac{18}{5} = B\left(\dfrac{2}{5} - 1\right) \Rightarrow B = 1$.

So, from (1) :

$$G(x) = 4(1-2x)^{-1} - (1-5x)^{-1} = 4\sum_{n=0}^{\infty} 2^n x^n - \sum_{n=0}^{\infty} 5^n x^n$$

$$= \sum_{n=0}^{\infty} (2^{n+2} - 5^n)\, x^n$$

Consequently, $a_n = 2^{n+2} - 5^n;\ n = 0, 1, 2, 3, 4, 5, \ldots$

Example 4: *Using generating function, solve the recurrence relation:*

$a_n - 7a_{n-1} + 10a_{n-2} = 2$, for all $n > 1$ with $a_0 = a_1 = 3$. [*WBUT 2006, 2008*]

Solution: Proceeding as in the previous example, we get

$$(10x^2 - 7x + 1)\, G(x) - 3 + 18x = 2\sum_{n=2}^{\infty} x^n = 2\left\{\sum_{n=0}^{\infty} x^n - 1 - x\right\}$$

$$= 2\{(1-x)^{-1} - 1 - x\}$$

$$\therefore \quad (10x^2 - 7x + 1)\, G(x) = 3 - 18x + \frac{2}{1-x} - 2 - 2x$$

$$= \frac{1 - 20x - x + 20x^2 + 2}{1-x} = \frac{20x^2 - 21x + 3}{1-x}$$

$$\therefore \qquad G(x) = \frac{20x^2 - 21x + 3}{(1-x)(1-2x)(1-5x)} \qquad \text{...(1)}$$

Let $$\frac{20x^2 - 21x + 3}{(1-x)(1-2x)(1-5x)} = \frac{A}{1-x} + \frac{B}{1-2x} + \frac{C}{1-5x}.$$

$$\therefore \qquad 20x^2 - 21x + 3 = A(1 - 2x)(1 - 5x) + B(1 - x)(1 - 5x) + C(1 - x)(1 - 2x) \qquad \text{...(2)}$$

This is an identity in x and hence it is true for all values of x.

Putting $x = 1$ in (2), we get $2 = 4A \Rightarrow A = \frac{1}{2}$.

Putting $x = \frac{1}{2}$ in (2), we get $5 - \frac{21}{2} + 3 = B\left(1 - \frac{1}{2}\right)\left(1 - \frac{5}{2}\right) \Rightarrow B = \frac{10}{3}$.

Putting $x = \frac{1}{5}$ in (2), we get $\frac{4}{5} - \frac{21}{5} + 3 = C\left(1 - \frac{1}{5}\right)\left(1 - \frac{2}{5}\right) \Rightarrow C = -\frac{5}{6}$.

So, from (1): $$G(x) = \frac{1}{2}(1 - x)^{-1} + \frac{10}{3}(1 - 2x)^{-1} - \frac{5}{6}(1 - 5x)^{-1}$$

$$= \frac{1}{2}\sum_{n=0}^{\infty} x^n + \frac{10}{3}\sum_{n=0}^{\infty} 2^n x^n - \frac{5}{6}\sum_{n=0}^{\infty} 5^n x^n$$

$$= \sum_{n=0}^{\infty} \left\{\frac{1}{2} + \frac{10}{3}2^n - \frac{5}{6}.5^n\right\}x^n$$

Consequently, $$a_n = \frac{1}{2} + \frac{10}{3}.2^n - \frac{5}{6}.5^n;\ n = 0, 1, 2, 3, 4, 5, \ldots$$

Example 5: *Suppose a valid code word is an n-digit number in decimal notation consisting of an even number of 0s and let a_n denote the number of valid code words of length n. In Example 10 of Article 5.3, it is deduced that the sequence $\{a_n\}$ satisfies the recurrence relation:*

$$a_n = 8\,a_{n-1} + 10^{n-1},\ n \geq 2$$

with initial condition $a_1 = 9$. Using generating functions method to find an explicit formula for a_n.

Solution: To apply generating functions method, let us extend the given sequence by setting $a_0 = 1$. When we assign this value to a_0 and use the given recurrence relation, we have $a_1 = 8a_0 + 10° = 9$, which is consistent with the original initial condition. Also, it makes sense since there is one code word of length 0 : the empty string.

Multiplying both sides of the given recurrence relation by x^n and summing from $n = 2$ to ∞, we get

$$\sum_{n=2}^{\infty} a_n x^n = 8\sum_{n=2}^{\infty} a_{n-1}x^n + \sum_{n=2}^{\infty} 10^{n-1}x^n$$

$$= 8x\sum_{n=2}^{\infty} a_{n-1}x^{n-1} + x\sum_{n=2}^{\infty} 10^{n-1}x^{n-1}$$

$$= 8x \sum_{k=1}^{\infty} a_k x^k + x \sum_{k=1}^{\infty} 10^k x^k \quad \text{(Put } k = n-1\text{)}$$

Let $G(x)$ be the generating function for the sequence $\{a_n\}$, that is, $G(x) = \sum_{n=0}^{\infty} a_n x^n$.

$$\therefore \quad \{G(x) - a_0 - a_1 x\} = 8x\{G(x) - a_0\} + x\left\{\sum_{k=0}^{\infty} 10^k x^k - 1\right\}$$

$$\Rightarrow \quad \{G(x) - 1 - 9x\} = 8x\{G(x) - 1\} + x(1 - 10x)^{-1} - x$$

$$\Rightarrow \quad (1 - 8x)\, G(x) = 1 + \frac{x}{1 - 10x}$$

$$\Rightarrow \quad G(x) = \frac{1 - 9x}{(1 - 8x)(1 - 10x)} = \frac{1}{2} \frac{(1 - 10x) + (1 - 8x)}{(1 - 8x)(1 - 10x)}$$

$$= \frac{1}{2}\{(1 - 8x)^{-1} + (1 - 10x)^{-1}\}$$

$$= \frac{1}{2}\left\{\sum_{n=0}^{\infty} 8^n x^n + \sum_{n=0}^{\infty} 10^n x^n\right\}$$

$$= \sum_{n=0}^{\infty} \frac{1}{2}(8^n + 10^n)\, x^n$$

Consequently, $$a_n = \frac{1}{2}(8^n + 10^n)\ ;\ n = 0, 1, 2, 3, 4, \ldots$$

Note: See Ex. 5, Art. 5.5.

Alternative Statement

Using generating functions method, solve:

$$a_n - 8a_{n-1} = 10^{n-1},\ a_0 = 1$$ [*WBUT 2005*]

Example 6: *Solve the following recurrence relation using generating function:*

$$a_n = 4(a_{n-1} - a_{n-2}) + 2^n\ ;\ n \geq 2,\ \text{with } a_0 = 1,\ a_1 = 4.$$ [*WBUT 2012*]

Solution: Let $G(x)$ be the generating function for the sequence $\{a_n\}$, that is, $G(x) = \sum_{n=0}^{\infty} a_n x^n$.

Multiplying both sides of the given recurrence relation by x^n and summing from $n = 2$ to ∞, we get

$$\sum_{n=2}^{\infty} a_n x^n = 4\sum_{n=2}^{\infty} a_{n-1} x^n - 4\sum_{n=2}^{\infty} a_{n-2} x^n + \sum_{n=2}^{\infty} 2^n x^n$$

$$= 4x\sum_{n=2}^{\infty} a_{n-1}x^{n-1} - 4x^2\sum_{n=2}^{\infty} a_{n-2}x^{n-2} + \sum_{n=2}^{\infty} 2^n x^n$$

$$= 4x\sum_{k=1}^{\infty} a_k x^k - 4x^2\sum_{r=0}^{\infty} a_r x^r + \sum_{n=2}^{\infty} 2^n x^n \quad [\text{Put } k = n-1, r = n-2]$$

$$\therefore \quad \{G(x) - a_0 - a_1x\} = 4x\{G(x) - a_0\} - 4x^2 G(x) + \sum_{n=0}^{\infty} 2^n x^n - 1 - 2x$$

or $$\{G(x) - 1 - 4x\} = 4x\{G(x) - 1\} - 4x^2 G(x) + (1-2x)^{-1} - 1 - 2x \quad [\because a_0 = 1, a_1 = 4]$$

or $$(4x^2 - 4x + 1)G(x) = \frac{1}{1-2x} - 2x$$

$$\therefore \quad G(x) = \frac{4x^2 - 2x + 1}{(1-2x)^3} = \frac{(1-2x)^2 - 1 + 2x + 1}{(1-2x)^3}$$

$$= \frac{1}{1-2x} - \frac{1}{(1-2x)^2} + \frac{1}{(1-2x)^3}$$

or $$G(x) = (1-2x)^{-1} - (1-2x)^{-2} + (1-2x)^{-3}$$

$$= \sum_{n=0}^{\infty} 2^n x^n - \sum_{n=0}^{\infty} (n+1)2^n x^n + \sum_{n=0}^{\infty} \frac{1}{2}(n+1)(n+2)2^n x^n$$

$$= \sum_{n=0}^{\infty} \left\{1 - (n+1) + \frac{1}{2}(n+1)(n+2)\right\} 2^n x^n$$

$$= \sum_{n=0}^{\infty} (n^2 + n + 2)\, 2^{n-1} x^n$$

Consequently, $a_n = (n^2 + n + 2)2^{n-1};\ n = 0, 1, 2, 3, 4, \ldots$

Note: See Ex. 5, Art. 5.6.2.

Example 7: *Using the generating functions method, solve the recurrence relation:*

$$a_n - 5a_{n-1} + 6a_{n-2} = 2^n - 6n, \textit{ for } n = 2, 3, 4, \ldots.$$

with initial conditions $a_0 = a_1 = 1$.

Solution: Let $G(x)$ be the generating function for the sequence $\{a_n\}$, that is, $G(x) = \sum_{n=0}^{\infty} a_n x^n$.

Multiplying both sides of the given recurrence relation by x^n and summing from $n = 2$ to ∞, we get

$$\sum_{n=2}^{\infty} a_n x^n - 5\sum_{n=2}^{\infty} a_{n-1}x^n + 6\sum_{n=2}^{\infty} a_{n-2}x^n = \sum_{n=2}^{\infty} 2^n x^n - 6\sum_{n=2}^{\infty} nx^n$$

or $$\sum_{n=2}^{\infty} a_n x^n - 5x \sum_{n=2}^{\infty} a_{n-1} x^{n-1} + 6x^2 \sum_{n=2}^{\infty} a_{n-2} x^{n-2} = \sum_{n=2}^{\infty} 2^n x^n - 6x \sum_{n=2}^{\infty} n x^{n-1}$$

or $$\sum_{n=2}^{\infty} a_n x^n - 5x \sum_{k=1}^{\infty} a_k x^k + 6x^2 \sum_{r=0}^{\infty} a_r x^r = \sum_{n=2}^{\infty} 2^n x^n - 6x \sum_{k=1}^{\infty} (k+1) x^k$$

(Put $k = n - 1$ and $r = n - 2$)

or $$\{G(x) - a_0 - a_1 x\} - 5x\{G(x) - a_0\} + 6x^2 G(x) = \left\{\sum_{n=0}^{\infty} 2^n x^n - 1 - 2x\right\} - 6x\left\{\sum_{k=0}^{\infty} (k+1) x^k - 1\right\}$$

or $$\{G(x) - 1 - x\} - 5x\{G(x) - 1\} + 6x^2 G(x) = \{1 - 2x\}^{-1} - 1 - 2x - 6x\{(1 - x)^{-2} - 1\} \qquad (\because \ a_0 = a_1 = 1).$$

or $$(6x^2 - 5x + 1)\, G(x) = \frac{1}{1-2x} - \frac{6x}{(1-x)^2}$$

$$\therefore \qquad G(x) = \frac{1 - 8x + 13x^2}{(1-x)^2(1-2x)^2(1-3x)} \qquad \text{...(1)}$$

Let $$\frac{1 - 8x + 13x^2}{(1-x)^2(1-2x)^2(1-3x)} = \frac{A}{1-x} + \frac{B}{(1-x)^2} + \frac{C}{1-2x} + \frac{D}{(1-2x)^2} + \frac{E}{1-3x}.$$

$$\therefore \quad 1 - 8x + 13x^2 = A(1-x)(1-2x)^2(1-3x) + B(1-2x)^2(1-3x) + C(1-x)^2(1-2x)(1-3x) + D(1-x)^2(1-3x) + E(1-x)^2(1-2x)^2 \qquad \text{...(2)}$$

This is an identity in x and hence it is true for all values of x.

Putting $x = 1,\ 1 - 8 + 13 = B(1-2)^2(1-3) \quad \Rightarrow \quad B = -3$

Putting $x = \frac{1}{2},\ 1 - 4 + \frac{13}{4} = D\left(1 - \frac{1}{2}\right)^2\left(1 - \frac{3}{2}\right) \quad \Rightarrow \quad D = -2$

Putting $x = \frac{1}{3},\ 1 - \frac{8}{3} + \frac{13}{9} = E\left(1 - \frac{1}{3}\right)^2\left(1 - \frac{2}{3}\right)^2 \quad \Rightarrow \quad E = -\frac{9}{2}$

Equating the constant terms from both sides of (2), we get

$$A + B + C + D + E = 1.$$

Equating the coefficients of x^4 from both sides of (2), we get

$$12A + 6C + 4E = 0.$$

Solving, we get $A = -\frac{15}{2}$, $C = 18$

So, from (1), we get

$$G(x) = -\frac{15}{2}(1-x)^{-1} - 3(1-x)^{-2} + 18(1-2x)^{-1} - 2(1-2x)^{-2} - \frac{9}{2}(1-3x)^{-1}$$

$$= -\frac{15}{2}\sum_{n=0}^{\infty} x^n - 3\sum_{n=0}^{\infty}(n+1)x^n + 18\sum_{n=0}^{\infty} 2^n x^n - 2\sum_{n=0}^{\infty}(n+1)2^n x^n - \frac{9}{2}\sum_{n=0}^{\infty} 3^n x^n$$

$$= \sum_{n=0}^{\infty}\left\{-\frac{15}{2} - 3(n+1) + 18.2^n - (n+1)\,2^{n+1} - \frac{1}{2}.3^{n+2}\right\}x^n$$

Consequently, $a_n = -\frac{21}{2} - 3n + (8-n)2^{n+1} - \frac{1}{2}\,.\,3^{n+2}, n = 0, 1, 2, 3, 4, ...$

MULTIPLE CHOICE QUESTIONS

1. The recurrence relation that determines the sequence $3, \frac{6}{5}, \frac{12}{25}, \frac{24}{125}, ...$ is

(*a*) $a_n = a_{n-1} + \frac{2}{5}, a_0 = 3.$ (*b*) $a_n = a_{n-1} - \frac{2}{5}, a_0 = 3$

(*c*) $2a_n - 5a_{n-1} = 0, a_0 = 3$ (*d*) $a_n = \left(\frac{2}{5}\right)a_{n-1}, a_0 = 3.$

2. If $a_{n+1} - ba_n = 0$, $a_3 = 161$ and $a_5 = 2576$, then $b =$

(*a*) ± 4 (*b*) 4 (*c*) -4 (*d*) 16

3. The first four terms of the sequence $\{t_n\}$ defined by $t_1 = 1, t_n = n^2 t_{n-1} + n - 2$ are

(*a*) 1, 4, 46, 750 (*b*) 1, 4, 46, 754

(*c*) 1, 4, 46, 752 (*d*) 1, 4, 46, 756.

4. The solution of the recurrence relation $S_n = 2S_{n-1}$ with $S_0 = 1$ is $S_n =$

(*a*) 2^n (*b*) 2^{n-1}

(*c*) 2^{n+1} (*d*) none of these. [*WBUT 2006, 2008*]

5. The solution of the recurrence relation

$a_{n+1} - 2a_n = 5, n \geq 0, a_0 = 1$ is $a_n =$

(*a*) $6 . 2^n - 5$ (*b*) $5 - 6 . 2^n$ (*c*) $2^{n+1} - 1$ (*d*) none of these.

[*WBUT 2007*]

6. $\{a_n\}$ is a solution of the recurrence relation

$$a_n = 4a_{n-1} - 3a_{n-2} \text{ if } a_n =$$

(*a*) $A(-3)^n + B$ (*b*) $A(-3)^n + B(-1)^n$ (*c*) $A3^n + B$ (*d*) $A3^n + B(-1)^n$.

7. $\{a_n\}$ is a solution of the recurrence relation

$$a_n = 10a_{n-1} - 25a_{n-2} \text{ if } a_n \text{ is equal to}$$

(*a*) 5^n (*b*) $n5^n$ (*c*) $n^2 5^n$ (*d*) $(A + Bn)5^n$.

8. The general solution of the recurrence relation

$$a_n - 4a_{n-1} + 4a_{n-2} = 0 \text{ is}$$

(*a*) $A2^n$ (*b*) $(A + Bn)2^n$ (*c*) $A2^n + B(-2)^n$ (*d*) $(A + Bn)(-2)^n$.

9. If $a_n = a_{n-1} + n$ and $a_1 = 1$, then $a_n =$

(*a*) n^2 (*b*) $n^2 + 1$ (*c*) $\frac{n}{2}(n+1)$ (*d*) $n^2 + n$.

10. If $a_n = a_{n-1} + n^2$ and $a_1 = 1$, then a_n is equal to

(a) $\frac{1}{6} n(n+1)(2n+1)$ (b) $(n-1)^2 + n^2$ (c) $n^2 + 1$ (d) $n(n+1)(2n+1)$.

11. The homogeneous solution of the recurrence relation $a_n + 2a_{n-1} + a_{n-2} = 7^n$ is

(a) $A + Bn$ (b) $(An + B)(-1)^n$ (c) $A(-1)^n$ (d) $n(-1)^n$.

12. A particular solution of the recurrence relation $a_n - 6a_{n-1} + 8a_{n-2} = 7^n$ is

(a) 7^n (b) $\frac{15}{49} 7^n$ (c) 49.7^n (d) $\frac{49}{15} 7^n$.

13. If $G(x)$ is the generating function for the sequence $a_0, a_1, a_2, \ldots$, then $xG(x)$ is the generating function for the sequence

(a) $a_1, a_2, a_3, \ldots$ (b) $0, a_1, a_2, \ldots$ (c) $0, a_0, a_1, \ldots$ (d) $a_2, a_3, a_4, \ldots$.

14. $\frac{1}{1-x} - x^3$ generates the sequence

(a) 1, 1, 1, 0, 1, 1, ... (b) 1, 1, 0, 1, 1, 1, ... (c) 1, 0, 1, 1, ... (d) 0, 1, 1, ...

15. The generating function for the sequence 1, 0, – 1, 0, 1, 0, – 1, ... is

(a) $\frac{1}{1-x^2}$ (b) $\frac{1}{1+x^2}$ (c) $\frac{1}{1+x} + \frac{1}{1-x}$ (d) $\frac{1}{1+x} - \frac{1}{1-x}$.

16. The generating function for the sequence 0, 0, 1, 1, 1, ... is

(a) $\frac{1}{1-x}$ (b) $\frac{1}{(1-x)^2}$ (c) $\frac{x^2}{(1-x)^2}$ (d) $\frac{x^2}{1-x}$.

17. The generating function for the sequence 2, – 2, 2, – 2, ... is

(a) $\frac{1}{1+x}$ (b) $\frac{2}{1-x}$ (c) $\frac{2}{1+x}$ (d) none of these.

18. The generating function for the sequence 1, 2, 3, 4, 5, 6, ... is

(a) e^x (b) $\frac{1}{(1-x)^2}$ (c) $\frac{1}{(1+x)^2}$ (d) $\frac{1}{1-x}$.

19. The generating function for the finite sequence 1, 0, – 2, 3, 4, – 5 is $1 - 2x^2 + 3x^3 + 4x^4 - 5x^5$.

(a) True (b) False

20. The generating function for the finite sequence $\binom{n}{0}, \binom{n}{1}, \binom{n}{2}, \ldots, \binom{n}{n}$ is

(a) $(1-x)^n$ (b) $(1+x)^{-n}$ (c) $1 + x$ (d) $(1+x)^n$.

[WBUT 2007]

21. The generating function for the sequence $\left\{1, 1, \frac{1}{2!}, \frac{1}{3!}, \frac{1}{4!}, \ldots\right\}$ is

(a) e^x (b) e^{-x} (c) $\log(1+x)$ (d) $(1-x)^{-1}$.

22. The generating function for the sequence $1, -\frac{1}{2}, \frac{1}{3}, -\frac{1}{4}, \frac{1}{5}, -\frac{1}{6}, \ldots$ is

(*a*) $\log(1+x)$ (*b*) $\frac{1}{x}\log(1-x)$ (*c*) $\frac{1}{x}\log(1+x)$ (*d*) e^x.

23. The order of the recurrence relation $a_n = a_{n-1} + a_{n-2}$, $n \geq 2$, is two.

(*a*) True (*b*) False

24. The solution to the recurrence equation $T(2^k) = 3T(2^{k-1}) + 1$, $T(1) = 1$ is

(*a*) 2^k (*b*) $\frac{1}{2}(3^{k+1} - 1)$ (*c*) $3^{\log_2 k}$ (*d*) $2^{\log_3 k}$

[*GATE 2002*]

ANSWERS

1. (*d*)	**2.** (*a*)	**3.** (*b*)	**4.** (*a*)
5. (*a*)	**6.** (*c*)	**7.** (*d*)	**8.** (*b*)
9. (*c*)	**10.** (*a*)	**11.** (*b*)	**12.** (*d*)
13. (*c*)	**14.** (*a*)	**15.** (*b*)	**16.** (*d*)
17. (*c*)	**18.** (*b*)	**19.** (*a*)	**20.** (*d*)
21. (*a*)	**22.** (*c*)	**23.** (*a*)	**24.** (*b*)

[**Hint of 24:**

$$T(2^k) = 3\{3T(2^{k-2}) + 1\} + 1 = 3^2\{3T(2^{k-3}) + 1\} + 3 + 1$$
$$= \ldots = 3^{k-1}\{3T(2^{k-k}) + 1\} + 3^{k-2} + 3^{k-3} + \ldots + 1$$
$$= 3^k + 3^{k-1} + 3^{k-2} + \ldots + 1 = \frac{3^{k+1} - 1}{3 - 1}.$$

PROBLEMS

1. Define a recurrence relation (or difference equation). What do you mean by its solution? Prove that the sequence $\{a_n\}$ defined by $a_n = 4.2^n + 7.3^n$ is a solution of the recurrence relation $a_n - 5a_{n-1} + 6a_{n-2} = 0$.

2. Define generating functions of an infinite and a finite sequence with examples.

3. Show that

(*i*) the sequence 2, 3, 4, 5, ..., $2r + 1$, ... satisfies the recurrence relation $a_n = 2a_{n-1} - a_{n-2}$, $n \geq 2$,

(*ii*) the sequence 0, 1, 3, 7, ..., $2^r - 1$, ... satisfies the recurrence relation $a_n = 3a_{n-1} - 2a_{n-2}$, $n \geq 2$.

4. Solve the recurrence relation $x_n = x_{n-1} + 4$, for all $n \geq 2$, where $x_1 = 2$ by iterative method.

[*WBUT 2004*]

5. Find the recurrence relation from the sequence 0, 1, 1, 2, 3, 5, 8, 13, ...

6. Solve: $a_{n+1} - 3a_n = 0$, $a_0 = 1$.

7. Find the recurrence relation for the solution $a_n = 5.2^n$.

8. Find the generating function corresponding to each of the following sequences:

(*i*) 2, 4, 8, 16, 32, 64, ...

(*ii*) 2, – 2, 2, – 2, 2, – 2, ...

(*iii*) 1, 0, 1, 0, 1, 0, ...

(*iv*) 0, 1, – 2, 4, – 8, 16, ...

9. Determine the generating function corresponding to the sequence $\{a_n\}$ defined by $a_n = 2^n + 3^n, n \geq 0$.

10. Find the generating function for the sequence $\{a_n\}$ defined by

(*i*) $a_n = \begin{cases} 2^n, & \text{if } n \text{ is even} \\ -2^n, & \text{if } n \text{ is odd} \end{cases}$ (*ii*) $a_n = \begin{cases} 2, & \text{for } n = 0 \\ 7, & \text{for } n = 1 \\ 2^{n+1}, & \text{for } n \geq 2 \end{cases}$

11. Find the sequence corresponding to each of the following generating functions:

(*i*) $\dfrac{1}{x^2 - 6x + 5}$ (*ii*) $\dfrac{2}{1 - 4x^2}$

(*iii*) $\dfrac{x^4}{1 - 2x}$ (*iv*) $g(z) = 5(z^5 - 1)/(z - 1)$ [*WBUT 2012*]

12. A computer virus (that spreads through e-mail message) is planted in 4 computers on the 1st day and each infected computer infects 3 new computers each day. On the 3rd day a software solution is found to counteract the virus and 5 computers are cleared off the virus. From the 4th day onwards the number of computers cleared off the virus is five times the number of computers cleared off on the previous day. In how many days is the computer virus entirely eradicated?

13. An investor deposits ₹ 10,000/- on 1st January every year in a savings account at a bank that offers 8% interest per annum compounded on a quarterly basis. Find the balance on 1st January of the nth year.

14. Any gossip in a large organisation is spread among the people in the organisation through telephone. When a telephone conversation occurs between any two persons, each person tells the other all the gossip he/she has heard. If t_n denotes the minimum number of telephone calls among n persons so that all gossip is known to each of the n persons, prove that

(*i*) $t_2 = 1$, $t_3 = 3$ and $t_4 = 4$

(*ii*) $t_n \leq t_{n-1} + 2$

(*iii*) $t_n \leq 2n - 4$, for $n \geq 4$.

15. The number of bacteria in a colony triples every hour.

(*i*) Construct a recurrence relation (difference equation) for the number of bacteria after n hours have elapsed.

(*ii*) If 100 bacteria are assumed to begin a new colony, how many bacteria will be in the colony in 10 hours?

16. A factory manufactures custom sports cars at an increasing rate. In the first month only one car is manufactured, in the second month two cars are made and so on with n cars are made in the nth month.

(*i*) Construct a recurrence relation for the number of cars manufactured in the first n months by this factory.

(*ii*) How many cars are made in the first year?

(*iii*) Determine an explicit formula for the number of cars produced in the first n months by this factory.

17. Solve each of the following recurrence relations using iterative method.

(*i*) $a_n = a_{n-1} + 2, a_0 = 1$ (*ii*) $a_n = a_{n-1} + 3, a_0 = 2$

(*iii*) $a_n = 2a_{n-1}, a_0 = 3$ (*iv*) $a_n = 2a_{n-1} + 1, a_0 = 3$

(*v*) $a_n = na_{n-1}, a_0 = 5$ (*vi*) $a_n - 4a_{n-1} + 4a_{n-2} = 0, a_0 = 1, a_1 = 4$

(*vii*) $y_n = y_{n-1} + 2y_{n-2}, a_0 = 2, a_1 = 7$

18. Solve, by characteristic roots method, each of the following recurrence relations together with the given initial conditions.

(*i*) $a_n = 5a_{n-1} - 6a_{n-2}$, for $n \geq 2$, $a_0 = 1$, $a_1 = 0$

(*ii*) $a_n = 4a_{n-1} - 4a_{n-2}$, for $n \geq 2$, $a_0 = 6$, $a_1 = 8$

(*iii*) $t_n = 2t_{n-1} + 8t_{n-2}$, for $n \geq 2$, $t_0 = 4$, $t_1 = 10$

(*iv*) $a_n = 2a_{n-1} + 15a_{n-2}$, for $n \geq 2$, $a_0 = 0$, $a_1 = -1$

(*v*) $a_n + 4(a_{n-1} + a_{n-2}) = 0$, for $n \geq 2$, $a_0 = 0$, $a_1 = 1$

19. (*i*) Find all solutions of the recurrence relation $a_n - 2a_{n-1} = 3^n$

(*ii*) Find the solution of the recurrence relation in (*i*) subject to the initial condition $a_1 = 5$.

20. Find the solution of each of the following recurrence relations:

(*i*) $a_n + 6a_{n-2} = 5a_{n-1} + 2^n + 3n$

(*ii*) $a_n - 4(a_{n-1} - a_{n-2}) = (n+1)2^n$

(*iii*) $a_n - 4a_{n-1} + 3a_{n-2} = 2^n + n + 3$
with initial conditions $a_0 = 1$ and $a_1 = 4$.

(*iv*) $t_n - 6t_{n-1} + 8t_{n-2} = n4^n$ with $t_0 = 8$, $t_1 = 22$.

21. Solve each of the following recurrence relations using generating functions method:

(*i*) $a_n = 4a_{n-2}$ with $a_0 = 0$, $a_1 = 1$

(*ii*) $a_n - 3a_{n-1} = 2$ with initial condition $a_0 = 1$

(*iii*) $a_n = 5a_{n-1} - 6a_{n-2}$ with initial conditions $a_0 = 6$ and $a_1 = 30$.

(*iv*) $a_n = 4a_{n-1} - 4a_{n-2} + n^2$ with initial conditions $a_0 = 2$ and $a_1 = 5$.

ANSWERS

4. $x_n = 4n - 2$

5. $a_n = a_{n-1} + a_{n-2}$, $n \geq 2$ with $a_0 = 0$, $a_1 = 1$.

6. $a_n = 3^n$

7. $a_n - 2a_{n-1} = 0$, with $a_0 = 5$.

8. (*i*) **Hint:** $G(x) = 2 + 4x + 8x^2 + 16x^3 + 32x^4 + \ldots$

$$= 2\{1 + 2x + (2x)^2 + (2x)^3 + (2x)^4 + \ldots\}$$

$$= \frac{2}{1-2x} \; ; |2x| < 1$$

(*ii*) $\frac{2}{1+x} \; ; |x| < 1$

(*iii*) $\frac{1}{1-x^2} \; ; |x^2| < 1$

(*iv*) $\frac{x}{1+2x} \; ; |2x| < 1.$

9. **Hint:** $G(x) = \sum_{n=0}^{\infty} a_n x^n = \sum_{n=0}^{\infty} (2^n + 3^n)x^n = \sum_{n=0}^{\infty} 2^n x^n + \sum_{n=0}^{\infty} 3^n x^n$

$$= (1-2x)^{-1} + (1-3x)^{-1} = \frac{1}{1-2x} + \frac{1}{1-3x} = \frac{2-5x}{(1-2x)(1-3x)}$$

10. (*i*) $\frac{1}{1+2x}$

(*ii*) **Hint:** $G(x) = \sum_{n=0}^{\infty} a_n x^n = a_0 + a_1 x + \sum_{n=2}^{\infty} a_n x^n$

$$= 2 + 7x + \sum_{n=2}^{\infty} 2^{n+1} x^n$$

$$= 3x + 2\left\{1 + 2x + \sum_{n=2}^{\infty} 2^n x^n\right\} = 3x + 2\sum_{n=0}^{\infty} (2x)^n$$

$$= 3x + \frac{2}{1-2x}.$$

11. (*i*) **Hint:** $$\frac{1}{x^2 - 6x + 5} = \frac{1}{(5-x)(1-x)} = \frac{1}{4}\,\frac{(5-x)-(1-x)}{(5-x)(1-x)}$$

$$= \frac{1}{4}\left\{(1-x)^{-1} - \frac{1}{5}\left(1-\frac{x}{5}\right)^{-1}\right\} = \frac{1}{4}\sum_{n=0}^{\infty} x^n - \frac{1}{20}\sum_{n=0}^{\infty}\left(\frac{x}{5}\right)^n$$

$$= \sum_{n=0}^{\infty} \frac{1}{4}\left\{1 - \frac{1}{5^{n+1}}\right\} x^n$$

So, the required sequence is $\{a_n\}$ defined by $a_n = \frac{1}{4}\left\{1 - \frac{1}{5^{n+1}}\right\}$, for all $n \geq 0$.

(*ii*) **Hint:** $$\frac{2}{1-4x^2} = \frac{2}{(1-2x)(1+2x)} = \frac{(1+2x)+(1-2x)}{(1-2x)(1+2x)}$$

$$= (1-2x)^{-1} + (1+2x)^{-1} = \sum_{n=0}^{\infty} 2^n x^n + \sum_{n=0}^{\infty} (-1)^n 2^n x^n$$

$$= \sum_{n=0}^{\infty} \{2^n + (-1)^n 2^n\} x^n$$

So, the required sequence is $\{a_n\}$ defined by

$$a_n = \{1 + (-1)^n\}\, 2^n, \text{ for all } n \geq 0, \text{ i.e.,}$$

$$a_n = \begin{cases} 0, & n \text{ is odd} \\ 2^{n+1}, & n \text{ is even} \end{cases}$$

(*iii*) $$\frac{x^4}{1-2x} = x^4(1-2x)^{-1} = x^4 \sum_{n=0}^{\infty} 2^n x^n = \sum_{n=0}^{\infty} 2^n x^{n+4}$$

$$= x^4 + 2x^5 + 2^2 x^6 + 2^3 x^7 + 2^4 x^8 + \ldots \text{ to } \infty$$

So, the required sequence is $\{a_n\}$ defined by

$$a_n = \begin{cases} 0, & n = 1, 2, 3 \\ 2^{n-4}, & n \geq 4 \end{cases}$$

(*iv*) $g(z) = 5(1 + z + z^2 + z^3 + z^4)$. Thus the required sequence is $\{a_n\}$ defined by

$$a_n = \begin{cases} 5z^n, & \text{for } n = 0, 1, 2, 3, 4 \\ 0, & \text{for } n \geq 5 \end{cases}.$$

12. **Hint:** Let I_n denote the number of infected computers at the end of the nth day. Then $I_1 = 4, I_2 = 16, I_3 = 4(I_2 - 5) = 44, I_4 = 4(I_3 - 25) = 76$. On the 5th day, 125 computers can be cleared off the virus and so the required answer is 5 days.

13. ₹ 10, 000 $(1.02)^{4n-4}$

15. (*i*) $a_n = 3a_{n-1}$ (*ii*) 5,904,900

16. (*i*) $a_n = a_{n-1} + n;\ a_0 = 0$ (*ii*) $a_{12} = 78$

(*iii*) $a_n = \frac{1}{2} n(n+1)$.

17. (*i*) $a_n = 2n + 1, n \geq 0$ (*ii*) $a_n = 3n + 2, n \geq 0$

(*iii*) $a_n = 3 \cdot 2^n, n \geq 0$ (*iv*) $a_n = 2^{n+2} - 1, n \geq 0$

(*v*) $a_n = 5(n!), n \geq 0$ (*vi*) $a_n = (n+1)2^n, n \geq 0$

(*vii*) $y_n = 3 \cdot 2^n - (-1)^n, n \geq 0$

18. (*i*) $a_n = 3 \cdot 2^n - 2 \cdot 3^n$ (*ii*) $a_n = 6 \cdot 2^n - 2n \cdot 2^n$

(*iii*) $t_n = 3 \cdot 4^n + (-2)^n$ (*iv*) $a_n = \frac{1}{2}\{3^n + (-5)^n\}$

(*v*) $a_n = n(-2)^{n-1}$

19. (*i*) $a_n = A\,2^n + 3^{n+1}$ (*ii*) $a_n = 3^{n+1} - 2^{n+1}$

20. (*i*) $a_n = A\,2^n + B3^n - n \cdot 2^{n+1} + \frac{3n}{2} + \frac{21}{4}$. (*ii*) $a_n = \left(A + Bn + n^2 + \frac{n^3}{6}\right)2^n$

(*iii*) $a_n = -2^{n+2} - \frac{n^2}{4} - \frac{5n}{2} + \frac{1}{8} + \frac{39}{8} \cdot 3^n$ (*iv*) $t_n = 3.4^n + 5.2^n + n(n-1)4^n$

21. (*i*) $a_n = \{1 - (-1)^n\}\,2^{n-2}$ (*ii*) $a_n = 2 \cdot 3^n - 1$

(*iii*) $a_n = 18 \cdot 3^n - 12 \cdot 2^n$ (*iv*) $a_n = n^2 + 8n + 20 + (6n - 18)2^n$.

Module IV

CHAPTER **6**

Graph Colouring

6.1 CHROMATIC NUMBER

Properly coloured graph: Painting all the vertices of a graph with colours so that no two adjacent vertices have the same colour is said to be the **proper colouring** (or sometimes simply **colouring**) of the graph.

A graph in which every vertex has been assigned a colour according to proper colouring is known as a **properly coloured graph.**

Note: Proper colouring is also known as **vertex colouring.**

K-chromatic graph and chromatic number: Usually a given graph can be properly coloured in many different ways. A graph G that needs K different colours for its proper colouring, and no less, is said to be a **K-chromatic graph** and the number K is called the **chromatic number** of G.

Note: The chromatic number of a graph G is denoted by $\chi(G)$.

K-colourable graph: A graph G is said to be **K-colourable** if it has a proper K-colouring. The **chromatic number** $\chi(G)$ is the least K such that G is K-colourable.

Example 1: The following graph is 3-chromatic:

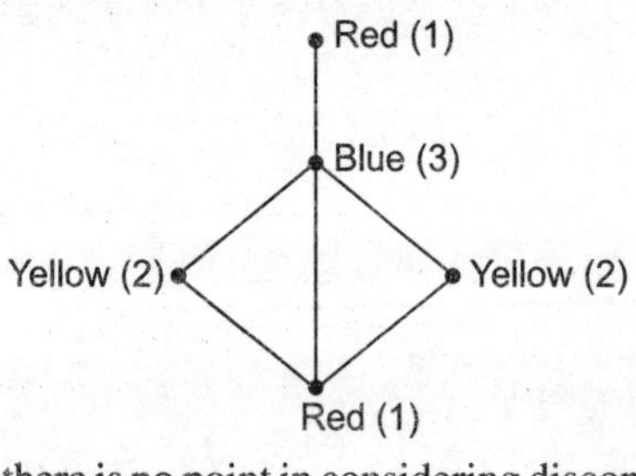

Notes: (*i*) In colouring graphs, there is no point in considering disconnected graphs. It is usual to investigate colouring of connected graphs only, because how we colour vertices in one component of a disconnected graph has no effect on the colouring of the other components.

(*ii*) All parallel edges between two vertices can be replaced by a single edge without affecting adjacency of vertices.

(*iii*) Self-loops must be disregarded.

Remark: For colouring problems, we need to consider only simple connected graphs.

Observations:

(*i*) A graph is 1-chromatic if it consists of only isolated vertices.

(*ii*) A graph with one or more edges (not a self-loop, of course) is at least **2-chromatic** (also known as **bichromatic**).

(*iii*) A complete graph of n vertices is n-chromatic, because any two of its vertices are adjacent. Therefore, a graph containing a complete graph of m vertices is at least m-chromatic. For example, every graph containing a triangle is at least 3-chromatic.

(*iv*) A path containing n (≥ 2) vertices is 2-chromatic.

(*v*) A graph consisting of simply one circuit with n (≥ 3) vertices is

(*a*) 2-chromatic if n is even, for instance:

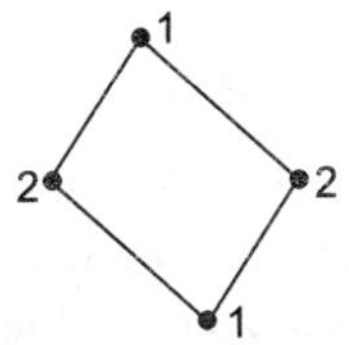

(*b*) 3-chromatic if n is odd, for instance:

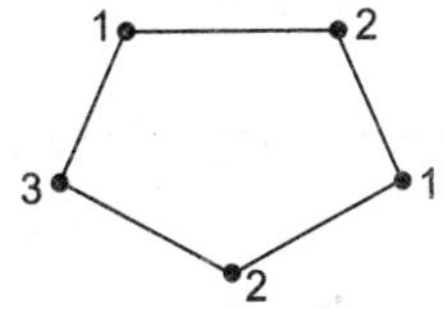

Note: Proper colouring of a given graph is simple enough. But in general, a proper colouring with the least number of colours is a difficult task.

Theorem 1: *Every tree with at least two vertices is 2-chromatic.*

Proof: Choose any vertex v in the given tree T. Consider T as a rooted tree at vertex v and paint v with colour 1. Paint all vertices adjacent to v using colour 2. Next, paint the vertices adjacent to these (that just have been coloured with 2) using colour 1. Continue this colouring process till every vertex in T has been painted.

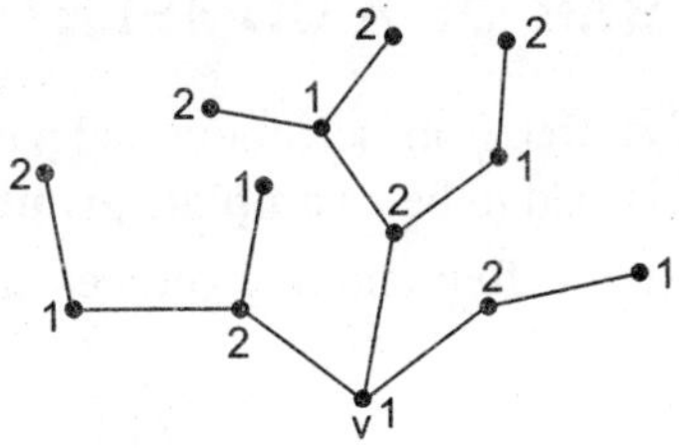

Observe that in T, all vertices at odd distances from v have colour 2, while v and vertices at even distances from v have colour 1.

Now along any path in T, the vertices are of alternating colours. We know that there is one and only one path between any two vertices in a tree, so no two adjacent vertices of T have the same colour. Thus, T has been properly coloured using two colours. Also, it is observed that a graph with one or more edges (not a self-loop, of course) is at least 2-chromatic. Hence the result.

Note: Though a tree (with at least two vertices) is 2-chromatic, not every 2-chromatic graph is a tree. For instance, the following graph is 2-chromatic but it is not a tree.

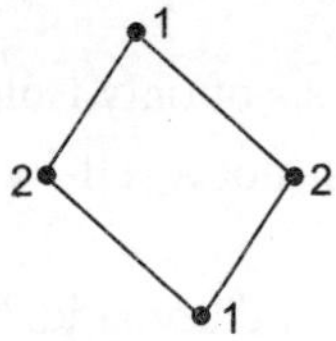

The following theorem characterize all 2-chromatic graphs.

Theorem 2: *A graph with at least one edge is 2-chromatic if and only if it has no circuits with odd length.*

Proof: Let G be a connected graph.

G has no circuits with odd length

In this case there are two possibilities.

Either G is a tree or G contains circuits of only even length.

If G is a tree then using Theorem 1, G is 2-chromatic.

If G contains circuits of only even length, consider a spanning tree T in G. Using the colouring procedure and the result of Theorem 1, let us properly colour T with two colours. Now add the chords to T one by one and since G has no circuits of odd length, the end vertices of every replaced chord are differently coloured. So, G is properly coloured with two colours. Thus G is 2-chromatic.

G is 2-chromatic

If G has a circuit of odd length, we require at least three colours for proper colouring of this circuit. For instance,

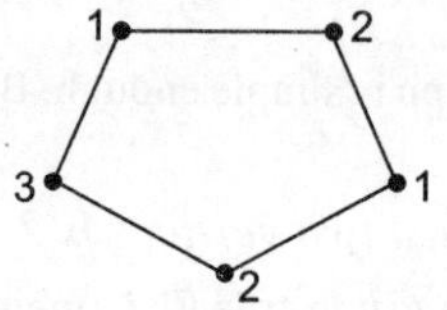

Hence G has no circuit with odd length. Thus the theorem.

Note: This theorem is due to **König** and is also known as **König's Theorem.**

6.2 CHROMATIC NUMBER OF A COMPLETE GRAPH (K_n)

Complete graph: A simple graph (*i.e.*, free from self-loops and parallel edges) in which there exists an edge between every pair of vertices is said to be a **complete graph** or a **universal graph** or a **clique.**

Complete graph of one, two, three, four and five vertices are shown below:

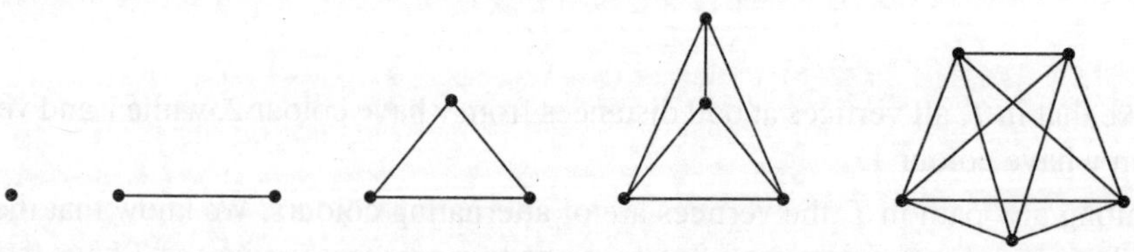

In a complete graph of n vertices the degree of every vertex is $(n-1)$ and the number of edges is $^nC_2 = \frac{1}{2}n(n-1)$, since it is free from self-loops, parallel edges and every vertex is joined with every other vertex through exactly one edge.

Notes: (*i*) A complete graph of *n* vertices is denoted by K_n.

(*ii*) The vertices of a complete graph are pairwise adjacent.

(*iii*) A complete graph is always a regular graph, *i.e.*, all vertices are of same degree.

(*iv*) K_n is not planar when $n \geq 5$.

Theorem 3: *The chromatic number of a complete graph K_n (with n vertices) is n.*

Proof: Let $v_1, v_2, ..., v_n$ are *n* vertices of a complete graph K_n with *n* vertices. Let a colour C_1 be assigned to v_1 and since v_2 is adjacent to v_1, a differently colour C_2 is required to be assigned to v_2. Next, since v_3 is adjacent to both v_1 and v_2, so another colour C_3 (different from C_1, C_2) is required to be assigned to v_3. In this way the different colours $C_1, C_2, ..., C_n$ are required to be assigned to the vertices $v_1, v_2, ..., v_n$ respectively because no two vertices can be assigned the same colour as every two vertices of K_n are adjacent. Thus at least *n* colours are required for proper colouring of K_n. Hence the chromatic number of K_n is *n*.

Theorem 4: *The chromatic number of the graph $K_n - v_i$, obtained by removing a vertex v_i from K_n, is $n - 1$.*

Proof: If a vertex is removed (along with the edges incident on it) from a complete graph then the graph remains to be complete. So, $K_n - v_i$ becomes a complete graph with $(n - 1)$ vertices. Using Theorem 3, the chromatic number of $K_n - v_i$ is $n - 1$.

Note: If K_n is a subgraph of a graph *G*, then $\chi(G) \geq \chi(K_n) \Rightarrow \chi(G) \geq n$.

Example 2: *Suppose final examinations of seven courses numbered 1 through 7 are to be scheduled at a Technological Institution, where the following pairs of courses have common students: 1 and 2 ; 1 and 3 ; 1 and 4 ; 1 and 7 ; 2 and 3 ; 2 and 4 ; 2 and 5 ; 2 and 7 ; 3 and 4; 3 and 6 ; 3 and 7 ; 4 and 5 ; 4 and 6 ; 5 and 6 ; 5 and 7 ; 6 and 7. Draw the graph associated with the set of courses. How can the final examination to be scheduled (in least slots) so that no student has two examinations at the same time?*

Solution: This scheduling problem can be solved using a graph model with vertices representing courses. There is an edge between two vertices if there is a common student in the courses (represented by the vertices). Each time slot for the final examination is represented by a different colour and a scheduling of the examinations corresponds to a colouring of the associated graph. In Fig. 6.1, the graph associated with the set of courses having common students is shown.

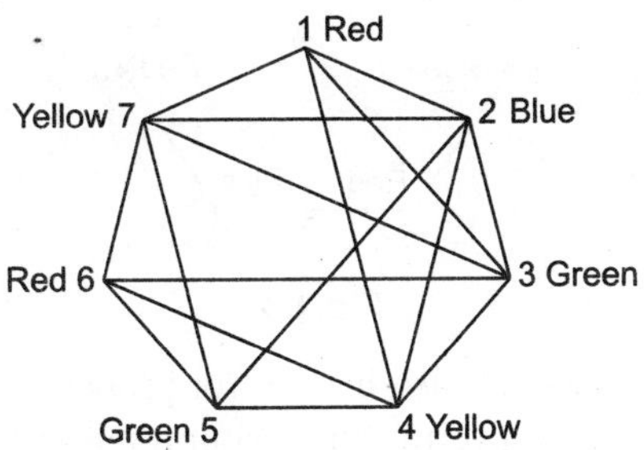

Fig. 6.1

A scheduling consists of a colouring of this graph. This graph has K_4 as a subgraph, induced by the vertices 1, 2, 3 and 7. So, the chromatic number of the graph (in Fig. 6.1) is ≥ 4. Observe that this graph is 4-colourable and so its chromatic number is 4. Thus, the seven courses can be scheduled in minimum four time slots as shown below:

Time slot	Courses
I	1, 6
II	2
III	3, 5
IV	4, 7

6.3 CHROMATIC NUMBER OF A CIRCUIT (C_n)

Circuit: A **circuit** in a graph is defined as a finite alternating sequence of vertices and edges, beginning and ending at the same vertex, such that each edge is incident with vertices preceding and following it. In a circuit no edge appears or is covered or traversed more than once and also no vertex (except the initial and the final vertex) appears more than once.

The number of edges in a circuit is known as the **length of the circuit.** A circuit of length n is denoted by C_n.

Notes: (*i*) Every self-loop is a circuit, but the converse is not true in general.

(*ii*) A circuit is also known as a **cycle, elementary cycle, circular path, polygon.**

(*iii*) In electrical engineering, a circuit is sometimes referred to as a loop (not to be confused with self-loop).

Theorem 5: *The chromatic number of a circuit C_n (with n vertices) is*

(*i*) *2 if n is even;* (*ii*) *3 if n is odd and* $n \geq 3$.

Proof: (*i*) To begin, let us take $n = 6$. Pick a vertex and colour it red and then proceed clockwise in the planar depiction of C_6 as shown in Fig. 6.2. It is now necessary to assign a second colour, say blue, to the next vertex reached. Continue in the clockwise direction; the third vertex can be coloured red, the fourth vertex blue, the fifth vertex red and finally, the sixth vertex, which is adjacent to the first, can be coloured blue. Hence, the chromatic number of C_6 is 2. Note that at least two colours are also needed if we move in the anticlockwise direction.

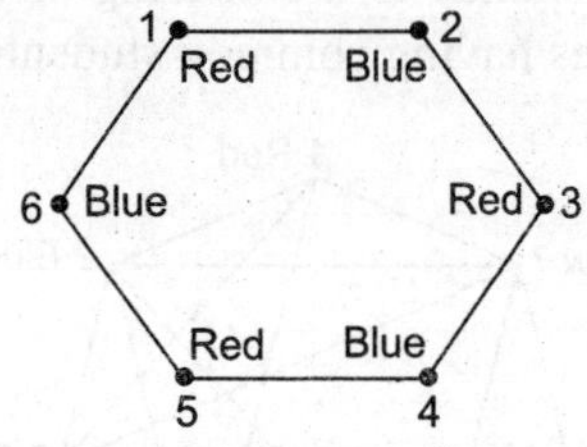

Fig. 6.2

In general, at least two colours are also needed for proper colouring of C_n when n is even. To construct such a colouring, simply pick a vertex and colour it red and then proceed along the graph in a clockwise (or, anticlockwise) direction (using a planar representation of the graph). Colour the second vertex blue, the third vertex red, and so on. The nth vertex can be coloured blue, since the two vertices adjacent to it, namely the $(n - 1)^{th}$ and the first vertices, are both coloured red as $(n - 1)$ is odd.

Hence, the chromatic number of C_n is 2 when n is even.

(*ii*) Let $n = 5$ and consider C_5. Pick a vertex and colour it red and then proceed clockwise in the planar depiction of C_5 as shown in Fig. 6.3. It is now necessary to assign a second colour, say blue, to the next vertex reached. Continue in the clockwise direction; the third vertex can be coloured red, the fourth vertex blue and finally, the fifth vertex which cannot be coloured either red or blue, because it is adjacent to the fourth and the first vertices. Consequently for proper colouring, a third colour is needed for the fifth vertex. Note that at least three colours are also required for proper colouring if we move in the anticlockwise direction. Hence, the chromatic number of C_5 is 3.

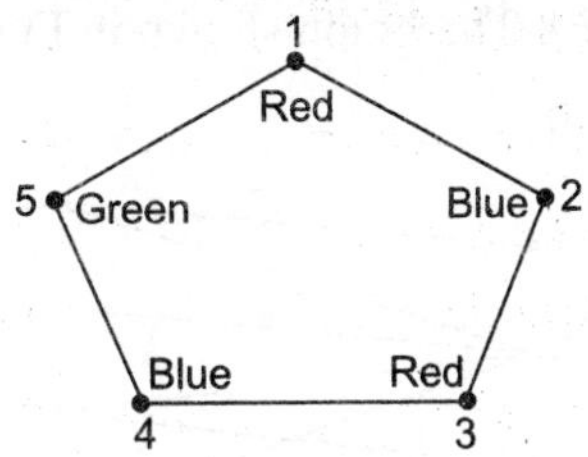

Fig. 6.3

In general, at least three colours are also required for proper colouring of C_n when n (> 1) is odd. To see this, pick an initial vertex and colour it red. To use only two colours, it is necessary to alternate colours as the graph is traversed in a clockwise (or, anticlockwise) direction. However, the nth vertex reached is adjacent to two vertices of different colours, namely the first and $(n-1)^{th}$, as n is odd. Hence, a third colour must be used for proper colouring.

Thus, the chromatic number of C_n is 3 when n is odd.

Notes: (*i*) If C_n is a subgraph of a graph G, then

$$\chi(G) \geq \chi(C_n) \quad \Rightarrow \quad \begin{cases} \chi(G) \geq 2 & \text{if } n \text{ is even} \\ \chi(G) \geq 3 & \text{if } n \text{ is odd} \end{cases}$$

(*ii*) The converse of Theorem 5 is not true as shown in the following example:

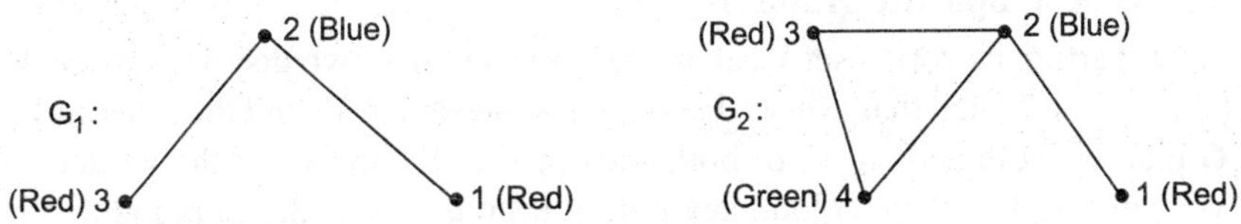

Observe that none of the graphs G_1 and G_2 is a circuit. But $\chi(G_1) = 2$ and $\chi(G_2) = 3$.

Wheel: **A wheel** of order n is defined as a graph obtained by joining a single new vertex (the 'hub') to each vertex of a circuit of length $n - 1$. Wheels of order n are denoted by W_n. The graph shown in Fig. 6.4 is a wheel of order 6, *i.e.*, W_6.

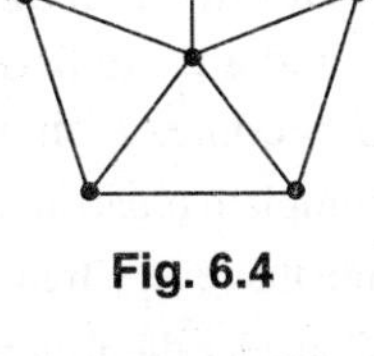

Fig. 6.4

6.4 CHROMATIC NUMBER OF A BIPARTITE GRAPH ($K_{m, n}$)

Bipartite graph: A graph $G = (V, E)$ is called **bipartite** if its vertex set can be decomposed into two disjoint (means without common elements) subsets V_1 and V_2 such that every edge of G has one end in V_1 and the other in V_2. Hence, there is no edge in G that has both ends in V_1 or both ends in V_2. Such a graph is also denoted as $G = (V_1, V_2, E)$. The graph shown in Fig. 6.5 is a bipartite graph.

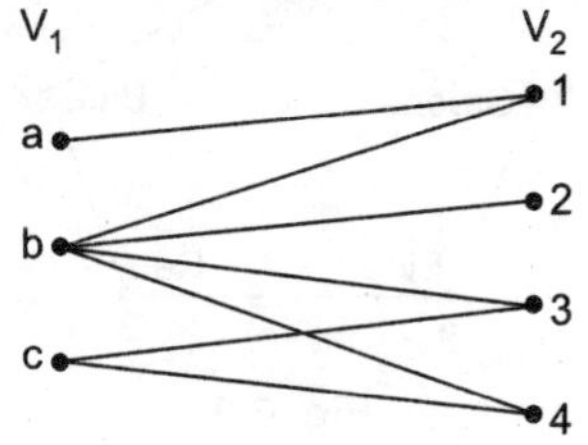

Fig. 6.5 Bipartite graph

Here the vertex set V is partitioned into two disjoint subsets $V_1 = \{a, b, c\}$ and $V_2 = \{1, 2, 3, 4\}$. It may represent the assignment of a set V_1 of workers to a set V_2 of jobs. The edges indicate that worker a can do the job 1, worker b the jobs 1, 2, 3, 4 and worker c the jobs 3, 4.

Complete bipartite graph: A bipartite graph $G = (V_1, V_2, E)$ is said to be a **complete bipartite graph** if there is an edge between every vertex of V_1 and every vertex of V_2.

A bipartite graph is usually denoted by $K_{m, n}$ where m and n $(m \leq n)$ are number of vertices in V_1 and V_2 respectively. Observe that a complete bipartite graph $K_{n, n}$ is a regular graph (*i.e.*, all vertices are of equal degree).

Theorem 6: *The chromatic number of a non-null graph is 2 if and only if it is bipartite.*

Proof: Let G be a bipartite graph ($K_{m, n}$)

Since G is bipartite, its vertex set V can be decomposed into two disjoint subsets $V_1 = \{u_1, u_2, ..., u_m\}$ and $V_2 = \{v_1, v_2, ..., v_n\}$ such that every edge of G has one end in V_1 and the other in V_2. Hence, there is no edge in G that has both ends in V_1 or both ends in V_2, *i.e.*, no two of the vertices in V_1 or V_2 are adjacent. So, we can assign a same colour, say red, to each u_i $(i = 1, 2, .., m)$ and another colour, say blue, to each $v_j = (j = 1, 2, ..., n)$. Thus, the graph G can be properly coloured by using at least two different colours. Hence, the chromatic number of G is 2, *i.e.*, $\chi(G) = 2$.

Let G be a graph such that $\chi(G) = 2$

So, G can be properly coloured by using two different colours, say red and blue. Assign the colour red to each of the vertices $u_1, u_2, ..., u_m$ (say) and colour blue to each of the vertices $v_1, v_2, ..., v_n$ (say). Let $V_1 = \{u_1, u_2, ..., u_m\}$ and $V_2 = \{v_1, v_2, ..., v_n\}$, then V_1, V_2 are disjoint and $V = V_1 \cup V_2$, where V is the vertex set of G.

Since all u_i $(i = 1, 2, ..., m)$ are assigned same colour, so no two u_i are adjacent and similarly, no two v_j $(j = 1, 2, ..., n)$ are adjacent. Also a u_i and a v_j are assigned with distinct colours, so each edge of G has joined some u_i with some v_j. Thus, G becomes a bipartite graph.

Notes: (*i*) The chromatic number of a complete bipartite graph is 2, *i.e.*, $\chi(K_{n, n}) = 2$.

(*ii*) From Theorems 1 and 6, we conclude that every tree with at least two vertices is bipartite.

(*iii*) Every bipartite graph is not a tree. Consider the following graphs:

The graphs shown in Fig. 6.6 (*a*), (*b*) are isomorphic to each other. So, the graph shown in Fig. 6.6(*a*) is a bipartite graph but it is not a tree.

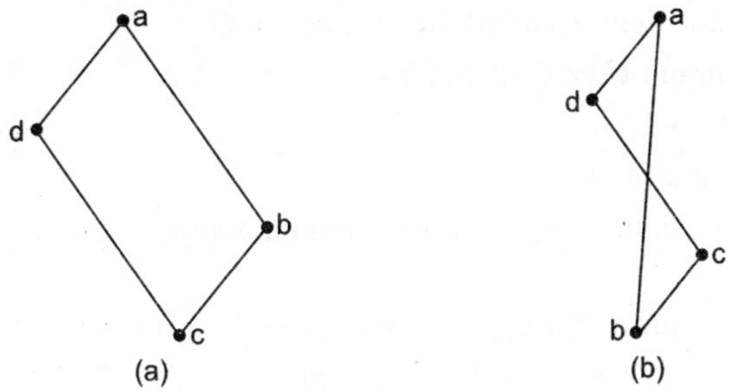

Fig. 6.6

(*iv*) A **null graph** of order *n* is defined as a graph with *n* vertices and no edges.

Question: Show that every bipartite graph is 2-chromatic. [*WBUT 2012*]

Using Theorems 1, 2 and 6, we obtain the following theorem:

Theorem 7: For a non-null graph *G*, the following statements are equivalent:

(*i*) $\chi(G) = 2$

(*ii*) *G* is a bipartite graph

(*iii*) *G* has no circuits of odd length.

Example 3: *Find the chromatic number of the following graph:*

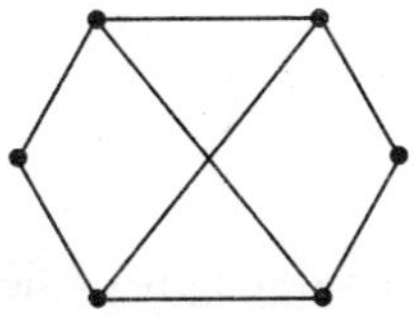

Solution: The given graph is isomorphic (*i.e.*, equivalent) to the following graph:

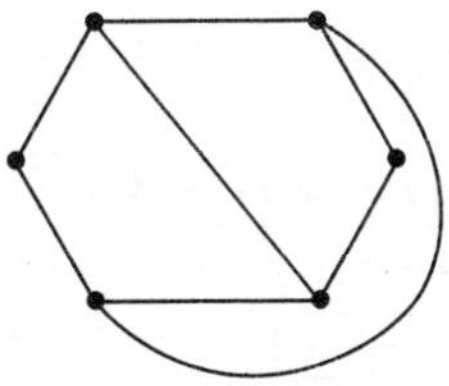

Observe that this graph has no circuits of odd length. Using Theorem 7, the chromatic number of this graph is 2. Hence, the chromatic number of the given graph is also 2.

6.5 UPPER BOUNDS OF CHROMATIC NUMBERS

Though we are able to find the exact chromatic numbers for K_n, C_n and $K_{m,n}$, but there is no efficient and convenient procedure for finding the chromatic number of any arbitrary graph. In general, it is extremely difficult to find the chromatic number of a graph, however there are various results which provide upper bounds for the chromatic number of an arbitrary graph if the degrees of their vertices are known.

Before stating these results, let us define some notations.

In a graph G:

$\Delta(G) \equiv$ maximum of the degrees of all the vertices of G.

$\delta(G) \equiv$ minimum of the degrees of all the vertices of G.

Theorem 8: In any graph G, $\chi(G) \leq \Delta(G) + 1$.

Notes: (*i*) $\chi(K_n) = \Delta(K_n) + 1$

(*ii*) $\chi(C_n) = \Delta(C_n) + 1$, if n is odd.

Theorem 9: If G is a connected graph other than a complete graph with $\Delta(G) \geq 3$, then $\chi(G) \leq \Delta(G)$.

Notes: (*i*) This theorem reduces the upper bound of $\chi(G)$ by 1 for a class of graphs.

(*ii*) This theorem is also known as **Brook's Theorem** and is due to Brooks (1941).

Example 4: *Consider the following graph G:*

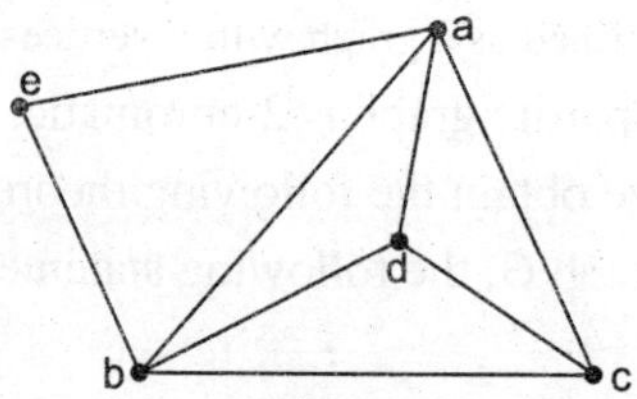

Fig. 6.7

Determine the chromatic number of G.

Solution: Observe that the graph induced by the four vertices a, b, c and d is a complete subgraph of G.

$$\therefore \qquad \chi(G) \geq 4 \qquad \text{...(1)}$$

On the other hand in G:

$$\deg(a) = \deg(b) = 4;\ \deg(c) = \deg(d) = 3;\ \deg(e) = 2.$$

So, $$\Delta(G) = \max_{v \in V(G)} \deg(v) = 4.$$

Since G is a connected graph other than a complete graph with $\Delta(G) = 4 > 3$, so by Brook's theorem

$$\chi(G) \leq \Delta(G), \text{ i.e., } \chi(G) \leq 4 \qquad \text{...(2)}$$

Using (1) and (2), we get $\chi(G) = 4$.

Theorem 10: For any graph G,

$$\chi(G) \leq 1 + \max_{H \subseteq G} \delta(H), \text{ where } \delta(H) = \min_{v \in V(H)} \deg(v)$$

for all subgraphs H of G.

Note: This theorem is also known as **Szekeres-Wilf Theorem** and is due to Szekeres and Wilf (1968).

Example 5: Consider the graph G shown in Fig. 6.7, Example 4. Observe that here, Max $\delta(H) = 3$, over all subgraphs of G. So, by Szekeres-Wilf Theorem: $\chi(G) \leq 1 + \max_{H \subseteq G} \delta(H) = 1 + 3 = 4$, *i.e.*, $\chi(G) \leq 4$. Also, by (1) of Example 4:

$$\chi(G) \geq 4.$$

$$\therefore \qquad 4 \leq \chi(G) \leq 4 \Rightarrow \chi(G) = 4.$$

6.6 RULES FOR FINDING CHROMATIC NUMBER OF A GRAPH G

Let us list few rules that may be helpful in finding the chromatic number $\chi(G)$ of a graph G.

Rule 1. If a graph G contains only isolated vertices, then $\chi(G) = 1$.

Rule 2. $\chi(G) \leq n$, where n is the number of vertices of G.

Rule 3. A triangle always requires 3 colours for proper colouring *i.e.*, $\chi(K_3) = 3$. More generally, $\chi(K_n) = n$, where K_n is the complete graph on n vertices.

Rule 4. If a subgraph of G requires m colours for proper colouring, then $\chi(G) \geq m$.

Rule 5. If deg $(v) = d$, for a vertex v in G, then at most d colours are needed for proper colouring of the vertices adjacent to v.

Rule 6. $\chi(G) = \text{Max}\ \{\chi(C)\}$, where Max. is taken over all connected components of G.

Rule 7. Every K-chromatic graph has at least K vertices, say $v_1, v_2, ..., v_K$ such that

$$\deg(v_i) \geq K - 1;\ i = 1, 2, ..., K.$$

Rule 8. For any graph G, $\chi(G) \leq \Delta(G) + 1$, where $\Delta(G)$ = maximum of the degrees of all the vertices of G.

Rule 9. For a non-null graph G, the following statements are equivalent:

(*i*) $\chi(G) = 2$

(*ii*) G is a bipartite graph

(*iii*) G has no circuits of odd lengths.

Rule 10. $\chi(G) \geq \dfrac{n}{n - \delta(G)}$, where n is the number of vertices of G and $\delta(G)$ = minimum of the degrees of all the vertices of G. See Miscellaneous Ex. 6.

Rule 11. If G is a tree with at least two vertices, then $\chi(G) = 2$.

Rule 12. $\chi(C_n) = \begin{cases} 2, & \text{if } n \text{ is even} \\ 3, & \text{if } n \text{ is odd} \end{cases}$

where C_n is a circuit with n vertices.

Rule 13. $\chi(K_{m,n}) = 2$, where $K_{m,n}$ is a bipartite graph.

Example 6: *Find the chromatic number of the wheel graph W_6.*

Solution: The graph shown in the following figure represents a wheel graph of order 6, *i.e.*, W_6.

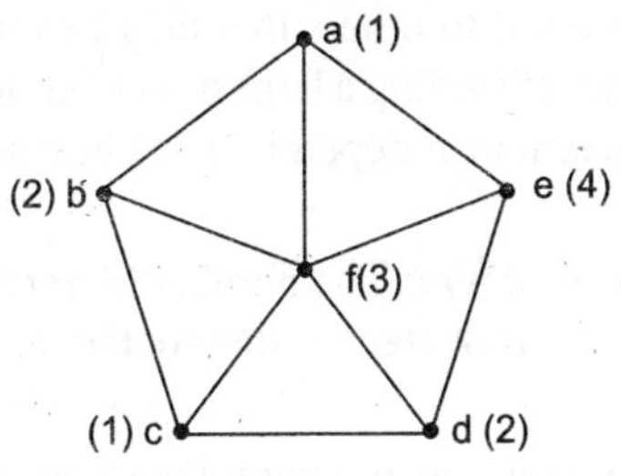

Since deg (a) = deg (b) = deg (c) = deg (d) = deg (e) = 3 and deg (f) = 5, we have $\Delta(G)$ = maximum of the degrees of all vertices = 5.

$$\therefore \qquad \chi(G) \leq 1 + \Delta(G) = 1 + 5 = 6 \text{ (By Rule 8)}.$$

Since there is a triangle subgraph of G, we have

$$\chi(G) \geq 3 \text{ (By Rule 4)}.$$

$$\therefore \qquad 3 \leq \chi(G) \leq 6.$$

Let us try to colour W_6 properly by using three colours, say 1, 2, 3. We start by colouring the vertices a, b, f with the colours 1, 2, 3 respectively. Now since c is adjacent to b and f having colours 2 and 3 respectively, c is forced to be coloured 1 and then d is forced to be coloured 2. Now, since the remaining vertex e is adjacent to the vertices a, d, f having colours 1, 2, 3 respectively, we require a fourth colour for e which yields $\chi(G) = 4$.

Note: Observe that by Brook's theorem, we get

$$\chi(G) \leq \Delta(G) = 5.$$

6.7 LOWER BOUNDS OF CHROMATIC NUMBERS

Definition 1: (***Independent set***)**:** A set of vertices in a graph G is said to be an **independent set of vertices** or simply an **independent set** or an **internally stable set** of G if no two vertices in the set are adjacent (in G). For example, $\{a, c, d\}$ is an independent set of the graph shown in Fig. 6.8. A single vertex in any graph constitutes an independent set of that graph.

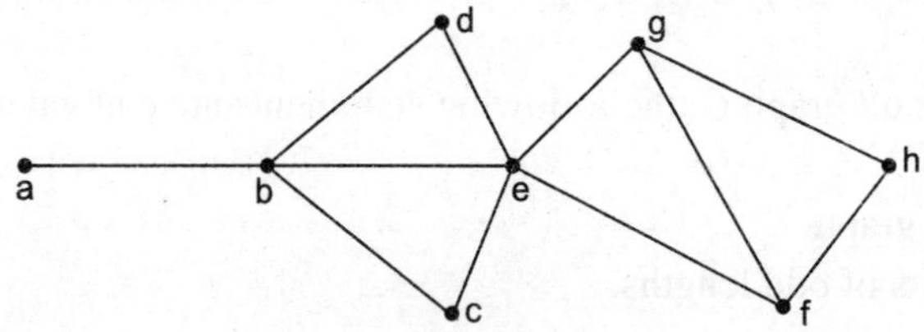

Fig. 6.8

Definition 2: (***Maximal independent set***)**:** An independent set in a graph G to which no other vertex of G can be included without destroying its independence property is said to be a **maximal independent set** or **maximal internally stable set** of G.

The sets $\{a, c, d, f\}$, $\{a, c, d, g\}$ and $\{a, c, d, h\}$ are maximal independent sets of the graph shown in Fig. 6.8. The sets $\{b, f\}$, $\{b, g\}$, $\{b, h\}$ are also maximal independent sets. In general, a graph may have more than one maximal independent set of different sizes. Among all maximal independent sets of a graph, one with the largest number of vertices is often of particular interest.

Illustration: Suppose that the graph shown in Fig. 6.8 describes the following problem:

Each of the 8 vertices is a possible code word to be used in some communication system. There are words that are so close (say, in sound) to others that may be confused for each other, pairs of such words are joined by edges. The problem of finding a largest set of code words for a reliable communication system is a problem of finding a maximal independent set having largest number of vertices. Here, $\{a, c, d, f\}$ is an answer.

Definition 3: (***Independence number***)**:** The number of vertices in the largest independent set of a graph G is called the **independence number** or the **coefficient of internal stability** of G and is usually denoted by $\beta(G)$.

Theorem 11: *If G be a K-chromatic graph with n vertices, then*

$$\beta(G) \geq \frac{n}{K}, \quad or \quad \chi(G) \geq \frac{n}{\beta(G)}$$

where $\chi(G) = K$ is the chromatic number of G.

Proof: Since G is a K-chromatic graph with n vertices, all the n vertices of G can be properly coloured with K different colours. Since the largest number of vertices in G having the same colour cannot exceed the independence number $\beta(G)$ and there are K distinct colours for proper colouring, we must have

$$K\beta(G) \geq n \quad \Rightarrow \quad \beta(G) \geq \frac{n}{K}, \quad \text{or} \quad \chi(G) \geq \frac{n}{\beta(G)}.$$

6.7.1 Boolean Algebra: Some Definitions and Properties

Definition 4: A class of elements B together with two binary operations denoted as + and · is said to be a **Boolean algebra** iff the following four postulates are satisfied:

P1: Both + and · are commutative, *i.e.*,

$$a + b = b + a \quad \text{and} \quad ab = ba, \; \forall \; a, b \in B.$$

($a \cdot b$ is written as ab).

P2: Each of + and · is distributive over the other *i.e.*,

$$a + bc = (a + b)(a + c) \text{ and } a(b + c) = ab + ac, \; \forall \; a, b, c \in B.$$

P3: There exist in B distinct identity elements 0 and 1 ($0 \neq 1$) for the operations + and · respectively, such that

$$a + 0 = 0 + a = a \text{ and } a1 = 1a = a, \; \forall \; a \in B.$$

P4: For every element a of B, there exists an element a' (called the complement of a) in B, such that

$$a + a' = 1 \quad \text{and } aa' = 0.$$

The term **constant** is used to represent any single symbol which is a specified element (such as 0 or 1) of the Boolean algebra, while the word **variable** will be used to describe any arbitrary or unspecified element of a Boolean algebra.

Few identities

I1: $a + a = a, \; \forall \; a \in B.$

Proof:

$$\begin{aligned} \text{RHS} = a &= a + 0 && \text{(by } P3) \\ &= a + aa' && \text{(by } P4) \\ &= (a + a)(a + a') && \text{(by } P2) \\ &= (a + a)1 && \text{(by } P4) \\ &= a + a && \text{(by } P3) \\ &= \text{LHS}. \end{aligned}$$

I2: $aa = a, \; \forall \; a \in B$

Proof:

$$\begin{aligned} \text{RHS} = a &= a1 && \text{(by } P3) \\ &= a(a + a') && \text{(by } P4) \\ &= aa + aa' && \text{(by } P2) \\ &= aa + 0 && \text{(by } P4) \\ &= aa && \text{(by } P3) \\ &= \text{LHS}. \end{aligned}$$

I3: (*i*) $a + 1 = 1, \; \forall \; a \in B.$

(*ii*) $a0 = 0, \; \forall \; a \in B$

Proof: (*i*)

$$\begin{aligned} \text{LHS} = a + 1 &= 1(a + 1) && \text{(by } P3) \\ &= (a + a')(a + 1) && \text{(by } P4) \\ &= a + a'1 && \text{(by } P2) \end{aligned}$$

$= a + a'$ (by $P3$)

$= 1$ (by $P4$)

= RHS.

Similarly (*ii*) can be proved.

Note: $0' = 1$ and $1' = 0$.

I4: $a + ab = a, \forall\, a, b \in B$

Proof: LHS $= a + ab = a1 + ab$ (by $P3$)

$= a(1 + b)$ (by $P2$)

$= a1$ (by $P1$ and $I3$ (*i*))

$= a$ (by $P3$)

= RHS.

I5: De Morgan's laws:

$$(a + b)' = a'b' \text{ and } (ab)' = a' + b', \forall a, b \in B.$$

Proof is omitted.

6.7.2 Finding all Maximal Independent Sets of a Graph

A reasonable (but not very efficient for large graphs) method for obtaining all maximal independent sets in any graph can be developed based on Boolean arithmetic on the vertices as follows:

Let $G = (V, E)$ be a simple connected graph and let each vertex of the vertex set $V = \{v_1, v_2, ..., v_n\}$ be treated as a Boolean variable. Let

(*i*) the logical (or Boolean) sum $v_i + v_j$ denote the operation of including vertex v_i or v_j or both,

(*ii*) the logical (or Boolean) multiplication v_iv_j denote the operation of including both vertices v_i and v_j,

(*iii*) the Boolean complement v_i' indicate that vertex v_i is not included.

To find a maximal independent set for G, we have to find a maximal subset of V that does not include the two end vertices of any edge in E.

Let us express an edge (v_i, v_j) as a Boolean product (v_iv_j) of its end vertices v_i and v_j, and sum all such products in G to get a Boolean expression:

$$\varphi = \Sigma v_iv_j, \text{ for all } (v_i, v_j) \text{ in } G.$$

Take the Boolean complement φ' of this expression and express it as a sum of Boolean products:

$$\varphi' = f_1 + f_2 + ... + f_m.$$

A vertex set is a maximal independent set iff

$\varphi = 0$ (logically false)

$\Leftrightarrow$ $\varphi' = 1$ (logically true)

$\Leftrightarrow$ at least one $f_i = 1$

$\Leftrightarrow$ each vertex appearing in f_i (in complemented form) is excluded from the vertex set V.

Thus each f_i will yield a maximal independent set and so every maximal independent set will be produced by this method.

Illustration: Consider the graph $G = (V, E)$ as shown in Fig. 6.9, where the vertex set $V = \{a, b, c, d, e, f, g\}$.

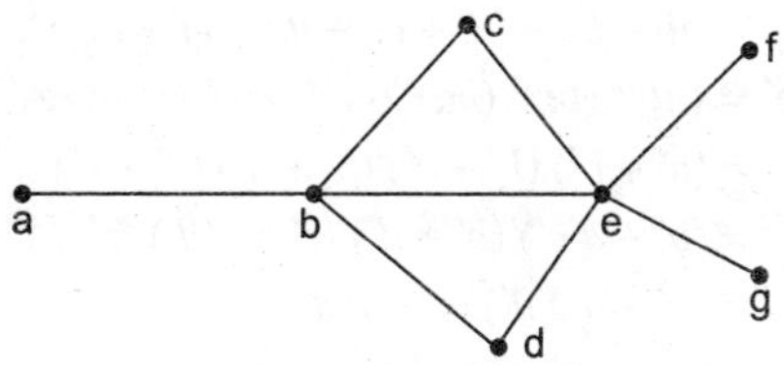

G

Fig. 6.9

Let us express each edge as a Boolean product of its end vertices and sum all such products to get a Boolean expression:

$$\varphi = ab + bc + bd + be + ce + de + ef + eg$$

$$\therefore \quad \varphi' = (ab)'\,(bc)'\,(bd)'\,(be)'\,(ce)'\,(de)'\,(ef)'\,(eg)'$$

$$= (a' + b')\,(b' + c')\,(b' + d')\,(b' + e')\,(c' + e')\,(d' + e')\,(e' + f')\,(e' + g')$$

(by De Morgan's law)

Using the distributive property and the identities:

$a + a = a$, $aa = a$, $a + ab = a$, we get

$$(a' + b')\,(b' + c') = a'b' + a'c' + b'b' + b'c'$$

$$= a'b' + a'c' + b' + b'c'$$

$$= a'b' + a'c' + b' = b' + a'c'$$

$$\therefore \quad \varphi' = (b' + a'c')\,(b' + d'e')\,(e' + c'd')\,(e' + f'g')$$

$$= (b' + a'c'd'e')\,(e' + c'd'f'g')$$

$$= b'e' + b'c'd'f'g' + a'c'd'e' + a'c'd'e'f'g'$$

$$= b'e' + b'c'd'f'g' + a'c'd'e'$$

Now, if we exclude from the vertex set V vertices appearing (in complemented form) in any one of these three terms, we get a maximal independent set. The three maximal independent sets are

$$\{a, c, d, f, g\},\ \{a, e\} \text{ and } \{b, f, g\}.$$

These are all the maximal independent sets of the graph shown in Fig. 6.9.

Independence number: The size of the maximal independent set having largest number of vertices is 5.

$\therefore$ Independence number is $\beta(G) = 5$.

Chromatic number: To find the chromatic number of G, we must find the minimum number of these maximal independent sets which collectively include all the vertices of G.

$\therefore \quad \chi(G) = 3$, *i.e.*, G is 3-chromatic.

Notes: (*i*) Consider the graph shown in Fig. 6.10:

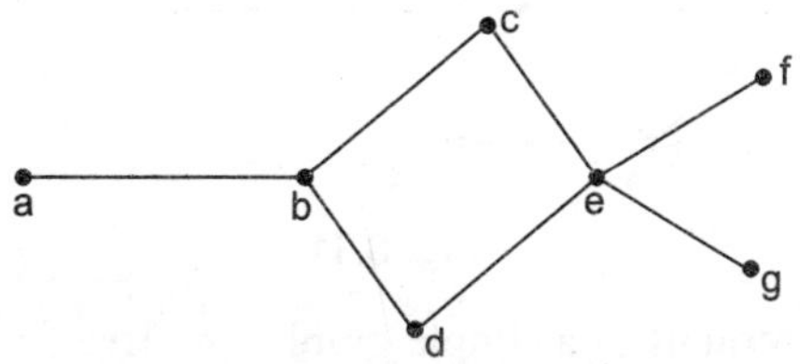

G

Fig. 6.10

Here, $\varphi = ab + bc + bd + ce + de + ef + eg$

$$\begin{aligned}\therefore \quad \varphi' &= (ab)'(bc)'(bd)'(ce)'(de)'(ef)'(eg)' \\ &= (a' + b')(b' + c')(b' + d')(c' + e')(d' + e')(e' + f')(e' + g') \\ &= (b' + a'c')(b' + d')(e' + c'd')(e' + f'g') \\ &= (b' + a'c'd')(e' + c'd'f'g') \\ &= b'e' + b'c'd'f'g' + a'c'd'e' + a'c'd'f'g'\end{aligned}$$

The four maximal independent sets are

$$\{a, c, d, f, g\}, \{a, e\}, \{b, f, g\} \text{ and } \{b, e\}$$

Here $\beta(G) = 5$ and $\chi(G) = 2$.

(*ii*) Theorem 11 holds for the graphs shown in Figs. 6.9 and 6.10.

Opposite to an independent set of vertices in a graph is a **clique** defined as follows:

Definition 5: A complete subgraph of a graph G is called a **clique** in G. In other words, a graph C is said to be a **clique** in a graph G if (*i*) all the vertices and all the edges of C are in G, (*ii*) each edge of C has the same end vertices in C as in G, (*iii*) C is free from self-loops and parallel edges and (*iv*) there exists an edge between every pair of vertices of C.

Definition 6: The order or cardinal number (*i.e.*, the number of vertices) of the largest clique in a graph G is called the **clique number** of G and is denoted by $\omega(G)$.

For example, the complete subgraph with vertex set $\{b, c, e\}$ is one of the largest cliques in G as shown in Fig. 6.8. So, the clique number of G is $\omega(G) = 3$.

Theorem 12: *For every graph G with chromatic number $\chi(G)$ and clique number $\omega(G)$,*

$$\chi(G) \geq \omega(G).$$

Proof: Let H be one of the largest cliques in G and so $\chi(H) = \omega(G)$. Since H is a subgraph of G, it follows that $\chi(G) \geq \chi(H) = \omega(G)$.

Note: Theorems 11 and 12 give lower bounds for the chromatic number of a graph G in terms of independence number $\beta(G)$ and clique number $\omega(G)$ of G.

6.8 PERFECT GRAPH

Definition 1. A graph G is said to be a **Perfect graph** or a **Berge graph** if $\chi(H) = \omega(H)$ for every induced subgraph H of G.

Equivalent, a graph G is called a **Perfect graph** or a **Berge graph** if the chromatic number of every induced subgraph H of G is equal to the clique number $\omega(H)$ of H.

Illustration 1: The graph shown in Fig. 6.11 is a perfect graph.

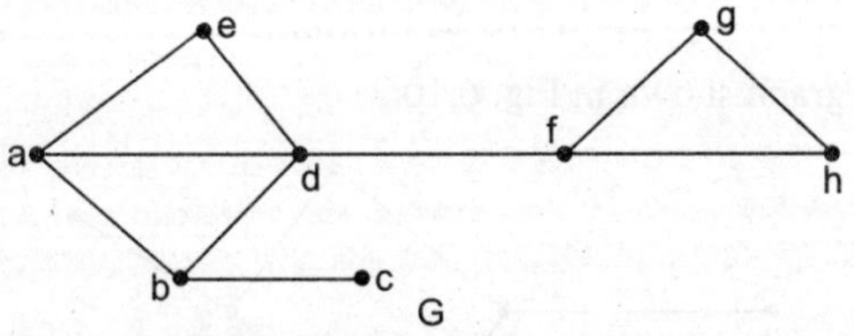

Fig. 6.11

Definition 2: The **complement** of a simple graph (*i.e.*, free from self-loops and parallel edges) G is a simple graph denoted as $\overline{G}$ which has same vertex set as G and two vertices in $\overline{G}$ are adjacent if they are not adjacent in G and vice versa.

Illustration 2: In Fig. 6.12, $\{a, b, c\}$ is a clique of size 3 and $\{c, e\}$ is an independent set of size 2, and these are the largest such sets. These values reverse in the complement $\overline{G}$ (shown in Fig. 6.12), since cliques become independent sets and vice versa under complementation.

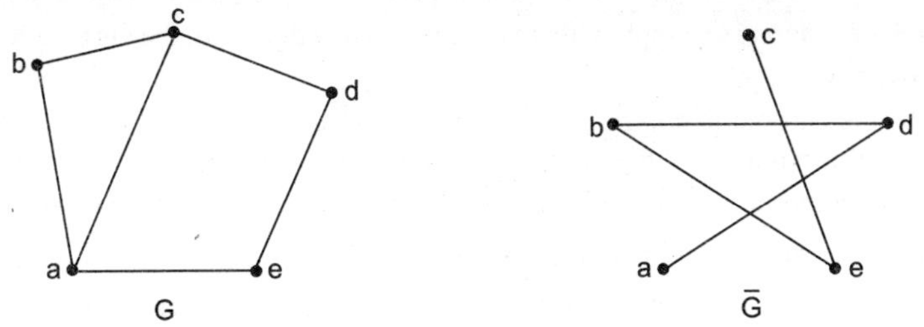

Fig. 6.12

Theorem 13: (The Perfect Graph Theorem (PGT))

A graph is perfect if and only if its complement is perfect.

Proof is omitted.

It can be show that bipartite graphs are perfect graphs.

Illustration 3: Both G and its complement $\overline{G}$ as shown in Fig. 6.13 are perfect graphs.

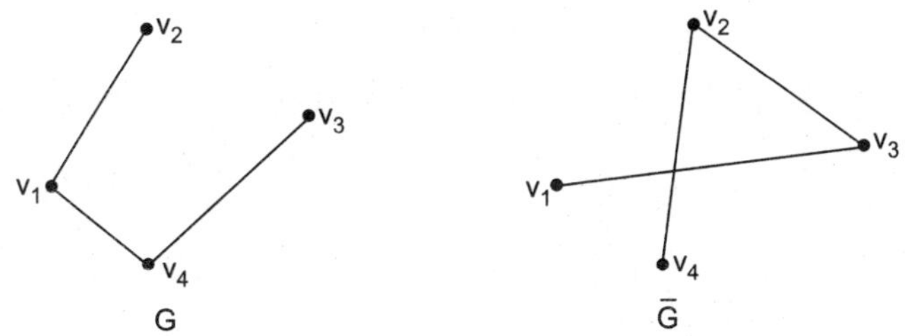

Fig. 6.13

6.9 CHROMATIC POLYNOMIAL

Definition 1: A given graph G with n vertices can be properly coloured in many different ways using a sufficiently large number of colours, say λ. This property of G can be expressed by means of a polynomial which is called **Chromatic polynomial** of G and is denoted as $P_n(\lambda)$.

Determination of $P_n(\lambda)$: Let a_i be the different ways of properly colouring a graph G using exactly i different colours. Now, i colours can be chosen out of λ colours in

$$^{\lambda}C_i \equiv \binom{\lambda}{i} \text{ different ways.}$$

So, there are $a_i \binom{\lambda}{i}$ different ways of properly colouring G using exactly i colours out of λ colours and the chromatic polynomial is a sum of these terms.

$$\therefore \quad P_n(\lambda) = \sum_{i=1}^{n} a_i \binom{\lambda}{i} = a_1 \frac{\lambda}{1!} + a_2 \frac{\lambda(\lambda-1)}{2!} + a_3 \frac{\lambda(\lambda-1)(\lambda-2)}{3!} + \ldots + a_n \frac{\lambda(\lambda-1)(\lambda-2)\ldots(\lambda-n+1)}{n!}.$$

Notes: (*i*) Here i is running from 1 to n, because it is not possible to use more than n colours on n vertices.

(*ii*) Each a_i has to be evaluated individually for a given graph. For example, any non-null graph requires at least two colours for proper colouring, and therefore $a_1 = 0$.

A graph with n vertices and using n different colours can be properly coloured in $n!$ ways and hence $a_n = n!$

(*iii*) A null graph of order n is a graph with n vertices and no edges. So, the chromatic polynomial of a null graph of order n is $P_n(\lambda) = \lambda^n$.

(*iv*) Let $P_n(\lambda)$ be the chromatic polynomial (using λ colours) of a graph G with n vertices. The smallest λ for which $P_n(\lambda) \neq 0$ is nothing but $\chi(G)$, the chromatic number of G. Because $\chi(G)$ is the minimum number of colours with which G can be properly coloured. In other words, $P_n(\lambda) = 0$ if $\lambda < \chi(G)$. See Example 2.

(*v*) $P_n(0) = 0$, for any graph G with n vertices.

Example 1: *Determine the chromatic polynomial of the following graph*:

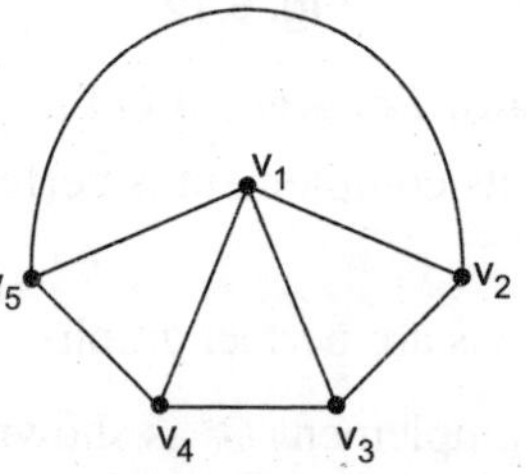

Fig. 6.14

Solution: The given graph has five vertices and so its chromatic polynomial is

$$P_5(\lambda) = a_1\lambda + a_2\frac{\lambda(\lambda-1)}{2!} + a_3\frac{\lambda(\lambda-1)(\lambda-2)}{3!}$$

$$+ a_4\frac{\lambda(\lambda-1)(\lambda-2)(\lambda-3)}{4!} + a_5\frac{\lambda(\lambda-1)(\lambda-2)(\lambda-3)(\lambda-4)}{5!}$$

Since the graph contains a triangle, it requires at least three different colours for proper colouring.

$\therefore \quad a_1 = a_2 = 0. \quad$ Also, $a_5 = 5!$

Moreover, to evaluate a_3, suppose that we have three colours, say x, y, z. The vertices v_1, v_2 and v_3 can be properly coloured by using these three colours in $3! = 6$ different ways. Having done that, we have no more choices left, because vertices v_4 and v_5 must have the same colours as v_2 and v_3 respectively.

$\therefore \quad a_3 = 6.$

Similarly, with four colours, vertices v_1, v_2 and v_3 can be properly coloured in ${}^4C_3 \times 3! = 4 \times 6 = 24$ different ways. The fourth colour can be assigned to v_4 or v_5, thus resulting two choices. The remaining vertex provides no additional choice.

$\therefore \quad a_4 = 24 \times 2 = 48.$

Substituting these values of a_i in $P_5(\lambda)$, we get

$$P_5(\lambda) = \lambda(\lambda-1)(\lambda-2) + 2\lambda(\lambda-1)(\lambda-2)(\lambda-3) + \lambda(\lambda-1)(\lambda-2)(\lambda-3)(\lambda-4)$$
$$= \lambda(\lambda-1)(\lambda-2)(1 + 2\lambda - 6 + \lambda^2 - 7\lambda + 12) = \lambda(\lambda-1)(\lambda-2)(\lambda^2 - 5\lambda + 7).$$

This is the required chromatic polynomial of the given graph.

Note: The graph shown in Fig. 6.14 is 3-chromatic.

Theorem 14. *A graph with n vertices is a complete graph if and only if its chromatic polynomial is*

$$P_n(\lambda) = \lambda(\lambda-1)(\lambda-2)\ldots(\lambda-n+1).$$

Proof: Using λ colours, there are λ different ways of colouring any selected vertex of a graph. A second vertex can be coloured properly in exactly $(\lambda - 1)$ ways, the third vertex in $(\lambda - 2)$ ways, the fourth in $(\lambda - 3)$ ways, ..., and the nth vertex (*i.e.*, the last one) in $(\lambda - n + 1)$ ways if and only if every vertex is adjacent to every other, *i.e.*, if and only if the graph is complete.

Hence, a graph with n vertices is a complete graph if and only if its chromatic polynomial is

$$P_n(\lambda) = \lambda(\lambda - 1)(\lambda - 2) \dots (\lambda - n + 1).$$

Theorem 15: *A graph with n vertices is a tree if and only if its chromatic polynomial is*

$$P_n(\lambda) = \lambda(\lambda - 1)^{n-1}.$$ [*WBUT 2012*]

Proof: Let a graph be a tree. Clearly, $P_1(\lambda) = \lambda$ and $P_2(\lambda) = \lambda(\lambda - 1)$. Thus, the theorem is true for $n = 1, 2$.

n = 1 n = 2

Let the theorem be true for all trees having less than m vertices. Let us consider a tree T with m vertices. Removal of a pendant vertex v (along with the edge incident on it) from T will yield a tree T' with $(m - 1)$ vertices. By induction hypothesis, T' can be coloured by using λ colours in $P_{m-1}(\lambda) = \lambda(\lambda - 1)^{m-2}$ ways. Now, putting the removed vertex v (along with the edge incident on it) back into T' will yield the tree T. For any kind of colouring of T', vertex v can be coloured in $(\lambda - 1)$ ways, hence T can be coloured in $(\lambda - 1) P_{m-1}(\lambda)$ ways, *i.e.*, $\lambda(\lambda - 1)^{m-1}$ ways. Therefore, if the result is true for a tree with $(m - 1)$ vertices, then it is true for a tree with m vertices. But we have already seen that the result is true for $n = 1, 2$, hence it is true for $2 + 1 = 3$, $3 + 1 = 4$, $4 + 1 = 5$ etc., *i.e.*, the result is true for a tree with n vertices, where n is any positive integer.

Theorem 16: (*Decomposition Theorem*)

Given:

(i) a and b be two non-adjacent vertices in a graph G with n vertices.

(ii) G′ (with n vertices) be a graph obtained by adding an edge between a and b.

(iii) G″ (with n – 1 vertices) be a graph obtained from G by fusing the vertices a and b together.

Then,

$$P_n(\lambda) \text{ of } G = P_n(\lambda) \text{ of } G' + P_{n-1}(\lambda) \text{ of } G''.$$

Proof: The number of ways of properly colouring G can be grouped into the following two cases:

Case 1: Vertices a and b are of the same colour

In this case, no. of ways of properly colouring G = no. of ways of properly colouring G''

$$= P_{n-1}(\lambda) \text{ of } G''.$$

Case 2: Vertices a and b are of different colours

In this case, no. of ways of properly colouring G = no. of ways of properly colouring G'

$$= P_n(\lambda) \text{ of } G'.$$

Combining these two cases, we conclude that

$$P_n(\lambda) \text{ of } G = P_n(\lambda) \text{ of } G' + P_{n-1}(\lambda) \text{ of } G''.$$

Notes: (*i*) This theorem is often used in evaluating the chromatic polynomial of a graph (see the following Example 2).

(*ii*) Theorems 14–16 provide a glimpse into the colourful world of chromatic polynomials.

Example 2: *Using decomposition theorem, find the chromatic polynomial and hence the chromatic number of the graph shown in Fig. 6.15.*

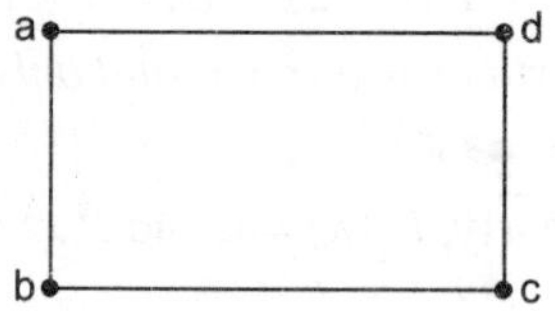

Fig. 6.15

Solution: Here the graph G' is obtained from G by adding the edge (a, d) and G'' is obtained from G by fusing (or merging) the vertices a and d together as shown in the following figures. Since G is a tree with 4 vertices, we have

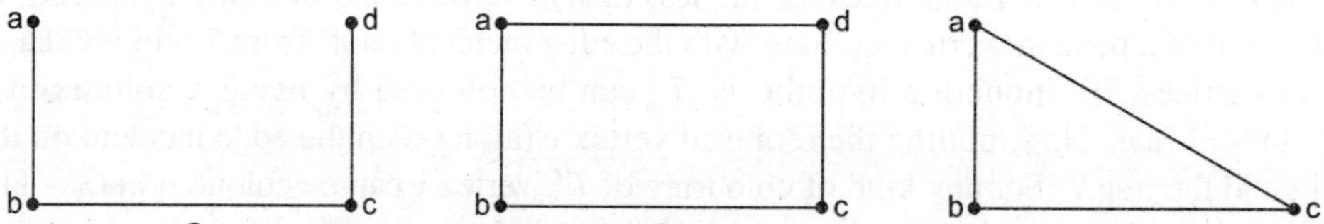

G G′ G″

$$P_4(\lambda) \text{ of } G = \lambda(\lambda - 1)^3.$$

Since G'' is a complete graph with 3 vertices, we have

$$P_3(\lambda) \text{ of } G'' = \lambda(\lambda - 1)(\lambda - 2).$$

So, by decomposition theorem, we have

$$P_4(\lambda) \text{ of } G = P_4(\lambda) \text{ of } G' + P_3(\lambda) \text{ of } G''$$

$$\Rightarrow \quad \lambda(\lambda - 1)^3 = P_4(\lambda) \text{ of } G' + \lambda(\lambda - 1)(\lambda - 2)$$

$$\Rightarrow \quad P_4(\lambda) \text{ of } G' = \lambda(\lambda - 1)^3 - \lambda(\lambda - 1)(\lambda - 2)$$

$$= \lambda(\lambda - 1)(\lambda^2 - 2\lambda + 1 - \lambda + 2)$$

$$= \lambda(\lambda - 1)(\lambda^2 - 3\lambda + 3)$$

$$= \lambda^4 - 4\lambda^3 + 6\lambda^2 - 3\lambda$$

This is the required chromatic polynomial of the given graph G'.

Observe that $P_4(1)$ of $G' = 0$,

but $P_4(2)$ of $G' = 2^4 - 4 \times 2^3 + 6 \times 2^2 - 3 \times 2 = 2 \neq 0$.

Hence, the chromatic number of the given graph is 2.

6.10 FOUR AND FIVE COLOUR THEOREMS

The **four colour theorem** was originally posed as a conjecture in 1852 by Francis Guthire, a post-graduate student at the University College, London. It was finally proved by the American Mathematicians Kenneth Appel and Wolfgang Haken in 1976 and they used 400 pages of arguments and about 1200 hours of computer time on some of the best computers in the world to arrive at the solution. The statement of the theorem (without proof) is given below:

Theorem 17: (***Four Colour Theorem***)**:** The chromatic number of every planar graph is four or less.

Let us state and proof **five colour theorem** which is due to Heawood (1890).

Theorem 18: (***Five Colour Theorem***)**:** *The vertices of every planar graph can be properly coloured with five colours.*

Proof: The theorem will be proved by induction on the number of vertices. Since the vertices of any planar graph with 1, 2, 3, 4 or 5 vertices can be properly coloured with five colours, let us assume that vertices of every planar graph with $(n - 1)$ vertices can be properly coloured with five colours. Then, if we prove that any planar graph G with n vertices will need no more than five colours for proper colouring, we shall have proved the theorem.

Consider a planar graph G with n vertices and so it has a vertex of degree 5 or less (see Ex. 3, Art. 10.2, Ch-10, Engg. Math-IV). Let this vertex be v.

Let G' be a graph with $(n - 1)$ vertices obtained from G by removing vertex v (along with all edges incident on it). According to the induction hypothesis, G' does not require more than five colours for proper colouring. Suppose that the vertices in G' have been properly coloured and now we put the removed vertex v along with all edges incident on v back to G'. If deg $(v) \leq 4$, we have no difficulty in assigning a proper colour to v.

Now consider the case in which deg $(v) = 5$ and all the five colours, say 1, 2, 3, 4, 5 have been used for proper colouring the vertices as shown in Fig. 6.16(a).

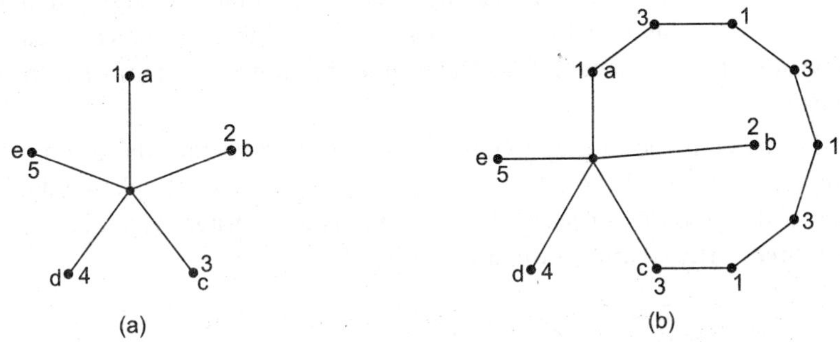

Fig. 6.16

Let H be the subgraph of G generated by the vertices coloured 1 and 3. If the vertices a and c belong to the different components of H, interchanging the colours 1 and 3 in the component containing a, the vertices a and c will be assigned the same colour 3. So, we get the proper colouring of G by assigning colour 1 to the vertex v.

If a and c are in the same component of H, then there exists on a–c path in G all of whose vertices are coloured alternately with colours 1 and 3 [see Fig. 6.16(b)]. Since G is planar, a similar path between the vertices b and d coloured alternately with colours 2 and 4 does not exist. Let K denotes the subgraph of G generated by the vertices coloured 2 and 4, then b and d belong to the different components of K. Let us interchange the colours 2 and 4 in the component containing b, the vertices b and d will be assigned the same colour 4. So, we get the proper colouring of G by assigning colour 2 to the vertex v. This completes the proof.

Note: Figure 6.16 is a part of the planar representation of G'.

Example 3: *Show that every planar graph is 6-colourable.* [*WBUT 2012*]

Solution: We shall use mathematical induction on the number of vertices. Since the vertices of any planar graph with 1, 2, 3, 4, 5 or 6 vertices can be properly coloured with six colours, let us assume that vertices of every planar graph with $(n-1)$ vertices can be properly coloured with six colours. Then, if we prove that any planar graph G with n vertices will need no more than six colours for proper colouring, we shall have proved the theorem.

Consider a planar graph G with n vertices and so it has a vertex, say v of degree 5 or less (see Ex. 3, Art. 10.2, Ch-10, Engg. Math-IV).

Let G' be a graph with $(n-1)$ vertices obtained from G by removing vertex v (along with all edges incident on it). According to the induction hypothesis, G' does not require more than six colours for proper colouring. Suppose that the vertices of G' have been properly coloured and now we put the removed vertex v along with all edges incident on v back to G'. Since deg $(v) \leq 5$, we have no difficulty in assigning a proper colour to v by using six different colours. Hence the result.

6.11 APPLICATIONS OF GRAPH COLOURINGS

Graph colouring has a variety of applications to problems involving scheduling and assignments as shown below:

I. Scheduling Final Examinations: How can the final examination to be scheduled (in least slots) so that no student has two examinations at the same time?

See the example of Art. 6.2.

II. Storage of Chemicals: A company has to store eight different chemicals. Some of these chemicals are incompatible with some other chemicals and so they have to be stored in different rooms. The company has numbered these chemicals in such a way that three chemicals bearing consecutive numbers cannot be stored in the same room. Determine the minimum number of rooms needed for storing the chemicals.

Solution: Let us construct a graph G for which the vertices are the chemicals and adjacent vertices are incompatible chemicals as shown in Fig. 6.17. So, the given problem reduces to finding the chromatic number of the graph G. Observe that G contains a subgraph which is isomorphic to K_3 (a complete graph of three vertices) and so we have $\chi(G) \geq 3$.

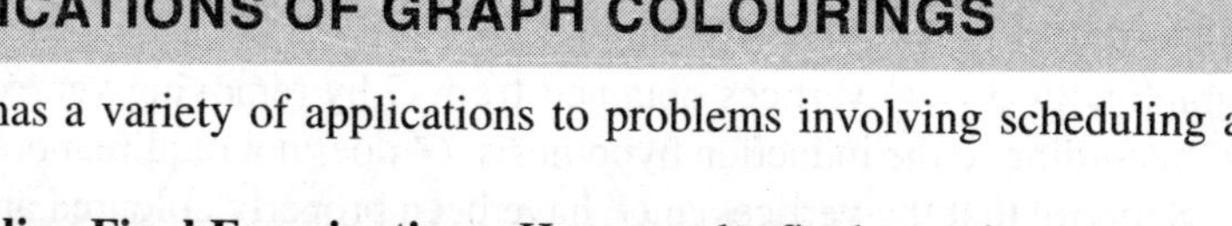

Fig. 6.17

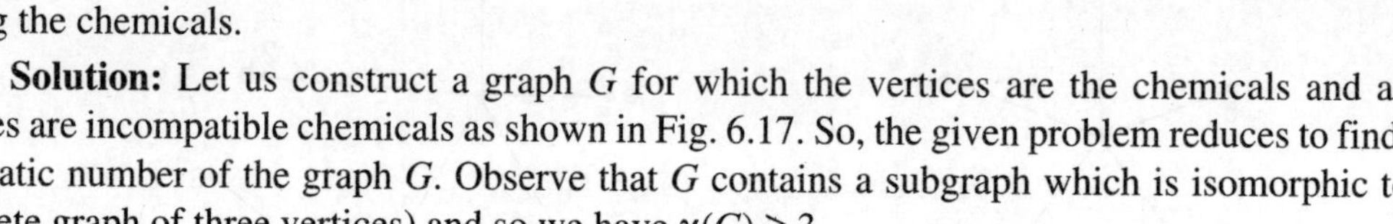

Fig. 6.18

A proper colouring of G is shown in Fig. 6.18 (b), where R, B and Gr denote red, blue and green respectively. Thus, minimum number of rooms needed for storing the chemicals is 3.

III. Index Registers: In efficient compilers of a computer, the execution of loops is speeded up when frequently used variables are stored temporarily in index registers in the central processing unit instead of in regular memory. For a given loop, how many least possible index registers are needed?

Let us illustrate this problem using a graph colouring model. Consider a program segment containing a loop as follows:

```
for     (i = 1 ; i ≤ 100)
    {
        x = y + i ;
        u = u + i ;
        v = v + i ;
        z = u + v ;
        y = z/i ;
    }
```

The loop consists of six variables i, x, y, z, u and v. We represent these variables as vertices of a graph G as shown in Fig. 6.19. Draw an edge between two vertices of G if the variables they represent must be stored in registers at the same time during the execution of the loop. Since different registers must be assigned to variables when the vertices representing these variables are adjacent in the graph G and so the chromatic number of G gives the number of registers required. Hence the minimum number of index registers needed for the execution of the loop is $\chi(G) = 3$.

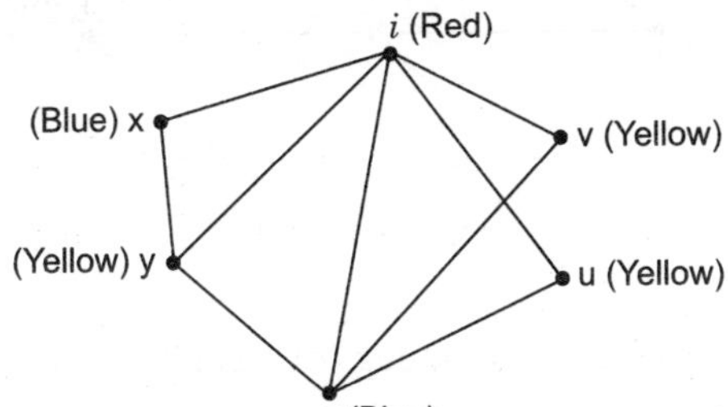

G

Fig. 6.19

MISCELLANEOUS EXAMPLES

Example 1: *Find the chromatic number of the following graph:*

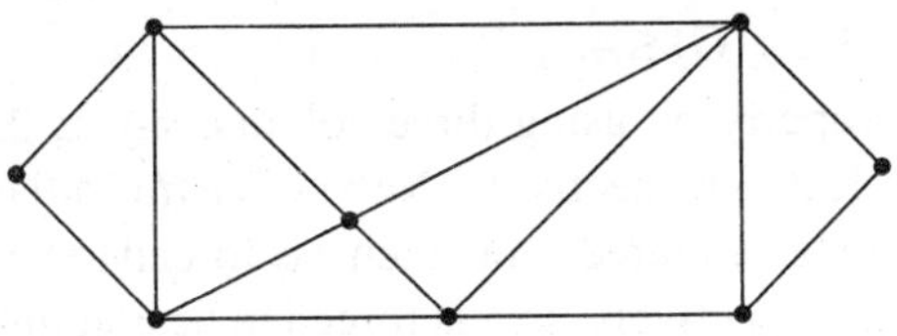

Solution: Since the given graph, say G contains a triangle, so $\chi(G) \geq 3$.

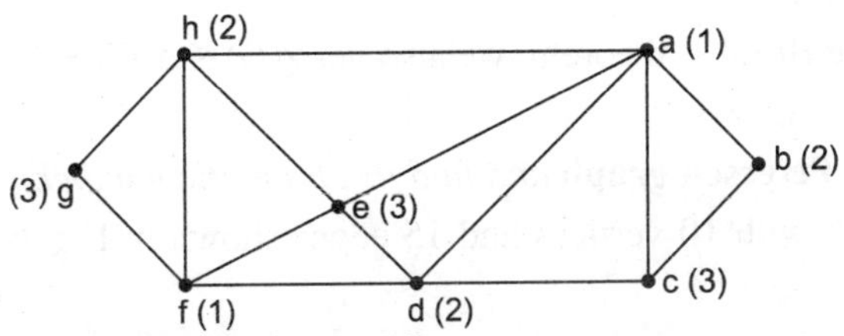

G

Let us try to colour G properly by using three colours, say 1, 2, 3. We start by colouring the vertices a, b, c with the colours 1, 2, 3 respectively. Now, since d is adjacent to a and c having colours 1 and 3 respectively, d is forced to be coloured 2 and then e is forced to be coloured 3. Also, f is adjacent to d and e having colours 2 and 3 respectively, f is forced to be coloured 1 and then h is forced to be coloured 2. Since g is adjacent to f and h having colours 1 and 2 respectively, g is forced to be coloured 3. Hence, the chromatic number of the given graph is 3.

Example 2: *Find the chromatic number of the following graph:*

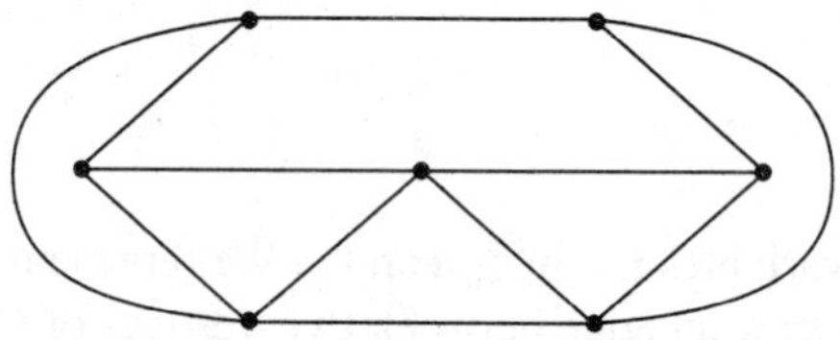

Solution: Since the given graph, say G contains a triangle, so $\chi(G) \geq 3$. ...(1)

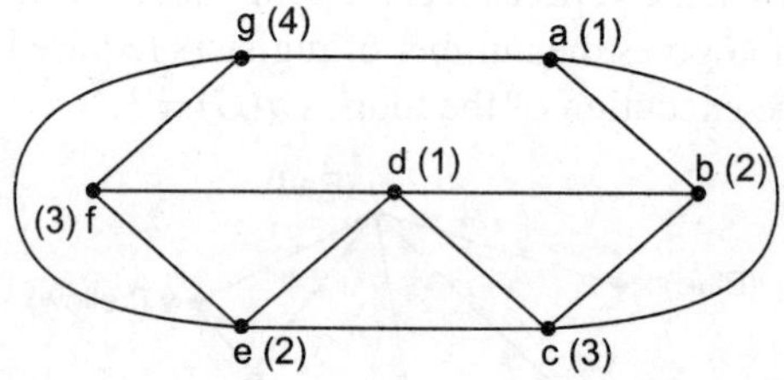

Observe that

$$\deg(a) = \deg(b) = 3, \deg(c) = \deg(d) = \deg(e) = 4, \deg(f) = \deg(g) = 3.$$

$$\therefore \quad \Delta(G) = \max_{v \in V(G)} \deg(v) = 4.$$

Using Theorem 8, we get

$$\chi(G) \leq \Delta(G) + 1 = 4 + 1 = 5 \quad ...(2)$$

From (1) and (2), we get $3 \leq \chi(G) \leq 5$.

If possible, let $\chi(G) = 5$, then G should have at least 5 vertices with degree ≥ 4 (using Rule 7, Art. 6.6). But there are only three such vertices here. So, $\chi(G) = 5$ is not possible and hence

$$3 \leq \chi(G) \leq 4.$$

Let us try to colour G properly by using three colours, say 1, 2, 3. We start by colouring the vertices a, b, c with the colours 1, 2, 3 respectively. Now, since d is adjacent to b and c having colours 2 and 3 respectively, d is forced to be coloured 1 and then e is forced to be coloured 2. Also, f is adjacent to d and e having colours 1 and 2 respectively, so f is forced to be coloured 3. Now, since the remaining vertex g is adjacent to the vertices a, e and f having colours 1, 2 and 3 respectively, we require a fourth colour for g which yields $\chi(G) = 4$.

Observation: Using Brooks' theorem, we also get $\chi(G) \leq \Delta(G) = 4$.

Note: Clique number $\omega(G) = 3$.

Example 3: *Define* **Petersen graph** *and find its chromatic number.*

Solution: The graph with 10 vertices and 15 edges shown in Fig. 6.20 is called the **Petersen graph.**

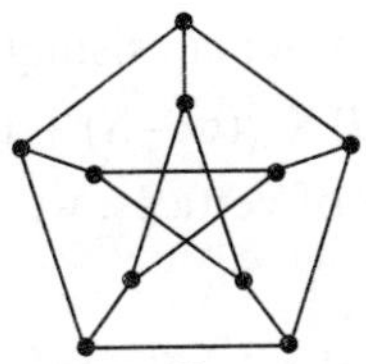

G

Fig. 6.20 Petersen graph

The graph G is neither complete nor a circuit. Also, it is not bipartite and does not contain circuits of odd length. So, $\chi(G) \neq 2$. Obviously $\chi(G) \neq 1$. Here, $\Delta(G) = \underset{v \in V(G)}{\text{Max}} \deg(v) = 3$.

Therefore, by Brook's theorem : $\chi(G) \leq 3$.

Hence, $\chi(G) = 3$.

Note: Clique number $\omega(G) = 2$.

Example 4: *Consider K_6, the complete graph on the six vertices a, b, c, d, e, f. The graph G_1 is obtained from K_6 by deleting the edge ab. The graph G_2 is obtained from G_1 by deleting the edge cd. What are the chromatic numbers of G_1 and G_2?* [WBUT 2012]

Solution: We know that the chromatic number of K_6 is $\chi(K_6) = 6$.

G_1:

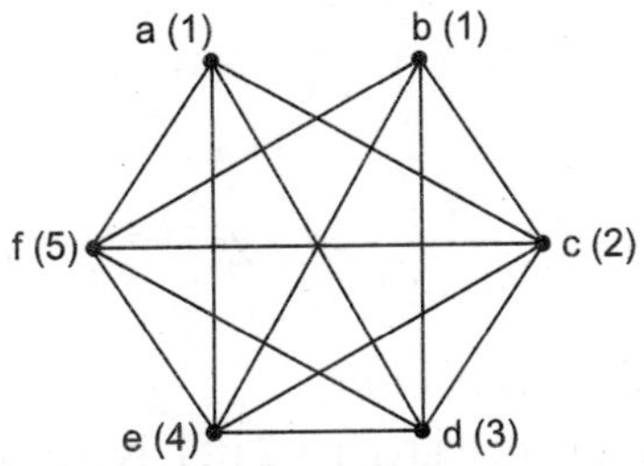

G_2:

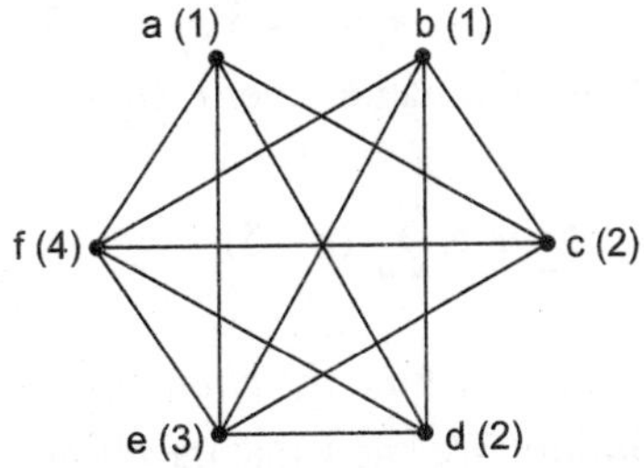

Since a minimum of 5 colours are required for proper colouring of the graph G_1, so chromatic number of G_1 is $\chi(G_1) = 5$.

A minimum of 4 colours are needed for proper colouring of the graph G_2, so chromatic number of G_2 is $\chi(G_2) = 4$.

Example 5: *Prove that for any graph G having n vertices,*

$$\frac{n}{\beta(G)} \leq \chi(G) \leq n - \beta(G) + 1$$

where $\chi(G)$ and $\beta(G)$ are respectively chromatic number and independence number of G.

Solution: Let k be the least number of colours with which G can be properly coloured, that is, $\chi(G) = k$. Also, let $c_1, c_2, ..., c_k$ be these distinct colours and let $n_1, n_2, ..., n_k$ be the number of vertices having colours $c_1, c_2, ..., c_k$ respectively.

Now, a set of vertices having the same colour consists of only those vertices which are non-adjacent. So, $n_1, n_2, ..., n_k$ are the orders or cardinal numbers of the maximal independent sets of G.

Since $\beta(G)$ is the order of the maximal independent set having largest number of vertices of G, so none of $n_1, n_2, ..., n_k$ can exceed $\beta(G)$.

$$\therefore \quad n_i \leq \beta(G), i = 1, 2, ..., k$$

$$\Rightarrow \quad \sum_{i=1}^{k} n_i \leq \sum_{i=1}^{k} \beta(G) \Rightarrow n \leq k\,\beta(G) \Rightarrow \chi(G) = k \geq \frac{n}{\beta(G)} \qquad ...(1)$$

Next, let S be a maximal independent set consisting of $\beta(G)$ number of vertices.

$\therefore \quad \chi(G-S) \geq \chi(G) - 1 \quad \Rightarrow \quad \chi(G) \leq \chi(G-S) + 1$ ...(2)

Since $G - S$ has $\{n - \beta(G)\}$ number of vertices, we have

$$\chi(G-S) \leq n - \beta(G) \qquad ...(3)$$

From (2) and (3): $\quad \chi(G) \leq \chi(G-S) + 1 \leq n - \beta(G) + 1$ ...(4)

From (1) and (4), we get $\dfrac{n}{\beta(G)} \leq \chi(G) \leq n - \beta(G) + 1$. **(Proved)**

Example 6: *If G is a graph with n vertices and* $\delta = \min\limits_{v \in V(G)} \deg(v)$, *then prove that the chromatic number* $\chi(G) \geq \dfrac{n}{n-\delta}$.

Solution: Given, δ is the minimum of the degrees of all vertices of G and so every vertex of G has at least δ number of adjacent vertices. Thus at most $(n - \delta)$ vertices can have the same colour.

Let k be the least number of colours with which G can be properly coloured, that is, $\chi(G) = k$.

Let $c_1, c_2, ..., c_k$ be these distinct colours. Let $n_1, n_2, ..., n_k$ be the number of vertices having colours $c_1, c_2, ..., c_k$ respectively and so

$$n_i \leq n - \delta \;;\; i = 1, 2, ..., k.$$

Taking summation over i, we get

$$\sum_{i=1}^{k} n_i \leq \sum_{i=1}^{k} (n-\delta) \;\Rightarrow\; n \leq k(n-\delta) \;\Rightarrow\; k \geq \frac{n}{n-\delta} \;\Rightarrow\; \chi(G) \geq \frac{n}{n-\delta}. \textbf{ (Proved)}$$

Example 7: *Find the clique number of a non-null bipartite graph.*

Solution: We know that the chromatic number of a non-null graph is 2 iff it is bipartite (see Th. 6, Art. 6.4). So, if G is a non-null bipartite graph, then $\chi(G) = 2$.

Also, we know that $\omega(G) \leq \chi(G)$ [see Theorem 12, Art. 6.7.2]

$\therefore \quad \omega(G) \leq 2$. But G is non-null and so $\omega(G) \neq 1$.

$\therefore \quad \omega(G) = 2$, this is the required clique number.

Note: A *null graph* has no edge, it has only vertices (or vertex).

Example 8: *Prove that every bipartite graph is a perfect graph.*

Solution: Let G be a bipartite graph. Let H be a subgraph of G and so H is either a null graph or a bipartite graph. Since clique number of a null graph is not defined, so we assure that H is a bipartite graph. The vertex set of G can be decomposed into two disjoint subsets V_1 and V_2 such that every clique of H consists of two vertices, one of V_1 and another of V_2. Thus every clique of H has two vertices and so $\omega(H) = 2$. Also, $\chi(H) = 2$ (see Th. 6, Art. 6.4).

$\therefore \qquad \chi(H) = \omega(H)$, for every induced subgraph H of G.

Hence G is a perfect graph.

Example 9: *Show that a complete graph* K_n *with n vertices is a perfect graph.*

Solution: We know that a complete graph K_n with n vertices has chromatic number $\chi(K_n) = n$ (see Th. 3, Art. 6.2).

Since every pair of vertices of K_n are adjacent, so K_n itself is the largest clique of K_n.

Hence $\omega(K_n) = n$ and so $\chi(K_n) = \omega(K_n) = n$.

Obviously, for every subgraph H_i of K_n,

$$\chi(H_i) = \omega(H_i).$$

Thus K_n is a perfect graph.

Example 10: *Show that*

(*i*) *a circuit with 3 vertices (i.e., length 3) is perfect.*

(*ii*) *a circuit with n vertices is not perfect if n is odd and n > 3,*

(*iii*) *a circuit with n vertices is a perfect graph if n is even.*

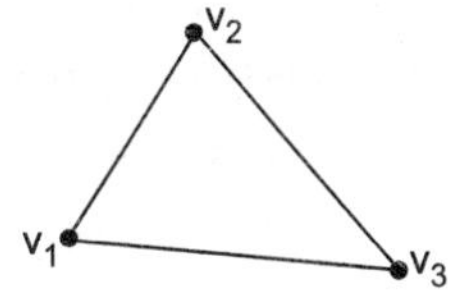

Solution: (*i*) Let C_3 be a circuit having three vertices and so $\chi(C_3) = 3$. If vertices v_1, v_2, v_3 form a circuit, then every pair of these vertices are adjacent.

Now, $C_3 \equiv \{v_1, v_2, v_3\}$ is the largest clique of C_3 and so its clique number $\omega(C_3) = 3$. For each subgraph $H_i \equiv \{v_i, v_j\}$, $\chi(H_i) = \omega(H_i) = 2$. Hence C_3 is a perfect graph.

(*ii*) Let C_n be a circuit formed by the successive vertices $v_1, v_2, v_3, ..., v_n, v_1$, where n is odd and $n > 3$.

$\therefore$ $\chi(C_n) = 3$, see Theorem 5, Art. 6.3.

In C_n, only two consecutive vertices are adjacent and so any clique of C_n has only two vertices (as $n > 3$), *i.e.*, $\omega(C_n) = 2$. Thus $\chi(C_n) \neq \omega(C_n)$ and hence C_n is not a perfect graph.

(*iii*) Let C_n be a circuit formed by the successive vertices $v_1, v_2, v_3, ..., v_n, v_1$, where n (> 2) is even. We know that $\chi(C_n) = 2$ (see Th. 5, Art 6.3). In C_n, only two consecutive vertices are adjacent and so any clique of C_n has only two vertices, *i.e.*, $\omega(C_n) = 2$.

For each subgraph H_i of C_n, obviously $\chi(H_i) = \omega(H_i) = 2$. Thus C_n is a perfect graph.

Note: A circuit is also known as a **cycle**.

Example 11: *Find the chromatic number of the following graph G. Determine whether this graph is perfect.*

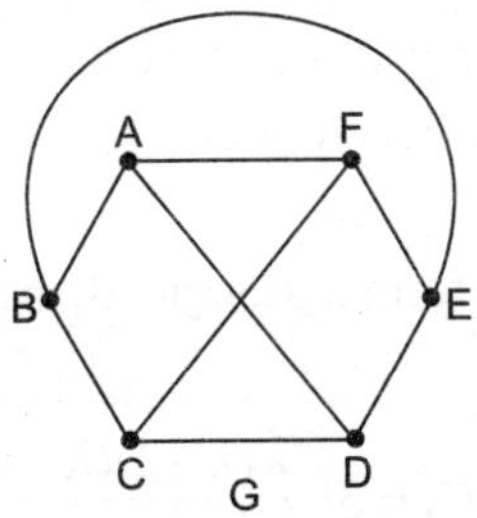

G

Solution: Let the vertex A be assigned a colour, say red. Since B, D and F are adjacent to A, so they should be assigned with a different colour, say blue. Since E and C are not adjacent and also not adjacent to A, they are assigned red colour. The labelling by colours is shown below:

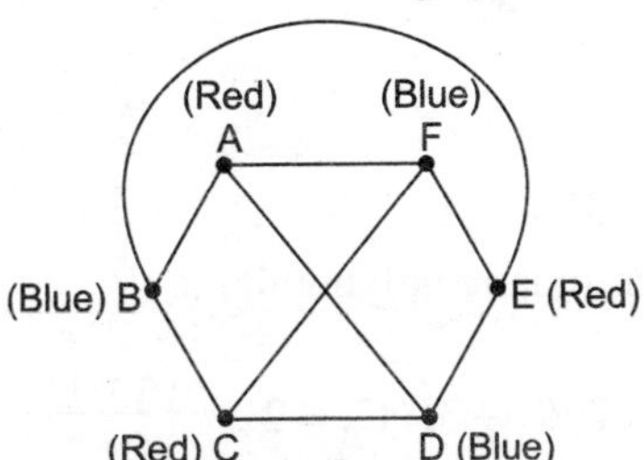

G

Hence the chromatic number of G is $\chi(G) = 2$.

Let H_1:

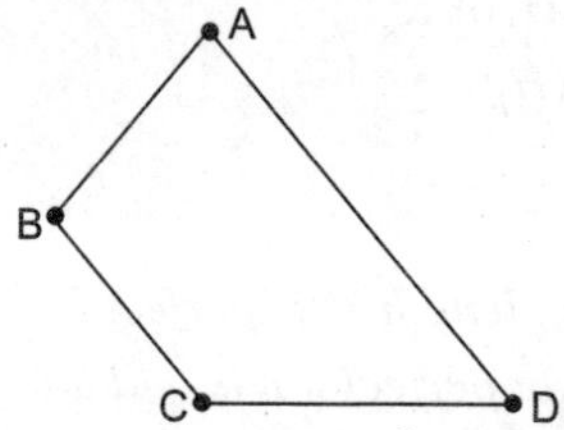

which is a subgraph of the given graph G. Obviously, $\omega(H_1) = \chi(H_1) = 2$. Similarly, for each subgraph H_i of G, $\omega(H_i) = \chi(H_i)$ and so the given graph G is a perfect graph.

Example 12: *Find whether the polynomial*

(*i*) $\lambda^4 - 4\lambda^3 + 7\lambda^2 - 2\lambda + 3$ *is a possible chromatic polynomial of some graph,*

(*ii*) $\lambda^3 - 5\lambda^2 + 3\lambda$ *is a possible chromatic polynomial of some non-null graph.*

Solution: (*i*) Let $P_4(\lambda) = \lambda^4 - 4\lambda^3 + 7\lambda^2 - 2\lambda + 3$.

We know that $P_n(0) = 0$, for any graph G with n vertices.

Here, $P_4(0) = 3 \neq 0$. So, the given polynomial is not a possible chromatic polynomial.

(*ii*) Let $P_3(\lambda) = \lambda^3 - 5\lambda^2 + 3\lambda$.

We know that a non-null graph G must have at least one edge, *i.e.*, at least two vertices must be adjacent. So, at least two colours will be needed for proper colouring of G, *i.e.*, $\chi(G) \geq 2 \Rightarrow P_n(1) = 0$, where $P_n(\lambda)$ be the chromatic polynomial of a non-null graph G with n vertices.

Here $P_3(1) = 1^3 - 5 \times 1^2 + 3 \times 1 = -1 \neq 0$.

Hence, the given polynomial can not be a possible chromatic polynomial of a non-null graph.

Example 13: *Determine the chromatic polynomial of a connected graph with three vertices.*

Solution: Since the given graph, say G has three vertices, so its chromatic polynomial is

$$P_3(\lambda) = a_1{}^{\lambda}C_1 + a_2{}^{\lambda}C_2 + a_3{}^{\lambda}C_3.$$

G must have at least one edge, since it is connected.

So, only one colour cannot properly colour G, *i.e.*, $a_1 = 0$.

There are two possibilities.

Case 1: If the three vertices of G form a triangle, then $a_2 = 0$ and $a_3 = 3! = 6$. So, in this case the chromatic polynomial becomes

$$P_3(\lambda) = 3!\,{}^{\lambda}C_3 = 3!\,\frac{\lambda(\lambda-1)(\lambda-2)}{3!} = \lambda(\lambda-1)(\lambda-2).$$

Case 2: If the three vertices does not make a triangle then it makes a path (*i.e.*, a tree). So, $a_2 = 2$.

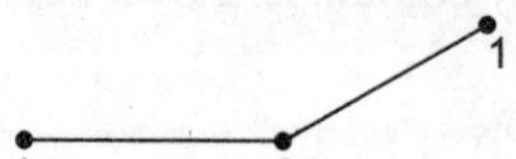

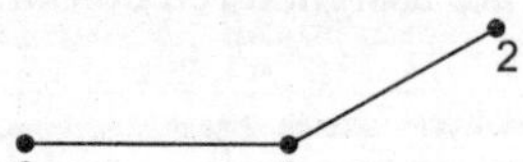

Also, $a_3 = 3!$

Therefore, in this case the chromatic polynomial is

$$P_3(\lambda) = 2{}^{\lambda}C_2 + 3!\,{}^{\lambda}C_3 = 2\,.\,\frac{\lambda(\lambda-1)}{2!} + 3!\frac{\lambda(\lambda-1)(\lambda-2)}{3!} = \lambda(\lambda-1)^2.$$

Example 14: *Using decomposition theorem, determine the chromatic polynomial of the following graph:*

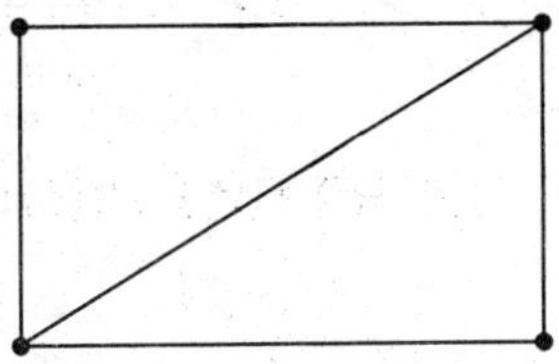

Hence find its chromatic number.

Solution: Let G be the given graph. Here the graph G' is obtained by adding the edge (a, c) and G'' is obtained from G by merging the vertices a and c together.

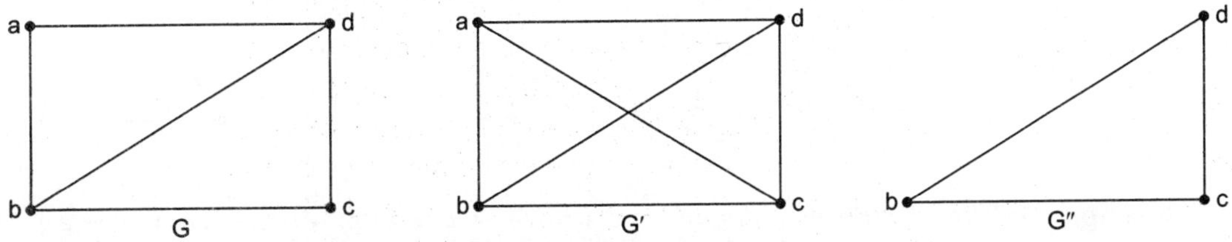

Since G' is a complete graph with 4 vertices, we have

$$P_4(\lambda) \text{ of } G' = \lambda(\lambda - 1)(\lambda - 2)(\lambda - 3) \quad \text{...(1)}$$

Since G'' is a complete graph with 3 vertices, we have

$$P_3(\lambda) \text{ of } G'' = \lambda(\lambda - 1)(\lambda - 2) \quad \text{...(2)}$$

So, by decomposition theorem, we have

$$P_4(\lambda) \text{ of } G = P_4(\lambda) \text{ of } G' + P_3(\lambda) \text{ of } G''$$

$$= \lambda(\lambda - 1)(\lambda - 2)(\lambda - 3) + \lambda(\lambda - 1)(\lambda - 2) \quad \text{[using (1) and (2)]}$$

$$= \lambda(\lambda - 1)(\lambda - 2)^2$$

Observe that, $P_4(1) = P_4(2) = 0$ and $P_4(3) = 6 \neq 0$.

$\therefore$ $\chi(G) = 3$.

Example 15: *Using decomposition theorem, find the chromatic polynomial of the following graph G.*

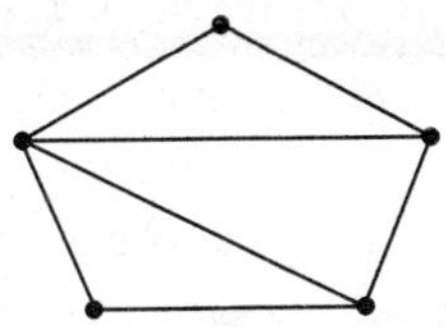

Solution: By successive application of addition of an edge between two non-adjacent vertices v_i, v_j and fusing these vertices, we get the graphs shown in Fig. 6.21.

Using decomposition theorem, we have

$$P_5(\lambda) \text{ of } G = P_5(\lambda) \text{ of } K_5 + 3P_4(\lambda) \text{ of } K_4 + P_3(\lambda) \text{ of } K_3$$

(where K_5, K_4 and K_3 are complete graphs with respective vertices 5, 4 and 3)

$$= \lambda(\lambda - 1)(\lambda - 2)(\lambda - 3)(\lambda - 4) + 3\lambda(\lambda - 1)(\lambda - 2)(\lambda - 3) + \lambda(\lambda - 1)(\lambda - 2)$$

$$= \lambda(\lambda - 1)(\lambda - 2)^3.$$

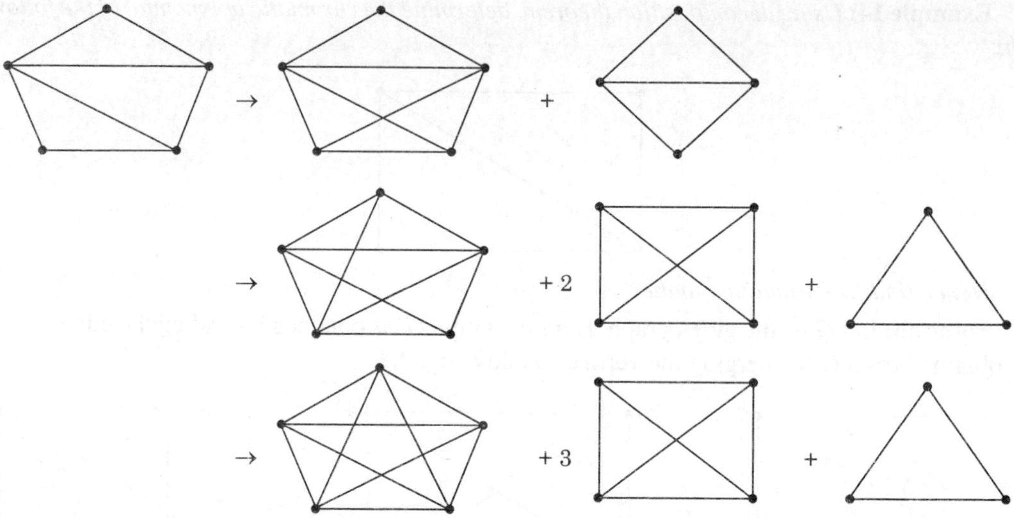

Fig. 6.21

Example 16: *Show that the non-planar graph with least number of vertices is a complete graph with five vertices.*

Solution: Let K_5 denotes the complete graph with five vertices. We know that the chromatic number of K_5 is five. But four colour theorem states that the chromatic number of every planar graph is four or less. Also, obviously a graph with four or less number of vertices is always planar. So, we come to the conclusion that the non-planar graph with smallest number of vertices is a complete graph with five vertices.

MULTIPLE CHOICE QUESTIONS

1. If the chromatic number of a graph G is K, then G is
   (*a*) K-colourable (*b*) $(K-1)$-colourable (*c*) $(K-2)$-colourable (*d*) $(K-3)$-colourable
2. If a graph G has chromatic number 7, then G is
   (*a*) 2-colourable (*b*) 3-colourable (*c*) 6-colourable (*d*) 8-colourable
3. If $\chi(G) = n$, then the graph G is K-colourable if
   (*a*) $K < n$ (*b*) $K = n$ (*c*) $K \geq n$ (*d*) none of these
4. If a graph is k-colourable, then it is $(k + m)$-colourable, $m \geq 0$. This statement is
   (*a*) True (*b*) False
5. If {1, 2, 3} be a set of three colours for proper colouring the graph:

   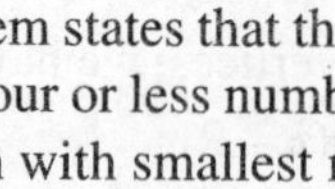

   If a is assigned the colour 1, then b is assigned the colour
   (*a*) 1 (*b*) 1 or 2 (*c*) 1 or 3 (*d*) 2 or 3
6. If $\chi(G) = 9$, then the minimum number of colours required for proper colouring of G is
   (*a*) 8 (*b*) 9 (*c*) 10 (*d*) none of these
7. If $\chi(G) = 4$, for a graph G with 6 vertices, then the maximum number of colours required for proper colouring of G is
   (*a*) 4 (*b*) 5 (*c*) 6 (*d*) 7

8. For proper colouring, a K-chromatic graph needs at least
(a) $K - 1$ colours (b) K colours (c) $K + 1$ colours (d) $K + 2$ colours
9. If G be a graph with n vertices, then
(a) $\chi(G) = n$ (b) $\chi(G) < n$ (c) $\chi(G) \leq n$ (d) none of these
10. A null graph with one vertex is
(a) 1-chromatic (b) 2-chromatic (c) 3-chromatic (d) none of these
11. A null graph with n vertices is
(a) 1-chromatic (b) $(n - 1)$-chromatic (c) n-chromatic (d) $(n + 1)$-chromatic
12. The chromatic number of a complete graph K_n (with n vertices) is n.
(a) True (b) False
13. The chromatic number of the graph $K_n - v$, obtained by removing a vertex v, is n.
(a) True (b) False
14. The chromatic number of the Kuratowski's first graph is
(a) 3 (b) 4 (c) 5 (d) none of these
15. The chromatic number of the Kuratowski's second graph is
(a) 1 (b) 2 (c) 3 (d) 6
16. The chromatic number of a circuit C_n (with n vertices) is 2.
(a) True (b) False
17. The chromatic number of a graph containing an odd circuit is
(a) 3 (b) 2
(c) greater than or equal to 3 (d) greater than or equal to 2. [*WBUT 2012*]
18. The chromatic number of a graph containing an even circuit is
(a) 2 (b) 3
(c) greater than or equal to 3 (d) greater than or equal to 2.
19. The chromatic number of a non-null graph is 2 if and only if it is bipartite.
(a) True (b) False
20. The chromatic number of a non-null graph is 2 if and only if it has no circuits of odd length.
(a) True (b) False
21. The chromatic number of a non-null graph is 2 if and only if it has no circuits of even length.
(a) True (b) False
22. A graph G is bipartite if and only if it has no circuits of odd length.
(a) True (b) False
23. Every tree with at least two vertices is
(a) 1-chromatic (b) 2-chromatic (c) 3-chromatic (d) 4-chromatic
24. In a chromatic polynomial $P_n(\lambda)$, λ is
(a) any real number (b) any rational number
(c) a non-negative integer (d) none of these
25. If $P(\lambda)$ be the chromatic polynomial of a graph G, then $P(10)$ is the number of different proper colouring of G with
(a) ten or more colours (b) exactly ten colours (c) less than ten colours (d) ten or fewer colours
26. The chromatic polynomial of a null graph of order n is $P_n(\lambda) =$
(a) λ^n (b) λ^{n-1} (c) $\lambda(\lambda - 1)^{n-1}$ (d) $\lambda(\lambda - 1)^n$
27. For a graph G, $P_n(\lambda) = 0$ if
(a) $\lambda = \chi(G)$ (b) $\lambda < \chi(G)$ (c) $\lambda > \chi(G)$ (d) none of these

28. $P_n(0) = 0$, for any graph G with n vertices.
(*a*) True (*b*) False

29. The chromatic polynomial of a complete graph with four vertices is
(*a*) λ (*b*) $\lambda(\lambda - 1)$ (*c*) $\lambda(\lambda - 1)(\lambda - 2)$ (*d*) $\lambda(\lambda - 1)(\lambda - 2)(\lambda - 3)$

30. The chromatic polynomial of a tree with n vertices is
(*a*) λ^n (*b*) λ^{n-1} (*c*) $\lambda(\lambda - 1)^{n-1}$ (*d*) $\lambda(\lambda - 1)^n$

31. If G be a graph with n vertices, then the degree of its chromatic polynomial $P_n(\lambda)$, where $\lambda \geq n$, is
(*a*) $< n$ (*b*) $= n$ (*c*) $> n$ (*d*) none of these

32. For a graph G, $\chi(G) \leq 4$.
(*a*) True (*b*) False

33. The chromatic number of every planar graph is four or less.
(*a*) True (*b*) False

34. Every planar graph is 4-colourable.
(*a*) True (*b*) False

35. Every planar graph is K-colourable, where
(*a*) $K \neq 4$ (*b*) $K < 4$ (*c*) $K \leq 4$ (*d*) none of these

36. Every planar graph is K-colourable, where $K \geq 4$. This statement is
(*a*) True (*b*) False

37. In any graph G, $\chi(G) \leq 1 + \max_{v \in V(G)} \deg(v)$.
(*a*) True (*b*) False

38. What is the chromatic number of the following graph with 7 vertices?

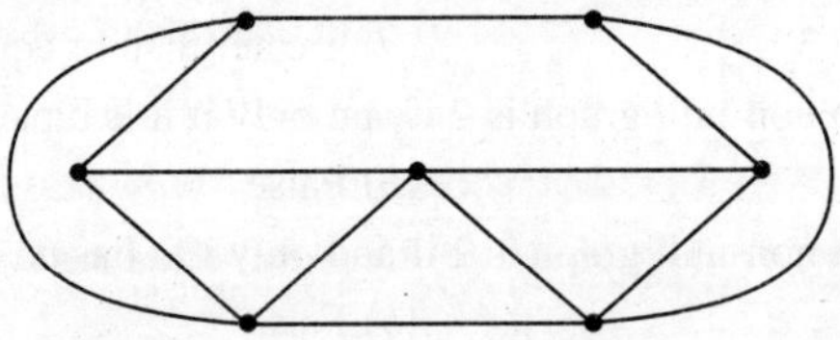

(*a*) 6 (*b*) 5 (*c*) 4 (*d*) 3 [*WBUT 2012*]

39. A sequence $d = (d_1, d_2, ..., d_n)$ is graphic if there is a simple undirected graph with degree sequence d. Which of the following degree sequences are graphic? Why?
P: (2, 3, 3, 4, 4, 5)
Q: (2, 3, 4, 4, 5)
(*a*) Neither P or Q (*b*) Both P and Q (*c*) P only (*d*) Q only.
[*WBUT 2012*]

40. If G be the following graph:

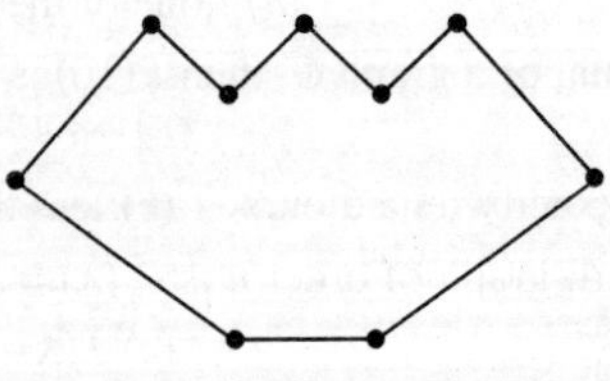

then
(*a*) $\chi(G) = 1$ (*b*) $\chi(G) = 2$ (*c*) $\chi(G) = 3$ (*d*) none of these

41. If C_{101} be a circuit of length 101, then $\chi(C_{101}) =$

(*a*) 2 (*b*) 3 (*c*) 101 (*d*) 102

42. If G be the following graph:

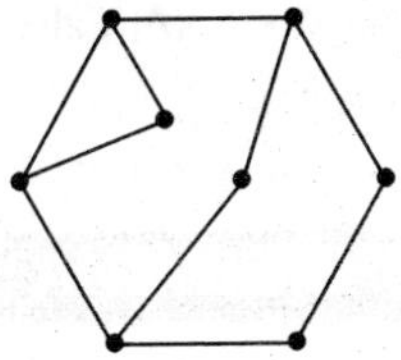

then

(*a*) $\chi(G) = 1$ (*b*) $\chi(G) = 2$ (*c*) $\chi(G) = 3$ (*d*) $\chi(G) = 4$

43. If G: 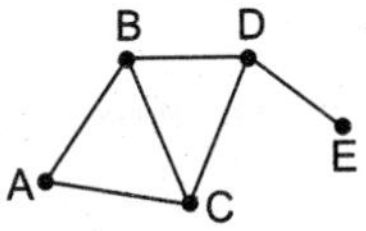

then which one of the following is a clique?

(*a*) $\{A, B, C\}$ (*b*) $\{A, B, D, C\}$ (*c*) $\{A, B, D\}$ (*d*) $\{A, B, E\}$

44. The clique number of the graph 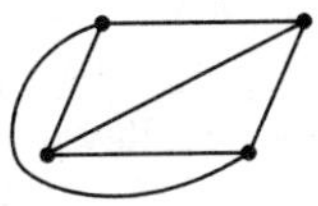 is

(*a*) 1 (*b*) 2 (*c*) 3 (*d*) 4

45. If G: 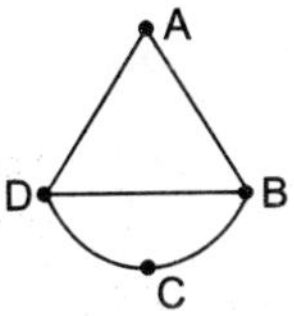

then which one of the following is not a clique of G?

(*a*) $\{A, B, C, D\}$ (*b*) $\{A, B, D\}$ (*c*) $\{B, C, D\}$ (*d*) $\{B, D\}$

46. If G: 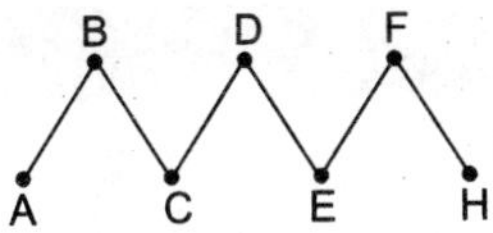

then which one of the following is independent set of vertices of G?

(*a*) $\{A, B, D\}$ (*b*) $\{B, D, F\}$ (*c*) $\{B, D, E, H\}$ (*d*) $\{A, C, D\}$

47. The independence number of the graph 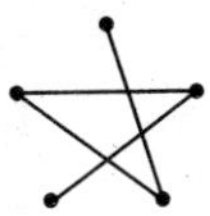 is

(*a*) 1 (*b*) 2 (*c*) 3 (*d*) 4

48. For a graph with 60 vertices and 15 chromatic number, the independence number ≥

(*a*) 4 (*b*) 5 (*c*) 6 (*d*) 7

49. If $\chi(G)$ and $\omega(G)$ are respectively chromatic number and clique number of a graph G, then

(*a*) $\chi(G) \leq \omega(G)$ (*b*) $\chi(G) = \omega(G)$ (*c*) $\chi(G) \geq \omega(G)$ (*d*) none of these

50. If G: 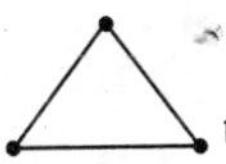 then G is a perfect graph.

(*a*) True (*b*) False

51. If G: 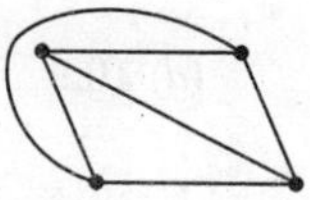 then G is a perfect graph.

(*a*) True (*b*) False

52. Every bipartite graph is a perfect graph.

(*a*) True (*b*) False

53. A complete graph is

(*a*) bipartite (*b*) perfect (*c*) 2-chromatic (*d*) 3-chromatic

54. The cycle C_{10} is perfect

(*a*) True (*b*) False

55. The cycle C_{13} is perfect

(*a*) True (*b*) False

56. The minimum number of colours required to colour the vertices of a cycle with n nodes in such a way that no two adjacent nodes have the same colour is

(*a*) 2 (*b*) 3 (*c*) 4 (*d*) $n-2\left[\frac{n}{2}\right]+2$

[*CS 2002*]

57. The minimum number of colours required to colour the following graph, such that no two adjacent vertices are assigned the same colour, is

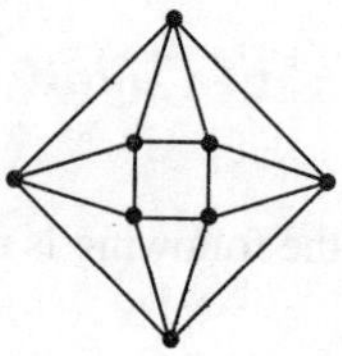

(*a*) 2 (*b*) 3 (*c*) 4 (*d*) 5 [*CS 2004*]

ANSWERS

1. (*a*)	**2.** (*d*)	**3.** (*c*)	**4.** (*a*)
5. (*d*)	**6.** (*b*)	**7.** (*c*)	**8.** (*b*)
9. (*c*)	**10.** (*a*)	**11.** (*a*)	**12.** (*a*)
13. (*b*)	**14.** (*c*)	**15.** (*b*)	**16.** (*b*)
17. (*c*)	**18.** (*d*)	**19.** (*a*)	**20.** (*a*)
21. (*b*)	**22.** (*a*)	**23.** (*b*)	**24.** (*c*)
25. (*d*)	**26.** (*a*)	**27.** (*b*)	**28.** (*a*)
29. (*d*)	**30.** (*c*)	**31.** (*b*)	**32.** (*b*)
33. (*a*)	**34.** (*a*)	**35.** (*c*)	**36.** (*a*)
37. (*a*)	**38.** (*c*)	**39.** (*d*)	($\because$ $2+3+4+4+5=18$, even)
40. (*c*)	**41.** (*b*)	**42.** (*c*)	**43.** (*a*)
44. (*d*)	**45.** (*a*)	**46.** (*b*)	**47.** (*c*)
48. (*a*)	**49.** (*c*)	**50.** (*a*)	**51.** (*a*)
52. (*a*)	**53.** (*b*)	**54.** (*a*)	**55.** (*b*)
56. (*d*)	**57.** (*c*)		

PROBLEMS

1. Find the chromatic number of each of the following graphs:

(i)

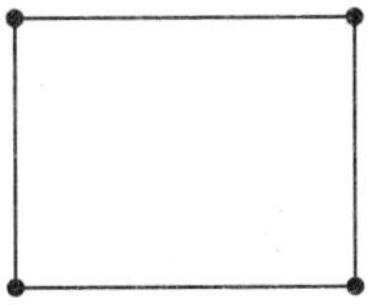

(ii)

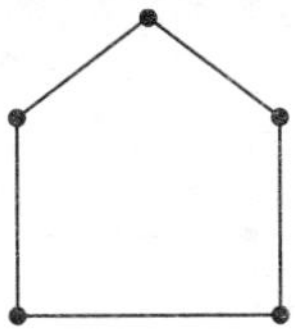

(iii)

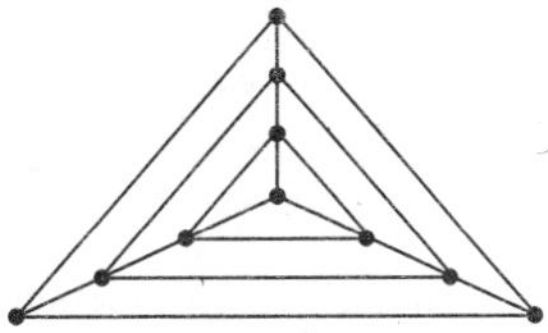

(iv)

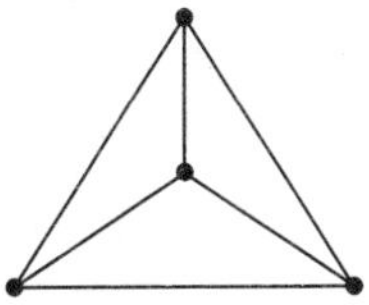

(v)

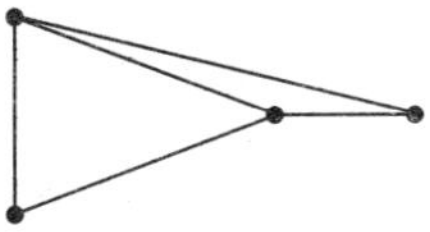

(vi)

(vii)

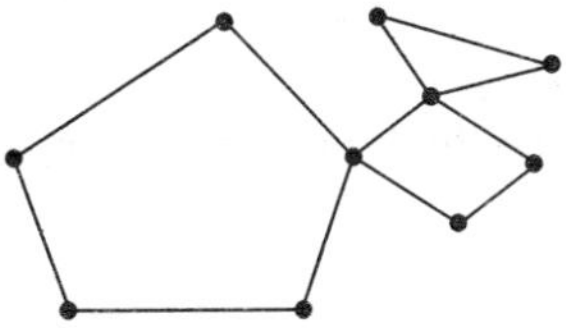

(viii)

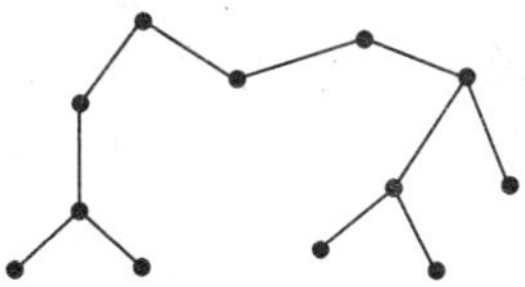

2. Find the chromatic number of the following graph G and verify Brook's theorem.

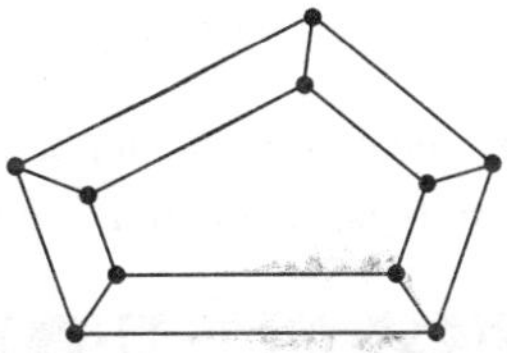

3. Using Decomposition Theorem, determine the chromatic polynomial and hence the chromatic number of each of the following graphs:

(i)

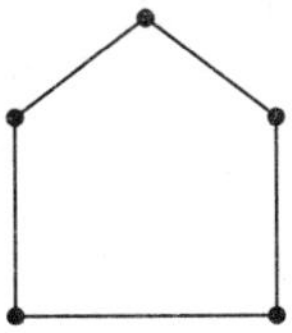

(ii)

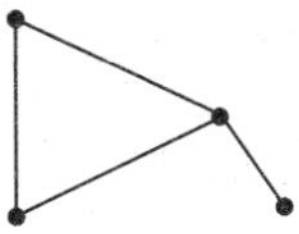

(iii)

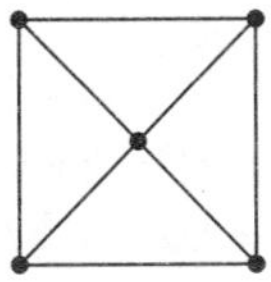

4. Show that the following two graphs have same chromatic polynomials though they are not isomorphic.

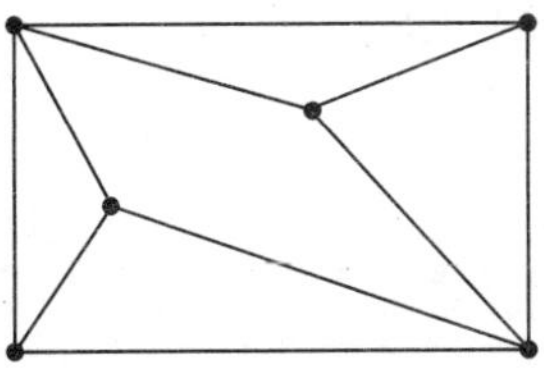

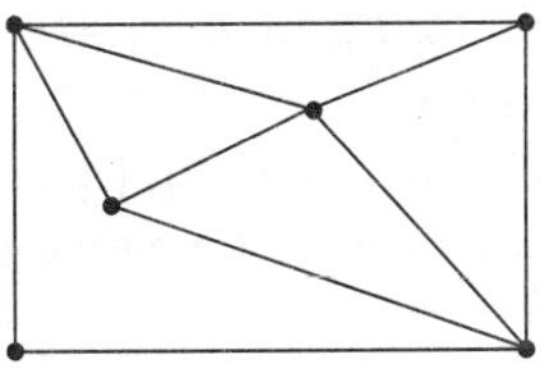

5. Prove that a simple connected planar graph with 17 edges and 10 vertices cannot be 2-colourable.

6. Find all the maximal independent sets of the following graph:

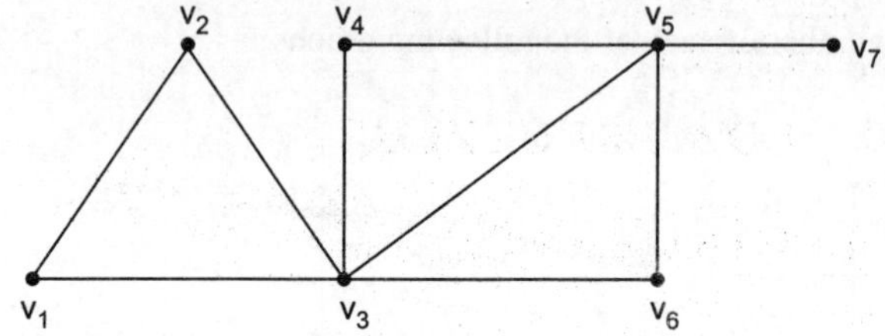

Hence find its independence number.

7. Find all the cliques and the clique number of the following graph:

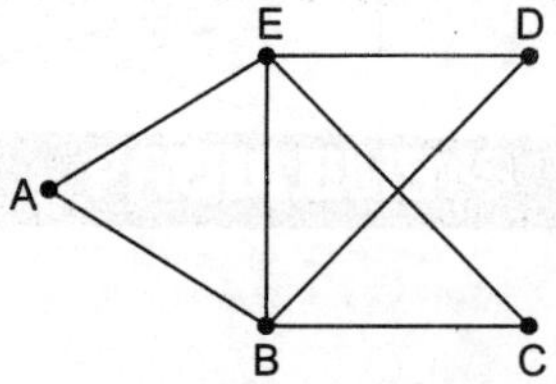

Show that it is a perfect graph.

8. Find all the maximal independent sets of the following graph using Boolean arithmetic on the vertices.

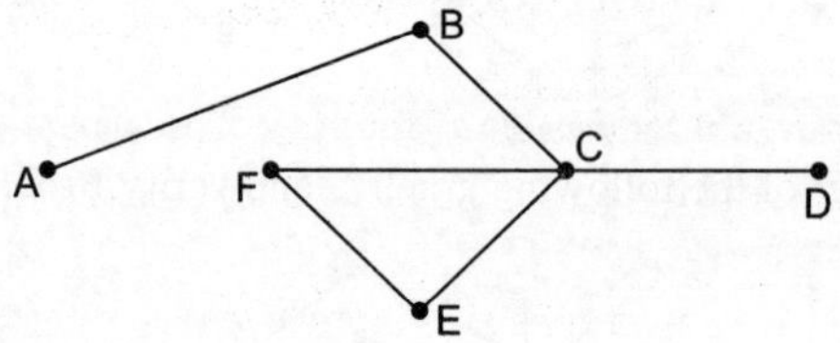

Hence find its independence number and chromatic number.

9. Define perfect graph. Show that every bipartite graph is a perfect graph.

ANSWERS

1. (*i*) 2 (*ii*) 3 (*iii*) 4 (*iv*) 4
(*v*) 3 (*vi*) 2 (*vii*) 3 (*viii*) 2
2. $\chi(G) = 3$
3. (*i*) $P_5(\lambda) = \lambda(\lambda - 1)(\lambda - 2)(\lambda^2 - 2\lambda + 2)$; $\chi(G) = 3$
(*ii*) $P_4(\lambda) = \lambda(\lambda - 1)^2(\lambda - 2)$; $\chi(G) = 3$
(*iii*) $P_5(\lambda) = \lambda(\lambda - 1)(\lambda - 2)(\lambda^2 - 5\lambda + 7)$; $\chi(G) = 3$.
6. $\{v_1, v_5\}, \{v_2, v_5\}, \{v_1, v_4, v_6, v_7\}, \{v_2, v_4, v_6, v_7\}$.
Independence number = 4.
7. $\{A, B\}, \{A, E\}, \{B, E\}, \{A, B, E\}, \{D, E\}, \{B, C\}, \{C, E\}, \{B, D\}$
Clique number = 3.
8. $\{A, F, D\}, \{B, F, D\}, \{B, E, D\}, \{A, E, D\}, \{A, C\}$.
Independence number = 3. Chromatic number = 3.

Module IV

CHAPTER 7

Matchings

7.1 INTRODUCTION

Let us consider the following problem:

Suppose there are seven vacant positions P_1, P_2, P_3, P_4, P_5, P_6 and P_7 in an organization. After screening, four applicants A_1, A_2, A_3 and A_4 have been chosen against these positions as per qualifications and experiences. It is observed that

(*i*) A_1 is suitable for P_2 and P_5, (*ii*) A_2 is suitable for P_2 and P_5,
(*iii*) A_3 is suitable for P_1, P_3, P_4, P_6 and P_7, (*iv*) A_4 is suitable for P_2 and P_5.

This situation is represented by the graph shown in Fig. 7.1 where the vertices correspond to vacant positions and applicants. The edges represent the suitability of each applicant against different positions. Clearly it is a bipartite graph, the vertex set is decomposed into two disjoint subsets $V_1 = \{A_1, A_2, A_3, A_4\}$ and $V_2 = \{P_1, P_2, P_3, P_4, P_5, P_6, P_7\}$.

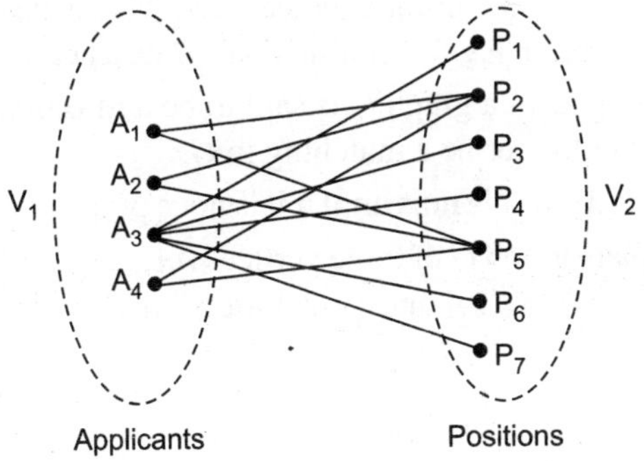

Fig. 7.1

In this situation, a reasonable question is: Is it possible to employ all the applicants and assign each a position for which he/she is suitable? If the answer is no, what is the maximum number of positions that can be filled up from the given set of applicants?

This problem is called a **matching** (or **assignment**) of one set of vertices (applicants) into another set of vertices (posts).

7.2 MATCHINGS: DEFINITIONS

Definition 1: A **matching** M in a graph $G = (V, E)$ is a set of pairwise non-adjacent edges of G.

If $u\,v \in M$, where $u, v \in V$, we say that the vertices u and v are **matched** under M. The vertices u and v are also called **M-satured.** Here uv represents the edge joining the vertices u and v.

Illustration 1: Consider the graph G_1 shown in Fig. 7.2. Here $M_1 = \{v_1v_6, v_2v_5, v_3v_4\}$ and $M_2 = \{v_1v_6, v_2v_4\}$ are two matchings in G_1. The vertices v_1 and v_2 are not M_1-saturated.

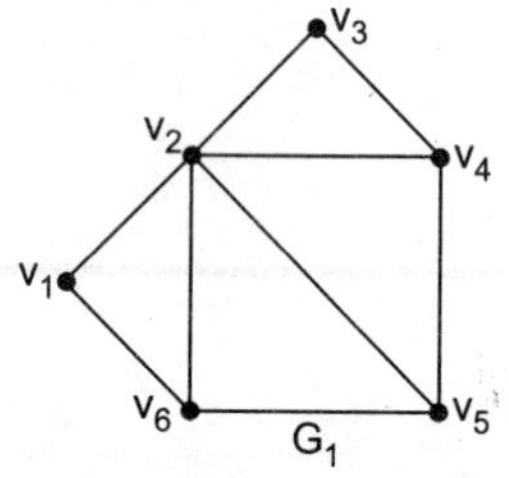

Fig. 7.2

Illustration 2: Consider the bipartite graph G_2 shown in Fig. 7.3. Here $M = \{v_2v_3, v_4v_5\}$ is a matching in G_2.

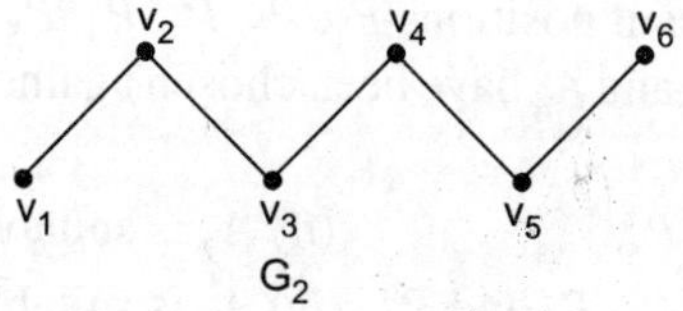

Fig. 7.3

Notes: (*i*) A single edge in a graph is obviously a matching.

(*ii*) Let $G = (V, E)$ be a bipartite graph whose vertex set V is decomposed into two disjoint subsets V_1 and V_2. So, a matching in G is a set $M = \{e_1, e_2, ..., e_r\}$ of edges, where $e_i = a_ib_i$ for $1 \le i \le r$ such that $a_1, a_2, ..., a_r$ are r distinct vertices of V_1 and $b_1, b_2, ..., b_r$ are r distinct vertices of V_2. We say that M **matches** the set $\{a_1, a_2, ..., a_r\}$ to the set $\{b_1, b_2, ..., b_r\}$. For example, in Fig. 7.3 (Illustration), M matches the set $\{v_2, v_4\}$ to the set $\{v_3, v_5\}$.

Definition 2: A matching M in a graph G is said to be a **maximal matching** if any edge (not in M) of G is added to M, it will no longer be a matching in G.

Equivalently, a matching M in G is **maximal** if it is not a proper subset of any other matching in G.

Illustration 3: A complete graph G of three vertices (*i.e.*, a triangle) any single edge is a maximal matching (see Fig. 7.4). So, $\{v_1v_2\}$, $\{v_2v_3\}$ and $\{v_3v_1\}$ are all the maximal matchings in G.

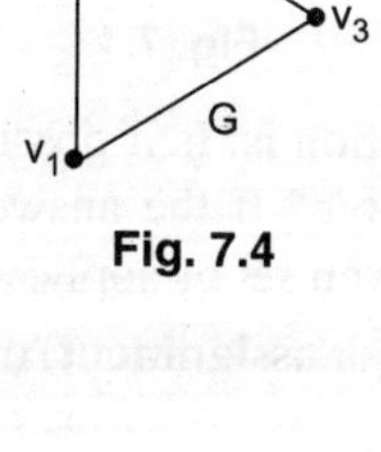

Fig. 7.4

Illustration 4: Consider the graph G shown in Fig. 7.5. Here $M_1 = \{v_1v_3\}$, $M_2 = \{v_1v_2, v_4v_3\}$ and $M_3 = \{v_1v_4, v_2v_3\}$ are the maximal matchings in G.

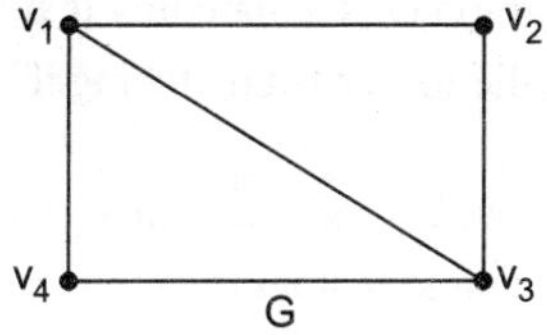

Fig. 7.5

Note: A graph may have many different maximal matchings, and of different sizes.

Definition 3: A **maximum matching** in a graph G is a maximal matching containing the largest number of edges.

For example, the graph G shown in Fig. 7.5 has two maximum matchings M_2 and M_3.

Notes: (*i*) Maximal ≠ maximum. See Illustration 4.

(*ii*) There may exist more than one maximum matching in a graph.

Illustration 5: *C_9 is a cycle (i.e., a circular chain) with the nine vertices a, b, c, d, e, f, g, h, i. How many distinct maximal matchings of size four in C_9 contain the edge ab?* [*WBUT 2012*]

Solution:

C_9:

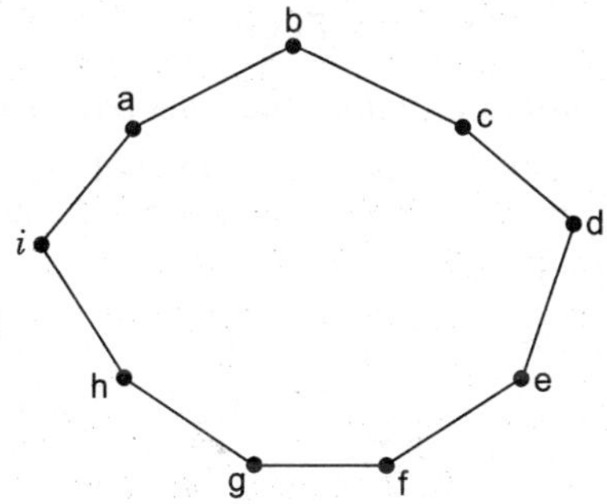

There are only four maximal matchings of size four in C_9 each containing the edge *ab* and these are

$M_1 = \{ab, cd, ef, gh\}$

$M_2 = \{ab, de, fg, hi\}$

$M_3 = \{ab, cd, fg, hi\}$

$M_4 = \{ab, cd, ef, hi\}$.

Definition 4: The number of edges of a maximum matching in a graph is called the **matching number** of the graph.

For examples, the matching number of the graph shown in Fig. 7.4 is 1; the matching number of the graph shown in Fig. 7.5 is 2 and the matching number of the graph shown in Illustration 5 is 4.

7.3 PERFECT MATCHINGS

Definition 5: A matching M in a graph G is said to be a **perfect matching** if every vertex of G is an end vertex of some edge in M.

A **perfect matching** is also called a **complete** matching.

Illustration 6: The matchings $M_1 = \{v_1v_6, v_2v_5, v_3v_4\}$ and $M_3 = \{v_5v_6, v_1v_2, v_3v_4\}$ are two perfect matchings in the graph G_1 shown in Fig. 7.2.

Notes: (*i*) A perfect matching in a graph *G* is a matching that saturates every vertex of *G*.

(*ii*) A matching *M* in a graph *G* is called a perfect matching if *M* matches every vertex of *G* to some vertex of *G*.

(*iii*) There may exist more than one perfect matching in a graph.

(*iv*) Every perfect matching is a maximum matching but the converse is not necessarily true. For example, in the following graph shown in Fig. 7.6:

$M = \{v_3v_4, v_5v_6\}$ is a maximum matching but it is not a perfect matching.

(*v*) There may not exist any perfect matching at all in a graph. For examples, the graph shown in Fig. 7.6 has no perfect matching, and also the graph shown in Illustration 5 has no perfect matching.

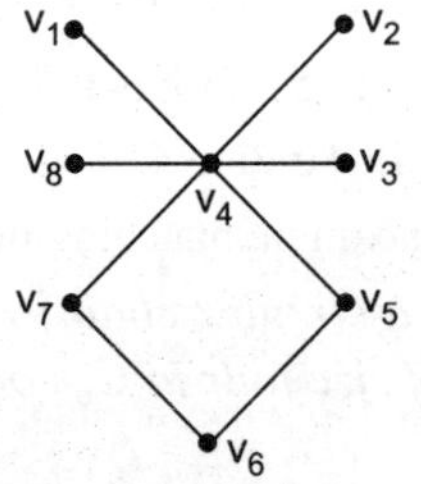

Fig. 7.6

Illustration 7: *For what values of n does the complete graph K_n has perfect matching.*

Solution: The complete graph K_n has a perfect matching if *n* is even. For example, if $V(K_n) = \{1, 2, ..., n\}$, then $\{(1, 2,), (3, 4), (5, 6)\}, ..., ((n-1), n)\}$ is a perfect matching of K_n. Therefore, K_n has a perfect matching if and only if *n* is even.

Illustration 8: *Determine the number of perfect matchings in the complete graph K_{2n}.*

Solution: Let $V(K_{2n}) = \{v_1, v_2, ..., v_{2n}\}$. The vertex v_1 can be saturated in $(2n - 1)$ ways, the next unsaturated vertex can be saturated in $(2n - 3)$ ways and so on. Hence the number of perfect matcings in K_{2n}.

$$= (2n-1)(2n-3)(2n-5) \ldots 5.3.1 = \frac{1.2.3.4.5. \ldots (2n-1)(2n)}{2.4.6. \ldots 2n} = \frac{(2n)!}{2^n (n!)}.$$

7.4 MATCHINGS IN BIPARTITE GRAPHS

Although matching is defined for any graph, it is mostly used in the context of bipartite graphs, as suggested by the problem given in Art. 7.1. In a bipartite graph with a vertex partition V_1 and V_2, the existence of a perfect (or complete) matching of V_1 to V_2 requires at least as many vertices in V_2 as there are in V_1 fulfilling the imposed restrictions. In the context of the given problem in Art. 7.1, there must be at least as many vacant positions as the number of applicants (as per suitability) if all the applicants are to be employed. This condition, however, is not sufficient. In this problem, this condition is violated because three applicants, A_1, A_2 and A_4, each qualifies only for the same two positions P_2 and P_5, and so one of the three applicants cannot be matched.

This leads to the following theorem for a perfect matching.

Theorem 1: Let $G = (V, E)$ be a bipartite graph and the vertex set *V* is decomposed into two disjoint subsets V_1 and V_2. Then a perfect matching of the set V_1 into the set V_2 exists if and only if every subset of r vertices in V_1 is collectively adjacent to at least r vertices in V_2 for all values of $r = 1, 2, ..., |V_1|$.

If $|V_1| = p$, then Theorem 1 requires that we have to take all $2^p - 1$ non-empty subsets of V_1. Then find the number of vertices of V_2 adjacent collectively to each of these subsets. The following theorem is a simplified version of Theorem 1 which is used for detection of a perfect matching in any large graph.

Theorem 2: *In a bipartite graph a perfect matching of V_1 to V_2 exists if (but not only if) there exists a positive integer m such that deg $(v_i) \geq m \geq$ deg (ω_i), $\forall\ v_i \in V_1$ and $\forall\ \omega_i \in V_2$.*

Proof: Let there exists a positive integer m such that the condition of the theorem is satisfied. Consider a subset of r vertices in V_1 and so these r vertices have at least mr edges incident on them (using the condition). Since the degree of every vertex in V_2 is not greater than m, these mr edges are incident on at least $\frac{mr}{m} = r$ vertices in V_2.

Thus any subset of r vertices in V_1 is collectively adjacent to at least r vertices in V_2 and so by Theorem 1, there exists a perfect matching of V_1 to V_2.

Notes: (*i*) In the bipartite graph of Art. 7.1 no such positive integer m exists, because deg $(P_2) = 3 > 2 =$ deg (A_1).

(*ii*) It must be emphasized that the condition of Theorem 2 is a sufficient condition but not necessary condition for the existence of a perfect (or, complete) matching. This means that if the condition of Theorem 2 is satisfied, then there exists a perfect matching in the bipartite graph; but there may exist perfect matching for which the condition of Theorem 2 is not satisfied (try to sketch such a bipartite graph).

(*iii*) In a **bipartite graph** $G = (V, E)$ where the vertex set V is decomposed into two disjoint subsets V_1 and V_2, a **perfect (or, complete) matching of vertices in set V_1 into those in V_2 is a matching in which there is one edge incident with every vertex in V_1**. Equivalently, in a perfect matching of vertices in V_1 into V_2, every vertex in V_1 is matched against some vertex in V_2.

(*iv*) **Complete bipartite graph:** A bipartite graph $G = (V_1, V_2, E)$ is said to be a **complete bipartite graph** if there is an edge between every vertex of V_1 and every vertex of V_2.

A bipartite graph is usually denoted by $K_{m,n}$ where m and n $(m \leq n)$ are the number of vertices in V_1 and V_2 respectively. The number of edges in a complete bipartite graph is mn because each of m vertices is connected to each of n vertices.

Illustration 9: *Determine the number of perfect matchings in the complete bipartite graph $K_{n,n}$.*

Solution: Let the vertex set of the complete bipartite graph $K_{n,n}$ is decomposed into two disjoint subsets $V_1 = \{v_1, v_2, ..., v_n\}$ and $V_2 = \{\omega_1, \omega_2, ..., \omega_n\}$.

We observe that any matching of $K_{n,n}$ that saturates every vertex of V_1 is a perfect matching. Now, the vertex v_1 can be saturated in n ways by choosing are one of the edges $(v_1, \omega_1), (v_1, \omega_2), ..., (v_1, \omega_n)$. Having saturated v_1, the vertex v_2 can be saturated in $(n - 1)$ ways. In general, having saturated $v_1, v_2, ..., v_i$, the next vertex v_{i+1} can be saturated in $(n - i)$ ways. Therefore, the number of perfect matchings in $K_{n,n}$ is $n(n - 1)(n - 2) ... 3.2.1 = n!$

7.5 MATRIX METHOD FOR FINDING PERFECT (OR, COMPLETE) MATCHING

Consider a bipartite graph G with disjoint sets of vertices $V_1 = \{v_1, v_2, ..., v_{n_1}\}$ and $V_2 = \{\omega_1, \omega_2 ..., \omega_{n_2}\}$ consist of n_1 and n_2 number of vertices respectively. Without loss of generality, let $n_1 \leq n_2$ where $n_1 + n_2 = n$, the number of vertices in G.

The adjacency matrix $A(G)$ of the bipartite graph G is of order $n \times n$ and can be written as

$$A(G) = \begin{array}{c|c|c} & v_1 v_2 \cdots v_{n1} & \omega_1 \omega_2 \cdots \omega_{n2} \\ \hline v_1, v_2, \ldots, v_{n_1} & O & A_{12} \\ \hline \omega_1, \omega_2, \ldots, \omega_{n2} & A_{12}^{T} & O \end{array}$$

where the submatrix A_{12} is an $n_1 \times n_2$ binary matrix (*i.e.*, each of its elements is either 0 or 1) containing information that which of the n_1 vertices of V_1 are connected to which of the n_2 vertices of V_2. Matrix A_{12}^T is the transpose of A_{12} and O is the null matrix.

$$\therefore \qquad A_{12} = \left[a_{ij}\right]_{n_1 \times n_2}$$

where
$$a_{ij} = \begin{cases} 1, \text{ if } i^{th} \text{ vertex of } V_1 \text{ is adjacent to } j^{th} \text{ vertex of } V_2 \\ 0, \text{ elsewhere.} \end{cases}$$

So, all the information regarding the adjacency and non-adjacency of the vertices of G are stored in the submatrix A_{12} of $A(G)$.

A matching of V_1 to V_2 corresponds to a selection of the 1's in A_{12} such that every row and every column must have at most one '1'.

A matching is perfect (or, complete) if we select exactly one '1' in each row in A_{12} such that each column must have at most one '1'.

Illustration 10: *Five senators S_1, S_2, S_3, S_4 and S_5 are members of three committees C_1, C_2 and C_3. The memberships are as follows:*

$$\{S_1, S_2\} \to C_1,\ \{S_1, S_3, S_4\} \to C_2 \quad and \quad \{S_3, S_4, S_5\} \to C_3.$$

Represent this situation by a graph.

One member from each committee is to be represented in a super-committee. Determine whether it is possible to send one distinct representative from each of the committees. If so, find an assignment using matrix method.

Solution: The situation of the problem is represented by the following bipartite graph G:

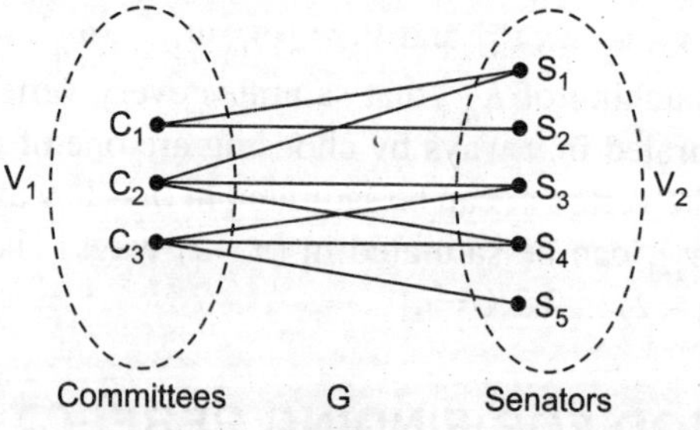

Fig. 7.7

The problem is to find a complete (or perfect) matching (if it exists) of a set $V_1 = \{C_1, C_2, C_3\}$ into a set $V_2 = \{S_1, S_2, S_3, S_4, S_5\}$ of the bipartite graph shown in Fig. 7.7.

Existence: In the bipartite graph of Fig. 7.7, degree of every vertex in $V_1 \geq 2 \geq$ degree of every vertex in V_2. Therefore, there exists a perfect matching of V_1 to V_2 according to Theorem 2.

To find perfect matching of G using matrix method: Adjacency matrix of G is

$$A(G)=\left[\begin{array}{c|c} O & A_{12} \\ \hline A_{12}^T & O \end{array}\right] \quad \text{where} \quad A_{12} = \begin{array}{c} \\ C_1 \\ C_2 \\ C_3 \end{array}\begin{array}{c} \begin{array}{ccccc} S_1 & S_2 & S_3 & S_4 & S_5 \end{array} \\ \begin{bmatrix} 1 & 1 & 0 & 0 & 0 \\ 1 & 0 & 1 & 1 & 0 \\ 0 & 0 & 1 & 1 & 1 \end{bmatrix} \end{array}$$

By inspection a perfect matching of V_1 into V_2 is given by

$$M = \begin{array}{c} \\ C_1 \\ C_2 \\ C_3 \end{array}\begin{array}{c} \begin{array}{ccccc} S_1 & S_2 & S_3 & S_4 & S_5 \end{array} \\ \begin{bmatrix} 1 & 0 & 0 & 0 & 0 \\ 0 & 0 & 1 & 0 & 0 \\ 0 & 0 & 0 & 1 & 0 \end{bmatrix} \end{array},$$

that is, $\{(C_1, S_1), (C_2, S_3), (C_3, S_4)\}$.

Notes: (*i*) There are several assignments (perfect matchings) of this problem.

(*ii*) This problem, called the problem of **distinct representatives**, was first formulated and studied by the English mathematician Philip Hall in 1935.

7.6 HALL'S MARRIAGE THEOREM

Hall's Marriage Theorem, also called **Hall's theorem,** is a combinatorial result that provides the condition for valid selection of a distinct element from each set of a collection of finite sets. It was stated and proved by the famous algebraist Philip Hall (born on April 11, 1904 in Hempstead, London) in his seminal paper published in 1935.

Let S be a finite or infinite family of finite sets where the individual sets may repeat multiple times in S.

Definition 6: A set T is said to be a **traversal** for S if there exists a bijective mapping $f : T \to S$ such that $t \in f(t)$, $\forall\, t \in T$.

A traversal is also called a **system of distinct representatives (SDR)**.

Definition 7: A collection S of sets is said to satisfy **marriage condition** (**MC**) if and only if for each subcollection $W \subseteq S$, we have $|W| \leq \left| \bigcup_{A \in W} A \right|$, where $|W|$ represents the number of elements of W.

Equivalently, a collection S of sets is said to satisfy **marriage condition (MC)** if the number of sets in each subcollection W of S is less than or equal to the number of distinct elements in the union over the subcollection W.

Question: Define SDR of a family of finite sets. What is Hall's Marriage Condition?

[*WBUT 2012*]

Theorem 3: ***(Hall's Marriage Theorem):*** A collection $S = \{S_1, S_2, ...\}$ of non-empty finite sets has a traversal (SDR) if and only if S satisfies the marriage condition, that is, if and only if for each integer k, the union of any k of these sets contains at least k elements.

Proof is omitted.

Illustration 11: Consider $S = \{A_1, A_2, A_3\}$ where $A_1 = \{1, 2, 3, 4\}$, $A_2 = \{4, 5, 6\}$ and $A_3 = \{3, 4, 5, 7\}$.

If $W = \{A_i\}$, $i = 1, 2, 3$, then $|W| = 1 < |A_i|$, $i = 1, 2, 3$.

If $W = \{A_1, A_2\}$, then $|W| = 2, |A_1 \cup A_2| = 6 \Rightarrow |W| < |A_1 \cup A_2|$.

If $W = \{A_1, A_3\}$, then $|W| = 2, |A_1 \cup A_3| = 6 \Rightarrow |W| < |A_1 \cup A_3|$.

If $W = \{A_2, A_3\}$, then $|W| = 2, |A_2 \cup A_3| = 5 \Rightarrow |W| < |A_2 \cup A_3|$.

If $W = \{A_1, A_2, A_3\}$, then $|W| = 3, |A_1 \cup A_2 \cup A_3| = 7 \Rightarrow |W| < |A_1 \cup A_2 \cup A_3|$.

Hence S satisfies marriage condition (MC) and so S has a system of distinct representatives (SDR).

Let $T = \{2, 3, 6\}$ and define a mapping $f : T \to S$ by $f(2) = A_1, f(3) = A_3, f(6) = A_2$. Then f is one-to-one and onto and so bijective. Also, $t \in f(t)$, for all $t \in T$. Hence $\{2, 3, 6\}$ is a valid SDR for S. Similarly, $\{2, 4, 7\}$, $\{2, 5, 7\}$, $\{2, 6, 7\}$ etc. are also valid SDR for S.

Illustration 12: *Consider the family of finite sets $S = \{A_1, A_2, A_3, A_4\}$ where $A_1 = \{a, b, d, e\}$, $A_2 = \{b, c, d, e, f\}$, $A_3 = \{c, f\}$ and $A_4 = \{b, c, f\}$. Show whether S satisfies the marriage condition. If yes, find two valid SDR of S.* [*WBUT 2012*]

Solution: Consider all possible subcollections W of S.

If $W = \{A_i\}$, then $|W| = 1 < |A_i|$, $i = 1, 2, 3, 4$.

If $W = \{A_1, A_2\}$, then $|W| = 2, |A_1 \cup A_2| = 6 \Rightarrow |W| < |A_1 \cup A_2|$.

If $W = \{A_1, A_3\}$, then $|W| = 2, |A_1 \cup A_3| = 6 \Rightarrow |W| < |A_1 \cup A_3|$.

If $W = \{A_1, A_4\}$, then $|W| = 2, |A_1 \cup A_4| = 6 \Rightarrow |W| < |A_1 \cup A_4|$.

If $W = \{A_2, A_3\}$, then $|W| = 2, |A_2 \cup A_3| = 5 \Rightarrow |W| < |A_2 \cup A_3|$.

If $W = \{A_2, A_4\}$, then $|W| = 2, |A_2 \cup A_4| = 5 \Rightarrow |W| < |A_2 \cup A_4|$.

If $W = \{A_3, A_4\}$, then $|W| = 2, |A_3 \cup A_4| = 3 \Rightarrow |W| < |A_3 \cup A_4|$.

If $W = \{A_1, A_2, A_3\}$, then $|W| = 3, |A_1 \cup A_2 \cup A_3| = 6$ and so $|W| < |A_1 \cup A_2 \cup A_3|$.

If $W = \{A_1, A_2, A_4\}$, then $|W| = 3, |A_1 \cup A_2 \cup A_4| = 6$ and so $|W| < |A_1 \cup A_2 \cup A_4|$.

If $W = \{A_1, A_3, A_4\}$, then $|W| = 3, |A_1 \cup A_3 \cup A_4| = 6$ and so $|W| < |A_1 \cup A_3 \cup A_4|$.

If $W = \{A_2, A_3, A_4\}$, then $|W| = 3, |A_2 \cup A_3 \cup A_4| = 5$ and so $|W| < |A_2 \cup A_3 \cup A_4|$.

If $W = \{A_1, A_2, A_3, A_4\}$, then $|W| = 4, |A_1 \cup A_2 \cup A_3 \cup A_4| = 6$ and so $|W| < |A_1 \cup A_2 \cup A_3 \cup A_4|$.

Thus for each subcollection W of S, we have

$$|W| < \left| \bigcup_{A_i \in W} A_i \right|.$$

Hence S satisfies the marriage condition. So, S possesses SDR by Hall's Marriage Theorem.

Let $T_1 = \{a, b, c, d\}$ and define a mapping $\varphi : T_1 \to S$ by $\varphi(a) = A_1$, $\varphi(b) = A_4$, $\varphi(c) = A_3$ and $\varphi(d) = A_2$. Then φ is one-to-one and onto and so bijective. Also, $t \in \varphi(t)$, $\forall\, t \in T_1$. Hence $\{a, b, c, d\}$ is a valid SDR for S.

Again consider the set $T_2 = \{b, c, d, e\}$ and define a mapping $\psi : T_2 \to S$ by $\psi(b) = A_4$, $\psi(c) = A_3$, $\psi(d) = A_2$ and $\psi(e) = A_1$. Then ψ is one-to-one and onto and so bijective. Also, $t \in \psi(t)$, $\forall\, t \in T_2$.

Hence $\{b, c, d, e\}$ is a valid SDR for S.

Illustration 13: Let $S = \{A_1, A_2, A_3, A_4, A_5\}$, where $A_1 = \{a, b\}$, $A_2 = \{b, c\}$, $A_3 = \{a, c\}$, $A_4 = \{a, b, c\}$ and $A_5 = \{a, c, d, e\}$.

In this case no valid SDR exists, because the marriage condition is violated as shown by the subcollection $W = \{A_1, A_2, A_3, A_4\}$, where $|W| = 4, |A_1 \cup A_2 \cup A_3 \cup A_4| = 3$

$\Rightarrow |W| > |A_1 \cup A_2 \cup A_3 \cup A_4|$.

Observation: Suppose there be two groups: (*i*) consists of n men and (*ii*) consists of n women. For each woman, there is a subset of the men such that any one of whom she would happily marry; and any man would be happy to marry a woman who wants to marry him. Let us investigate whether it is possible to pair up (marriage) the men and the women so that everyone is happy. It is assumed that polygamy and polyandry are prohibited, that is, every woman can marry only one man and every man can marry at most one woman.

Let A_i be the set of those men so that i-th woman would be happy if she marries any member of A_i. Then Hall's marriage theorem states that each woman can happily marry a man if and only if the collection $\{A_i\}$ satisfies marriage condition:

For any subset W of women, the number of men whom at least one of the woman would happy to marry $\left[i.e., \left| \bigcup_{A_i \in W} A_i \right|\right]$ must be at least the number of women in the set (*i.e.*, $|W|$).

Clearly this condition is **necessary** since if it does not hold, then there will be not enough men to share among $|W|$ women. It is interesting to note that this condition is also **sufficient**.

This observation leads to the following **graph theoretic formulation** of Hall's marriage theorem:

Let $G = (V, E)$ be a bipartite graph where the vertex set is decomposed into two disjoint subsets V_1 and V_2 representing the sets of women and men respectively and E consists of edges each of which represents the event that a specific woman prefers to marry a specific man. The problem is to determine whether G possesses a perfect matching.

The Hall's marriage theorem can be restated as follows:

Theorem 4: A bipartite graph $G = (V, E)$, where the vertex set V is decomposed into two disjoint subsets V_1 and V_2, has a perfect matching if and only if for every subset W of V_1,

$$|W| \leq |N_G(W)|. \qquad ...(1)$$

Here $N_G(W)$ denotes the neighbourhood of W in G, that is, $N_G(W)$ is the set of all vertices of G adjacent to some vertex of W.

Note: Condition (1) in Theorem 4 is called the **Hall's marriage condition.**

Illustration 14: *In a certain small village there are four boys and three girls who are of marriageable age. Let* $\{b_1, b_2, b_3, b_4\}$ *and* $\{g_1, g_2, g_3\}$ *be the sets of boys and girls respectively. It is the custom of this village that each girl must select a set of boys any one of whom she would happily marry and any boy must have to marry a girl who want to marry him. Also, polygamy and polyandry are strictly prohibited. The sets of boys selected by the girls are as follows:*

$$g_1 \to \{b_2, b_3, b_4\},\ g_2 \to \{b_1, b_3\} \text{ and } g_3 \to \{b_1, b_2\}.$$

(*i*) *Draw the bipartite graph corresponding to this problem.*

(*ii*) *Determine five different solutions (perfect matchings) of the corresponding marriage problem.*

(*iii*) *Verify the Hall's marriage condition for this problem.*

Solution: (*i*) Let us represent the given problem graphically as shown in Fig. 7.8 which is a bipartite graph where the vertex set is decomposed into two disjoint subsets V_1 and V_2 representing the sets of girls and boys respectively.

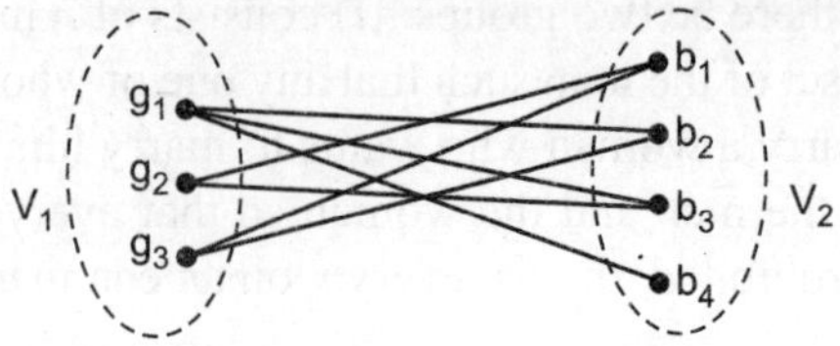

Fig. 7.8

(*ii*) Consider every subset W of V_1 (set of girls) and find whether each subset is collectively adjacent to equal or less vertices in V_2 (set of boys).

$\|W\|$	W	$N_G(W)$
0	Q	Q
1	$\{g_1\}$ $\{g_2\}$ $\{g_3\}$	$\{b_2, b_3, b_4\}$ $\{b_1, b_3\}$ $\{b_1, b_2\}$
2	$\{g_1, g_2\}$ $\{g_2, g_3\}$ $\{g_3, g_1\}$	$\{b_1, b_2, b_3, b_4\}$ $\{b_1, b_2, b_3\}$ $\{b_1, b_2, b_3, b_4\}$
3	$\{g_1, g_2, g_3\}$	$\{b_1, b_2, b_3, b_4\}$

From the above table the following five possible solutions are obtained:

1. $\{(g_1, b_4), (g_2, b_1), (g_3, b_2)\}$
2. $\{(g_1, b_4), (g_2, b_3), (g_3, b_1)\}$
3. $\{(g_1, b_4), (g_2, b_3), (g_3, b_2)\}$
4. $\{(g_1, b_2), (g_2, b_3), (g_3, b_1)\}$
5. $\{(g_1, b_3), (g_2, b_1), (g_3, b_2)\}$.

(*iii*) From the table in (*ii*), it is observed that for each $W \subseteq V_1$ the number of elements in $N_G(W)$ is greater than or equal to the number of elements in W. Therefore, the bipartite graph shown in Fig. 7.8 has a perfect (or, complete) matching as the Hall's marriage condition is satisfied.

EXERCISES

1. True/False

(*i*) A maximal matching in a graph is a matching that cannot be enlarged by adding an edge.

(*a*) True (*b*) False

(*ii*) A maximum matching is a matching of maximum size among all matchings in the graph.

(*a*) True (*b*) False

(*iii*) Any connected graph having an even number of vertices has a perfect (or complete) matching.

(*a*) True (*b*) False

(*iv*) K_6 (complete graph of 6 vertices) has a perfect matching.

(*a*) True (*b*) False

(v) $K_{4,7}$ has a perfect matching.

(a) True (b) False

(vi) If a non-bipartite graph G has a perfect matching, then G has an even number of vertices.

(a) True (b) False

(vii) K_n has a perfect matching iff n is even.

(a) True (b) False

(viii) A tree has atmost one perfect matching

(a) True (b) False

2. **Multiple Choice Questions**

(i) For a perfect matching, the corresponding graph in a matching problem must be

(a) a circuit with even number of vertices (b) a bipartite graph

(c) a tree (d) none of these.

(ii) The matching number of the graph 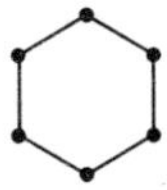 is

(a) 1 (b) 2

(c) 3 (d) 4

(iii) If G: 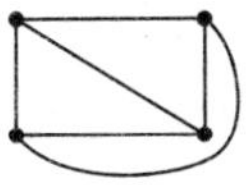, then the number of perfect matching in G is

(a) 3 (b) 2

(c) 1 (d) none of these

(iv) How many perfect matchings are there in a complete graph of 6 vertices?

(a) 15 (b) 24

(c) 30 (d) 60 [CS 2003]

(v) The number of perfect matchings in the complete bipartite graph $K_{n,n}$ is

(a) n (b) $2n$

(c) $n(n-1)$ (d) $n!$

3. Draw two graphs each of which has even order but no perfect matchings.
4. Draw the complete graph of six vertices (K_6) and write down all the perfect matchings in it.
5. In a certain village there are six girls and eight boys each of whom are of marriageable age. Let $\{g_1, g_2, ..., g_6\}$ and $\{b_1, b_2, ..., b_8\}$ be the sets of girls and boys respectively. It is the custom of the village that each girl must choose a set of boys any one of whom she would gladfully marry and any boy must have to marry a girl who want to marry him. Polygamy and polyandry are strictly prohibited in the village. The girls select the sets of boys as follows:

$$g_1 \to \{b_3, b_4\},\ g_2 \to \{b_1, b_5, b_8\},\ g_3 \to \{b_1, b_6\},$$
$$g_4 \to \{b_2, b_4, b_5\},\ g_5 \to \{b_2, b_3, b_7\},\ g_6 \to \{b_7, b_8\}.$$

Determine whether or not a perfect matching exists of the set of girls to the set of boys. If yes, find a possible perfect matching. Use matrix method.

6. Consider the family of finite sets $S = \{A_1, A_2, A_3\}$, where $A_1 = \{a, b, c, d\}$, $A_2 = \{a, e, f, g\}$ and $A_3 = \{c, d, f\}$. Determine whether or not S satisfies the marriage condition. If yes, find two valid system of distinct representatives (SDR) of S.

7. Define system of distinct representatives (SDR) for a collection S of finite sets and state marriage condition satisfied by S. State Hall's marriage theorem and give a graph theoretic formulation of this theorem. What is the corresponding marriage condition in graph theoretic formulation?

ANSWERS

1. *(i)* a *(ii)* a *(iii)* b *(iv)* a
   (v) b *(vi)* a *(vii)* a *(viii)* a
2. *(i)* b *(ii)* c

   (iii) a $\left[\textbf{Hint: } G \equiv K_4. \quad \therefore \text{Required no.} = \frac{4!}{2^2(2!)} = 3\right]$

   (iv) a $\left[\textbf{Hint: } \text{Required no. is } \frac{6!}{2^3(3!)} = 15. \text{ See Illustration 8}\right]$ *(v)* d

3. 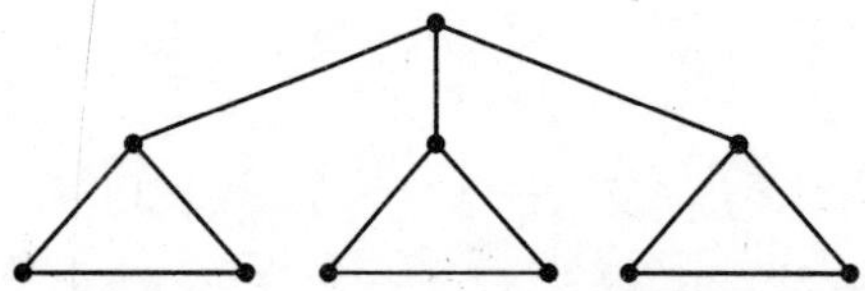 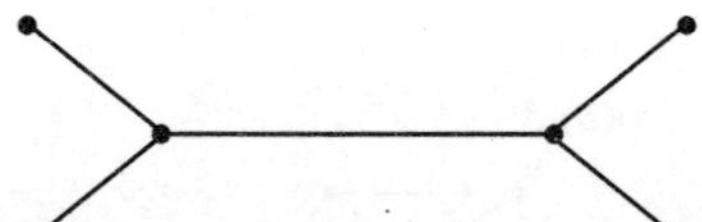

4. 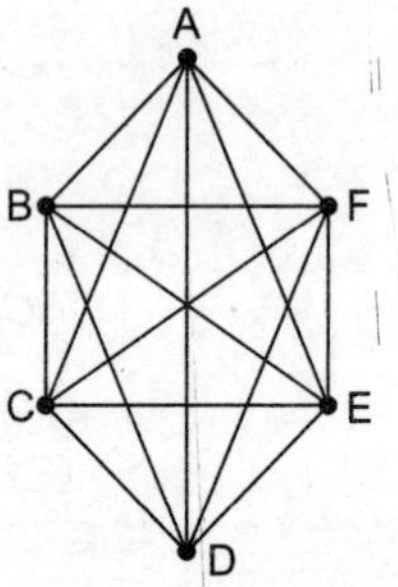

Perfect matchings: {AB, CF, DE}, {BC, AD, EF}, {CD, BE, AF}, {AB, CD, EF}, {BC, DE, AF}, {CD, EF, AB}, {DE, AF, BC}.

5. $\{(g_1, b_3), (g_2, b_1), (g_3, b_6), (g_4, b_2), (g_5, b_7), (g_6, b_8)\}$.
6. S satisfies MC. $\{b, c, e\}$, $\{c, f, g\}$ are two valid SDR of S.

EXAMINATION PAPER

Time Allotted : 3 Hours **Full marks : 70**

The figures in the margin indicate full marks.
Candidates are required to give their answers in their own words as far as practicable.

GROUP – A

(Multiple Choice Type Questions)

1. Choose the correct alternatives for the following: **10 × 1 = 10**

(*i*) What is the chromatic number of the following graph with 7 vertices?

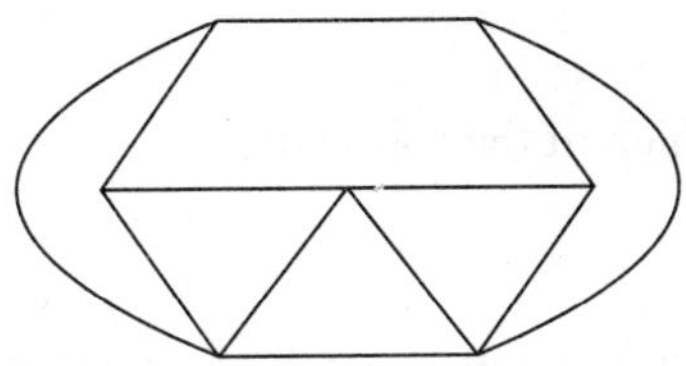

(*a*) 6 (*b*) 5
(*c*) 4 (*d*) 3

Ans. (*c*), see page 222.

(*ii*) If there are n^r arrangement of r objects and n bins, then

(*a*) the objects and bins are all distinguishable
(*b*) the objects are distinguishable and bins are indistinguishable
(*c*) the objects are indistinguishable and bins are distinguishable
(*d*) the objects and bins are indistinguishable.

(*iii*) Consider the set A of all integers greater than 1. Let D be a relation defined on A by $(x, y) \in D$ iff x divides y. Then which of the following is true?

(*a*) D is both a lattice and a partial ordering
(*b*) D is a lattice but not a partial ordering
(*c*) D is neither a lattice nor a partial ordering
(*d*) D is a partial ordering but not a lattice

Ans. (*d*)

(*iv*) If 12 distinct points are placed on the circumference of a circle and all the chords connecting these points are drawn, at how many points do the chords intersect? Assume that no three chords intersect at the same point.

(*a*) $C(12, 2)$ (*b*) $C(12, 4)$
(*c*) 2^{12} (*d*) $12!/2$.

(*v*) The set of natural numbers N with the relationship '|' (divides) is a poset. How many minimal and maximal elements does it have?

(*a*) 1 minimal and 1 maximal (*b*) 1 minimal and 0 maximal

(*c*) 1 minimal and more than 1 maximal (*d*) 0 minimal and 0 maximal.

Ans. (*b*)

(*vi*) What is the result of $(-3)\ X_8\ 5 +_8 (-3)\ X_8\ (-5)$ in $[Z_8, +_8, X_8]$, where Z_8 is the set of integers modulo 8, $+_8$ is the modulo 8 addition operation and X_8 is the modulo 8 multiplication operation?

(*a*) 0 (*b*) 7

(*c*) 8 (*d*) 2

Ans. (*a*)

(*vii*) How many ways are there to travel in *xyz* space from the origin (0, 0, 0) to the point (4, 3, 5) by taking unit steps in positive *x, y, z* directions only?

(*a*) 4!. 3!. 5! (*b*) 60

(*c*) 12!/ (5!4!3!) (*d*) 3^{12}

Ans. (*c*), see Example 19, page 130

(*viii*) $A \wedge B$ is equivalent to which of the following?

(*a*) $\neg A \to \neg B$ (*b*) $\neg A \to B$

(*c*) $\neg B \to A$ (*d*) $\neg (A \to \neg B)$

Ans. (*d*), [**Hint.** $\neg (A \to \neg B) \Leftrightarrow \neg(\neg A \vee \neg B) \Leftrightarrow \neg(\neg A) \wedge \neg(\neg B) \Leftrightarrow A \wedge B$.]

(*ix*) A sequence $d = (d_1, d_2, d_3, ..., d_n)$ is graphic if there is a simple undirected graph with degree sequence d. Which of the following degree sequences are graphic? Why?

P : (2, 3, 3, 4, 4, 5)

Q : (2, 3, 4, 4 5)

(*a*) Neither P or Q (*b*) Both P and Q

(*c*) P only (*d*) Q only

Ans. (*d*), since 2 + 3 + 4 + 4 + 5 = 18 is even

(*x*) A complemented, distributive lattice is also called a Boolean Algebra. Consider a set $S = \{a, b, c\}$ and let $M = \wp(S)$ be the power set of S. Consider the inclusion (subset) relation '$\subseteq$'. Then $(M, \subseteq)$ is

(*a*) not a partial ordering (*b*) a partial ordering but not a lattice

(*c*) a lattice but not a boolean algebra (*d*) a boolean algebra.

Ans. (*d*)

GROUP – B

(Short Answer Type Questions)

Answer any *three* of the following: **3 × 5 = 15**

2. C_9 is a cycle (*i.e.*, a circular chain) with the nine vertices *a, b, c, d, e, f, g, h, i*. How many distinct maximal matchings of size four in C_9 contain the edge *ab*?

Ans. See Illustration 5, page 237.

3. Consider K_6, the complete graph on the six vertices a, b, c, d, e, f. The graph G_1 is obtained from K_6 by deleting the edge ab. The graph G_2 is obtained from G_1 by deleting the edge cd. What are the chromatic numbers of G_1 and G_2 ?

Ans. See Example 4, page 223.

4. A new flag is to be designed with 6 vertical stripes using 4 colours. In how many ways can this be done so that no 2 adjacent strips have the same colour?

Ans. See Example 9, page 117.

5. Give the sequence whose generating function is $g(z) = 5\,(z^5 - 1) / (z - 1)$.

Ans. See answer of 11(*iv*), page 199

6. Consider the poset $S = \{2, 4, 6, 9, 12, 18, 27, 36, 48, 60, 72\}$ under the relation '|' (*i.e.* 'divides'). Find the following:

Maximum element, Minimal element, Greatest element, Least element, lub (2, 9), glb (60, 72).

Ans. See Example 21, page 98.

GROUP – C

(Long Answer Type Questions)

Answer any *three* of the following: **3 × 15 = 45**

7. (*a*) Show that s is a valid conclusion from the premises $p \rightarrow \sim q$, $q \vee r$, $\sim s \rightarrow p$.

Ans. See Example 36, page 18 ($\neg p$ is also denoted as $\sim p$)

(*b*) How many 10 bit binary strings are there none of which contains the patters '110'?

(*c*) Use theory of congruence to prove that for $n \geq 1$, $17 \mid (2^{3n+1} + 3.5^{2n+1})$. **5 + 5 + 5**

Ans. See Example 2, page 73

8. (*a*) Show that t is a valid conclusion from the premises $p \Rightarrow q$, $q \Rightarrow r$, $r \Rightarrow s$ and $p \vee t$.

Ans. See Example 37, page 18

(*b*) For any integer n, prove that the integer $8n + 3$ and $5n + 2$ are relatively prime. Hence find integers x, y such that $(8n + 3x) + (5n + 2) = \gcd(8n + 3, 5n + 2)$.

Ans. See Example 7, page 60

(*c*) Define CRS (mode m) (complete residue system modulo, m). Find all CRS (mod 5).

5 + 5 + 5

Ans. See Definition 9, page 79.

9. (*a*) Solve the recurrence relation:

$$a_{n+2} - 4a_{n+1} + 4a_{n-2} = (r + 1)\,2^r$$

Ans. See Example 7, page 182 and 20 (*ii*), page 197,

(*b*) Show that every bipartite graph is 2-chromatic. (See Theorem 6, page 206)

(*c*) A positive integer n is expressed in the form $10b + b$. Prove that n is divisible by 17 if $a - 5b$ is divisible by 17.

[**Hint.** $17 \mid (a - 5b) \Rightarrow a \equiv 5b \pmod{17} \Rightarrow 10a \equiv 50b \pmod{17} \Rightarrow 10a + b \equiv 51b \pmod{17}$ $\equiv 0 \pmod{17} \Rightarrow 17 \mid n$.] **3 + 5 + 7**

10. (*a*) Show that the poset given in the following Hasse diagram is a lattice. Is it distributive and complemented? Justify your answer.

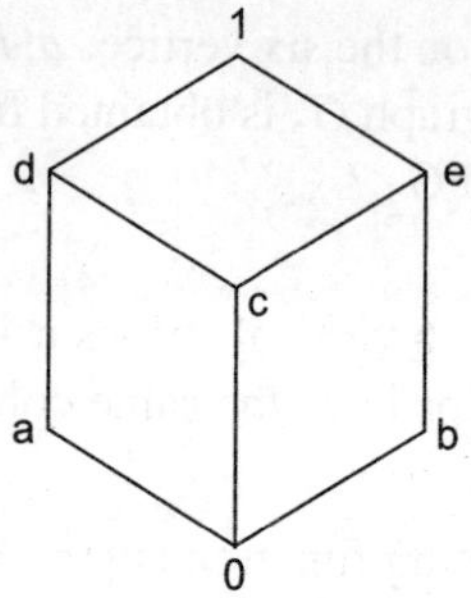

Ans. See Example 44, page 108

(*b*) Show that in a complemented distributive lattice $< L, \wedge, \vee >$

(*i*) $(a \wedge b)' = a' \vee b'$

(*ii*) $(a \vee b)' = a' \wedge b'$

Ans. See Theorem 15, page 109

(*c*) Solve the following recurrence relation using generating function: **6 + 4 + 5**

$$a_n = 4\,(a_{n-1} - a_{n-2}) + 2^n\ (n \geq 2);\ a_0 = 1,\ a_1 = 4.$$

Ans. See Example 6, page 190

11. (*a*) Check the validity of the following arguments:

"If my program runs successfully then I will submit my project. I can appear the examination only if 1 submit my project. Either my program runs successfully or the computer crashes then I cannot appear in examination."

(*b*) Define SDR of a family of finite sets. What is Hall's marriage condition? Consider the family of finite sets $S = \{A_1, A_2, A_3, A_4\}$ where $A_1 = \{a, b, d, e\}$, $A_2 = \{b, c, d, e, f\}$, $A_3 = \{c, f\}$ and $A_4 = \{b, c, f\}$. Show whether S satisfies the marriage condition. If yes, find two valid SDR of S.

Ans. See page 241 and Illustration 12, page 242.

(*c*) Write down the truth table for conditional and biconditional proposition. **5 + 5 + 5**

Ans. See pages 3 and 5.

Index